Differential Forms
Theory and Practice

T0348842

Differential Forms
Theory and Practice

Second edition

by
Steven H. Weintraub
Lehigh University
Bethlehem, PA, USA

AMSTERDAM • BOSTON • HEIDELBERG • LONDON • NEW YORK
OXFORD • PARIS • SAN DIEGO • SAN FRANCISCO • SINGAPORE
SYDNEY • TOKYO
Academic Press is an imprint of Elsevier

Academic Press is an imprint of Elsevier
The Boulevard, Langford Lane, Kidlington, Oxford OX5 1GB, UK
Radarweg 29, PO Box 211, 1000 AE Amsterdam, The Netherlands
225 Wyman Street, Waltham, MA 02451, USA
525 B Street, Suite 1800, San Diego, CA 92101-4495, USA

Second edition 2014

Library of Congress Cataloging-in-Publication Data
Weintraub, Steven H., author.
  Differential forms: theory and practice / by Steven H. Weintraub. -- [2] edition.
     pages cm
  Includes index.
  ISBN 978-0-12-394403-0 (alk. paper)
1. Differential forms. I. Title.
QA381.W45   2014
515'.37--dc23

                                                                    2013035820

British Library Cataloguing in Publication Data
A catalogue record for this book is available from the British Library

For information on all Academic Press publications
visit our web site at store.elsevier.com

Printed and bound in USA

14 15 16 17 10 9 8 7 6 5 4 3 2 1

ISBN: 978-0-12-394403-0

  Working together
to grow libraries in
developing countries

www.elsevier.com • www.bookaid.org

To my mother and the memory of my father

Contents

Preface

Differential forms are a powerful computational and theoretical tool. They play a central role in mathematics, in such areas as analysis on manifolds and differential geometry, and in physics as well, in such areas as electromagnetism and general relativity. In this book, we present a concrete and careful introduction to differential forms, at the upper-undergraduate or beginning graduate level, designed with the needs of both mathematicians and physicists (and other users of the theory) in mind.

On the one hand, our treatment is concrete. By that we mean that we present quite a bit of material on how to do computations with differential forms, so that the reader may effectively use them.

On the other hand, our treatment is careful. By that we mean that we present precise definitions and rigorous proofs of (almost) all of the results in this book.

We begin at the beginning, defining differential forms and showing how to manipulate them. First we show how to do algebra with them, and then we show how to find the exterior derivative $d\varphi$ of a differential form φ. We explain what differential forms really are: Roughly speaking, a k-form is a particular kind of function on k-tuples of tangent vectors. (Of course, in order to make sense of this we must first make sense of tangent vectors.) We carry on to our main goal, the Generalized Stokes's Theorem, one of the central theorems of mathematics. This theorem states:

THEOREM (*Generalized Stokes's Theorem (GST)*). *Let M be an oriented smooth k-manifold with boundary ∂M (possibly empty) and let ∂M be given the induced orientation. Let φ be a $(k-1)$-form on M with compact support. Then*

$$\int_M d\varphi = \int_{\partial M} \varphi.$$

This goal determines our path. We must develop the notion of an oriented smooth manifold and show how to integrate differential forms on these. Once we have done so, we can state and prove this theorem.

The theory of differential forms was first developed in the early twentieth century by Elie Cartan, and this theory naturally led to de Rham cohomology, which we consider in our last chapter.

One thing we call the reader's attention to here is the theme of "naturality" that pervades the book. That is, everything commutes with pull-backs–this cryptic statement will become clear upon reading the book—and this enables us to do all our calculations on subsets of $\mathbb{R}^n$, which is the only place we really know how to do calculus.

This book is an outgrowth of the author's earlier book *Differential Forms: A Complement to Vector Calculus*. In that book we introduced differential forms at a lower level, that of third semester calculus. The point there was to show how the theory of differential forms unified and clarified the material in multivariable calculus: the gradient of a function, and the curl and divergence of a vector field (in $\mathbb{R}^3$) are all "really" special cases of the exterior derivative of a differential form, and the classical theorems of Green, Stokes, and Gauss are all "really" special cases of the GST. By "really" we mean that we must first recast these results in terms of differential forms, and this is done by what we call the "Fundamental Correspondence."

However, in the (many) years since that book appeared, we have received a steady stream of emails from students and teachers who used this book, but almost invariably at a higher level. We have thus decided to rewrite it at a higher level, in order to address the needs of the actual readers of the book. Our previous book had minimal prerequisites, but for this book the reader will have to be familiar with the basics of point-set topology, and to have had a good undergraduate course in linear algebra. We use additional linear algebra material, often not covered in such a course, and we develop it when we need it.

We would like to take this opportunity to correct two historical errors we made in our earlier book. One of the motivations for developing vector calculus was, as we wrote, Maxwell's equations in electromagnetism. We wrote that Maxwell would have recognized vector calculus. In fact, the (common) expression of those equations in vector calculus terms was not due to him, but rather to Heaviside.

But it is indeed the case that this is a nineteenth century formulation, and there is an illuminating reformulation of Maxwell's equations in terms of differential forms (which we urge the interested reader to investigate). Also, Poincaré's work in celestial mechanics was another important precursor of the theory of differential forms, and in particular he proved a result now known as Poincaré's Lemma. However, there is considerable disagreement among modern authors as to what this lemma is (some say it is a given statement, others its converse). In our earlier book we wrote that the statement in one direction was Poincaré's Lemma, but we believe we got it backwards then (and correct now). See Remark 1.4.2.

We conclude with some remarks about notation and language. Results in this book have three-level numbering, so that, for example, Theorem 1.2.7 is the 7th numbered item in Chapter 1, Section 2. The ends of proofs are marked by the symbol $\square$. The statements of theorems, corollaries, etc., are in italics, so are clearly delineated. But the statements of definitions, remarks, etc., are in ordinary type, so there is nothing to delineate them. We thus mark their ends by the symbol $\diamond$. We use $A \subseteq B$ to mean that A is a subset of B, and $A \subset B$ to mean that A is a proper subset of B. We use the term "manifold" to mean precisely that, i.e., a manifold without boundary. The term "manifold with boundary" is a generalization of the term "manifold," i.e., it includes the case when the boundary is empty, in which case it is simply a manifold.

Steven H. Weintraub
Bethlehem, PA, USA
May, 2013

1 Differential Forms in $\mathbb{R}^n$, I

In this chapter we introduce differential forms in $\mathbb{R}^n$ as formal objects and show how to do algebra with them. We then introduce the operation of exterior differentiation and discuss closed and exact forms. Our discussion here is completely general, but for the sake of clarity and simplicity, we will be drawing most of our examples from $\mathbb{R}^1$, $\mathbb{R}^2$, or $\mathbb{R}^3$.

The special case $n = 3$ has some particularly interesting features mathematically, and corresponds to the world we live in physically. In Section 1.3 of this chapter, we show the correspondence between differential forms and functions/vector fields in $\mathbb{R}^3$.

1.0 Euclidean spaces, tangent spaces, and tangent vector fields

We begin by establishing some notation and by making some subtle but important distinctions that are often ignored.

We let $\mathbb{R}$ denote the set of real numbers.

DEFINITION 1.0.1. For a positive integer n, $\mathbb{R}^n$ is

$$\mathbb{R}^n = \{(x_1, \ldots, x_n) \mid x_i \in \mathbb{R}\}. \qquad \diamond$$

In dealing with $\mathbb{R}^1$, $\mathbb{R}^2$, and $\mathbb{R}^3$, we will often use (x), (x, y), and (x, y, z) rather than (x_1), (x_1, x_2), and (x_1, x_2, x_3), respectively, as coordinates.

We are about to introduce tangent spaces. For some purposes, it is convenient to have a single vector space that serves as a "model" for each tangent space, and we introduce that first.

Differential Forms, Second Edition. http://dx.doi.org/10.1016/B978-0-12-394403-0.00001-3

DEFINITION 1.0.2. For a positive integer n, $\mathbf{R}^n$ is the vector space

$$\left\{ v = \begin{bmatrix} a_1 \\ \vdots \\ a_n \end{bmatrix} \mid a_i \in \mathbb{R} \right\}.$$

with the operations

$$\begin{bmatrix} a_1 \\ \vdots \\ a_n \end{bmatrix} + \begin{bmatrix} b_1 \\ \vdots \\ b_n \end{bmatrix} = \begin{bmatrix} a_1 + b_1 \\ \vdots \\ a_n + b_n \end{bmatrix} \quad \text{and} \quad c \begin{bmatrix} a_1 \\ \vdots \\ a_n \end{bmatrix} = \begin{bmatrix} ca_1 \\ \vdots \\ ca_n \end{bmatrix}. \quad \diamond$$

DEFINITION 1.0.3. Let $p = (x_1, \ldots, x_n) \in \mathbb{R}^n$. The tangent space $\mathbf{R}^n_p = T_p\mathbb{R}^n$ to $\mathbb{R}^n$ at p is

$$\left\{ \mathbf{v}_p = \begin{bmatrix} a_1 \\ \vdots \\ a_n \end{bmatrix}_p \mid a_i \in \mathbb{R} \right\}. \quad \diamond$$

Let us carefully discuss the distinction here. Elements of $\mathbb{R}^n$ are points, while for each fixed point p of $\mathbb{R}^n$, elements $\mathbf{v}_p$ of $T_p\mathbb{R}^n$ are vectors. For a fixed point p of $\mathbb{R}^n$, if $\mathbf{v}_p \in T_p\mathbb{R}^n$, we say that $\mathbf{v}_p$ is a tangent vector based at that point, so that $T_p\mathbb{R}^n$ is the vector space of tangent vectors to $\mathbb{R}^n$ at p. This is indeed a vector space as we have the operations of vector addition and scalar multiplication given by

$$\begin{bmatrix} a_1 \\ \vdots \\ a_n \end{bmatrix}_p + \begin{bmatrix} b_1 \\ \vdots \\ b_n \end{bmatrix}_p = \begin{bmatrix} a_1 + b_1 \\ \vdots \\ a_n + b_n \end{bmatrix}_p \quad \text{and} \quad c \begin{bmatrix} a_1 \\ \vdots \\ a_n \end{bmatrix}_p = \begin{bmatrix} ca_1 \\ \vdots \\ ca_n \end{bmatrix}_p.$$

We say that $\mathbf{R}^n$ serves as a model for every tangent space $\mathbf{R}^n_p$ as we have an isomorphism from $\mathbf{R}^n$ to $\mathbf{R}^n_p$ given by

$$\begin{bmatrix} a_1 \\ \vdots \\ a_n \end{bmatrix} \mapsto \begin{bmatrix} a_1 \\ \vdots \\ a_n \end{bmatrix}_p.$$

Geometrically, it makes sense to add tangent vectors based at the same point of $\mathbb{R}^n$, and that is what we have done. But it does not make geometric sense to add tangent vectors based at different points of $\mathbb{R}^n$, nor does it make geometric sense to add points of $\mathbb{R}^n$, so we do not attempt to define such operations.

Let us emphasize the distinctions we have made. Often one sees points and vectors identified, but they are really different kinds of objects. Also, often one sees tangent vectors at different points identified but they are really different objects.

What we have just said is mathematically precise, but also makes sense physically. Consider, for example, a body under the gravitational influence of a star. The body has some position, i.e., it is located at some point p of $\mathbb{R}^3$, and the gravity of the star exerts a force on the body, this force being given by a vector $\mathbf{v}_p$ whose direction is toward the star and whose magnitude is given by Newton's law of universal gravitation. Furthermore $\mathbf{v}_p$ is indeed based at p as this is where the body is located, i.e., this is the point at which the force is acting. Thus we see a clear distinction between the position (point) and force (vector based at that point). If our body, located at the point p, is under the influence of a binary star system, with $\mathbf{v}_p$ the force vector for the gravitational attraction of the first star and $\mathbf{w}_p$ the force vector for the gravitational attraction of the second star, then their sum $\mathbf{v}_p + \mathbf{w}_p$ (given by the "parallelogram law") is the net gravitational force on the body, and this indeed makes sense. On the other hand, if $\mathbf{v}_p$ is a tangent vector based at p and $\mathbf{w}_q$ is a tangent vector based at some different point q, their sum $\mathbf{v}_p + \mathbf{w}_q$ is not defined, and would not make physical sense either, as what could this represent? (A body can't be in two different places at the same time!)

Next we come to the closely related notion of a tangent vector field.

DEFINITION 1.0.4. A *tangent vector field* $\mathbf{v}$ on $\mathbb{R}^n$ is a function $\mathbf{v}$ that associates to every point $p \in \mathbb{R}^n$ a tangent vector to $\mathbb{R}^n$ based at p, i.e., a tangent vector $\mathbf{v}_p \in T_p\mathbb{R}^n$. Thus we may write $\mathbf{v}(p) = \mathbf{v}_p$ for every $p \in \mathbb{R}^n$.   ◇

EXAMPLE 1.0.5. For any n-tuple of real numbers $(a_1, \ldots, a_n)$, we have the *constant vector field*

$$\mathbf{v} = \begin{bmatrix} a_1 \\ \vdots \\ a_n \end{bmatrix}$$

given by

$$\mathbf{v}(p) = \begin{bmatrix} a_1 \\ \vdots \\ a_n \end{bmatrix}_p$$

for any point $p = (x_1, \ldots, x_n) \in \mathbb{R}^n$.

In particular, for any i between 1 and n, if $(a_1, \ldots, a_n) = (0, \ldots, 0, 1, 0, \ldots, 0)$ with the entry of 1 in the ith position, we denote the corresponding constant tangent vector field by $\mathbf{e}^i$ and its value at the point p by $\mathbf{e}^i(p) = \mathbf{e}^i_p$.   ◇

The notation $\mathbf{e}^i$ is a bit ambiguous, as $\mathbf{e}^i$ is a vector field on $\mathbb{R}^n$ but the notation does not make clear what the value of n is. However, this will always be clear from the context.

In dealing with $\mathbb{R}^n$ for $n \leq 3$, we will often use $\mathbf{i}, \mathbf{j}, \mathbf{k}$ rather than $\mathbf{e}^1, \mathbf{e}^2, \mathbf{e}^3$.

Now, while it is improper to identify tangent vectors based at different points, it certainly is proper to consider constant tangent vector fields. (When people do identify tangent vectors based at different points, what they really should be doing is considering tangent vector fields.)

We observe that it makes sense to add two tangent vector fields: $(\mathbf{v} + \mathbf{w})_p = \mathbf{v}_p + \mathbf{w}_p$ for every point p; and to multiply a tangent vector field by a scalar: $(c\mathbf{v})_p = c\mathbf{v}_p$ for every point p. In particular, if $\mathbf{v}$ is the constant vector field in the above definition, then $\mathbf{v} = a_1\mathbf{e}^1 + a_2\mathbf{e}^2 + \cdots + a_n\mathbf{e}^n$.

As a matter of notation, we will maintain the distinction that $\mathbf{v}$ (no subscript) denotes a vector field and $\mathbf{v}_p$ denotes a tangent vector at the point p. However, often for emphasis we will use upper-case boldface letters to denote vector fields. Also, we remark that we use superscripts rather than subscripts (i.e., $\mathbf{e}^i$ rather than $\mathbf{e}_i$), as if we were to use subscripts, we would wind up using double subscripts (e.g., $\mathbf{e}_{i_p}$) and we wish to avoid that.

Now consider the physical situation of a particle moving in $\mathbb{R}^n$, say the earth orbiting the sun. If the particle is located at the point p, then we can consider its velocity vector as a tangent vector to $\mathbb{R}^n$ at p. Conversely, given a vector field $\mathbf{v}$, we may imagine that a point p represents the position of a particle and the tangent vector $\mathbf{v}_p$

represents its velocity. In this physical situation, given any point p, we may imagine that we start off at p at time $t = 0$, and move with velocity $\mathbf{v}$ for $0 \leq t \leq 1$, finishing up at position q at time $t = 1$. (For example, in the case of the earth orbiting the sun, with time measured in years, we would have q just about equal to p.) We can always find q by "integrating the vector field." This is something we do not want to discuss here, except to note that there is one very simple but very important case, that of a constant vector field. In this case, if p is the point $(x_1, \ldots, x_n)$ and $\mathbf{v}$ is the constant vector field $\begin{bmatrix} a_1 \\ \vdots \\ a_n \end{bmatrix}$, then q is the point

$$q = (x_1 + a_1, \ldots, x_n + a_n).$$

In this case we will write

$$q = p + \mathbf{v}.$$

(We emphasize that this notation is really an abuse of language. We are really not adding $\mathbf{v}$ to p; rather we are starting at p and following $\mathbf{v}$. But it is so concise and convenient that its virtues outweigh its vices.)

1.1   The algebra of differential forms

In this section we define differential forms and see how to do algebra with them. For the first part of our development, in the next few sections, we will be treating them as formal objects. Later on, we will of course see their true meaning.

DEFINITION 1.1.1. Let $\mathcal{R}$ be an open set in $\mathbb{R}^n$. Then

$$C^\infty(\mathcal{R}) = \{f : \mathcal{R} \longrightarrow \mathbb{R} \mid f \text{ has all partial derivatives of all}$$
$$\text{orders at every point } p \text{ of } \mathcal{R}\}.$$

A function $f \in C^\infty(\mathcal{R})$ is said to be *smooth* on $\mathcal{R}$.   ◇

For example, $f(x, y) = e^x(\sin(x + y^2))$ is a smooth function on $\mathbb{R}^2$, and $f(x, y) = 1/(x^2 + y^2)$ is a smooth function on $\mathbb{R}^2 - \{(0, 0)\}$.

DEFINITION 1.1.2. Let $\mathcal{R}$ be an open set in $\mathbb{R}^n$. Let k be a fixed nonnegative integer. A monomial k-form on $\mathcal{R}$ is an expression

$$f\,dx_{i_1}\cdots dx_{i_k},$$

where f is a smooth function on $\mathcal{R}$.

A k-form on $\mathcal{R}$ is a sum of monomial k-forms on $\mathcal{R}$.

A *differential form* φ on $\mathcal{R}$ is a k-form on $\mathcal{R}$ for some k. In this situation, k is the *degree* of φ.

We let $\Omega^k(\mathcal{R}) = \{k\text{-forms on } \mathcal{R}\}$ and $\Omega^*(\mathcal{R}) = \{\text{differential forms on } \mathcal{R}\}$.   ◇

We will be using lower-case Greek letters to denote differential forms.

When dealing with differential forms in $\mathbb{R}^1, \mathbb{R}^2$, or $\mathbb{R}^3$, we will often use dx, dy, and dz instead of dx_1, dx_2, and dx_3, respectively.

So far, $dx_1, \ldots, dx_n$ are just symbols. (That is what we mean by saying that these are formal objects.)

We let I be a *multi-index*, i.e., $I = (i_1, \ldots, i_k)$ is a sequence of positive integers. (We allow k to be 0.) We will adopt the notation dx_I to denote the string (possibly empty) $dx_I = dx_{i_1}\cdots dx_{i_k}$. Thus a general k-form φ can be written as $\varphi = A_1 dx_{I_1} + A_2 dx_{I_2} + \cdots + A_m dx_{I_m}$, where all of $dx_{I_1}, \ldots, dx_{I_m}$ have length k. In this case, we will also say that $A_1, A_2, \ldots, A_m$ are the functions involved in φ.

In case $\varphi = A_1 dx_{I_1} + A_2 dx_{I_2} + \cdots + A_m dx_{I_m}$ with each of the functions $A_1, \ldots, A_m$ constants, we will say that φ is a constant form.

EXAMPLE 1.1.3.

(0) $\varphi = x^2 y + e^z$ is a 0-form.

(1) $\varphi = x^2\,dx + (yz + 1)dz$ is a 1-form.

(2) $\varphi = xyz\,dy\,dz + xe^y\,dz\,dx + 2\,dx\,dy$ is a 2-form.

(3) $\varphi = (x^2 + xyz + z^3)dx\,dy\,dz$ is a 3-form.   ◇

Addition of differential forms is done term by term. The sum $\rho = \varphi + \psi$ is only defined when φ and ψ both have the same degree k, for some k, in which case ρ also has degree k. Also, addition is required to be commutative and associative.

Multiplication of differential forms is determined by:

The product of dx_{I_1} and dx_{I_2} is their concatenation $dx_{I_1}dx_{I_2}$.

This is extended to arbitrary differential forms by requiring that multiplication of forms commutes with multiplication by functions and distributes over addition of forms. Note that concatenation is associative and hence multiplication of arbitrary differential forms is associative. Note also that if φ is a k-form and ψ is an ℓ-form, then their product $\rho = \varphi\psi$ is a $(k + \ell)$-form.

So far, the algebra of differential forms does not seem to have any new features. But there is a very important new one, the definition of equality of differential forms. Equality of differential forms is determined by:

$$dx_i dx_j = -dx_j dx_i \text{ for any } i, j.$$

Note in particular the special case when $j = i$: $dx_i dx_i = -dx_i dx_i$ and hence:

$$dx_i^2 = 0 \text{ for any } i.$$

As a consequence, we also have:

LEMMA 1.1.4.

(1) If dx_I is any string with some dx_i appearing more than once, then $dx_I = 0$.

(2) If φ is a k-form on an open set in $\mathbb{R}^n$ with $k > n$, then $\varphi = 0$.

(3) Let $\{i_1, \ldots, i_k\}$ be distinct. If $p : \{i_1, \ldots, i_k\} \longrightarrow \{i_1, \ldots, i_k\}$ is any permutation, then

$$dx_{p(i_1)} \cdots dx_{p(i_k)} = \text{sign}(p)dx_{i_1} \cdots dx_{i_k}.$$

(4) Let φ be a k-form and ψ an ℓ-form. Then $\varphi\psi = (-1)^{k\ell}\psi\varphi$. (In other words, $\varphi\psi = \psi\varphi$ if at least one of k of ℓ is even, while $\varphi\psi = -\psi\varphi$ if both k and ℓ are odd.)

(5) Let φ be a k-form with k odd. Then $\varphi^2 = 0$.

Proof. We observe that if dx_I contains the substring $dx_i dx_j$ and dx_J is obtained from dx_I by transposing those two symbols (i.e., by replacing that substring by $dx_j dx_i$), then $dx_J = -dx_I$.

(1) If dx_i appears more than once in dx_I, then we may repeatedly transpose symbols to arrive at dx_J with $dx_J = \pm dx_I$ and with these two occurrences of dx_i adjacent. But $dx_i^2 = 0$ so $dx_J = 0$.

(2) If $k > n$, then in every term of φ, at least one of $dx_1, \ldots, dx_n$ must appear more than once.

(3) If $dx_I = dx_{i_1} \cdots dx_{i_k}$ and $dx_J = dx_{p(i_1)} \cdots dx_{p(i_k)}$, then $dx_J = \varepsilon dx_I$ where $\varepsilon = 1 \text{ or } -1$ according as dx_J is obtained from dx_I by transposing an even or odd number of adjacent symbols respectively. But this is one precisely one of the definitions of the sign of the permutation p.

(4) Let us compare $dx_I dx_J$ and $\pm dx_J dx_I$, where dx_I has degree k and dx_J has degree ℓ. To get each factor in dx_J from its original position behind dx_I to its new position in front of dx_I, we must move it across each of the k factors in dx_I. Every time we move across a factor, we multiply by -1, so in total we have multiplied by $(-1)^k$. But we must do this for each of the ℓ factors in dx_J, so in the end we have multiplied by $\big((-1)^k\big)^{\ell} = (-1)^{k\ell}$, as claimed.

(5) By (4), if k is odd, $\varphi^2 = (-1)^{k^2}\varphi^2 = -\varphi^2$. $\square$

For future reference, we record the following computation, which you can easily check:

$$dx \, dy \, dz = dy \, dz \, dx = dz \, dx \, dy = -dx \, dz \, dy$$
$$= -dz \, dy \, dx = -dy \, dx \, dz.$$

We now give some illustrative examples of the algebra of differential forms.

EXAMPLE 1.1.5. Let $\varphi = (x^2 + 2yz)dx + (xy - 1)dy + x^3 \, dz$ and $\psi = (x^2 - 2yz)dx + dy + x^3 \, dz$. Then $\varphi + \psi = 2x^2 \, dx + xy \, dy + 2x^3 \, dz$. $\diamond$

EXAMPLE 1.1.6. Let $\varphi = x^3 y\,dx + y\,dy$, $\psi = x^4\,dx + x\,dy + z^2\,dz$, and $\rho = xyz\,dz\,dx$.

Then

$$\begin{aligned}
\varphi\psi &= (x^3 y\,dx + y\,dy)(x^4\,dx + x\,dy + z^2\,dz) \\
&= x^7 y\,dx\,dx + x^4 y\,dx\,dy + x^3 yz^2\,dx\,dz \\
&\quad + x^4 y\,dy\,dx + xy\,dy\,dy + yz^2\,dy\,dz \\
&= 0 + x^4 y\,dx\,dy + x^3 yz^2\,dx\,dz \\
&\quad - x^4 y\,dx\,dy + 0 + yz^2\,dy\,dz \\
&= yz^2\,dy\,dz - x^3 yz^2\,dz\,dx.
\end{aligned}$$

Also,

$$\begin{aligned}
\varphi\rho &= (x^3 y\,dx + y\,dy)(xyz\,dz\,dx) \\
&= x^4 y^2 z\,dx\,dz\,dx + xy^2 z\,dy\,dz\,dx \\
&= xy^2 z\,dx\,dy\,dz.
\end{aligned}$$ ◇

REMARK 1.1.7. While the order of terms in a differential form does not matter, and the order of symbols within a given term of a differential only matters up to sign, when dealing with the special case of differential forms on a region in $\mathbb{R}^3$, we will always write them in the following form:

(0) A 0-form $\varphi = A$, where A is a function.

(1) A 1-form $\varphi = A\,dx + B\,dy + C\,dz$, where A, B, and C are functions.

(2) A 2-form $\varphi = A\,dy\,dz + B\,dz\,dx + C\,dx\,dy$, where A, B, and C are functions.

(3) A 3-form $\varphi = A\,dx\,dy\,dz$, where A is a function.

This is purely for mnemonic purposes. The reason for this will become clear in Section 1.3.   ◇

REMARK 1.1.8. The algebra of differential forms is called exterior algebra, and the product is sometimes called the exterior product. It is also sometimes called the wedge product, with the product of dx_i

and dx_j denoted by $dx_i \wedge dx_j$. But we find the wedge superfluous and do not use it.   ◇

We conclude this section by introducing the Hodge $*$-operator, which we will use later.

DEFINITION 1.1.9. The Hodge $*$-operator on $\mathbb{R}^n$ is a function that takes a k-form φ to an $(n - k)$-form φ^* defined as follows:

(1) Let $I = \{x_{i_1}, x_{i_2}, \ldots, x_{i_k}\}$ be an ordered subset of $\{x_1, x_2, \ldots, x_n\}$, and let $J = \{x_{j_1}, x_{j_2}, \ldots, x_{j_{n-k}}\}$ be the complement of I in $\{x_1, x_2, \ldots, x_n\}$, ordered so that
$dx_{i_1} \cdots dx_{i_k} dx_{j_1} \cdots dx_{j_{n-k}} = dx_1 dx_2 \cdots dx_n$. Then

$$(dx_{i_1} \cdots dx_{i_k})^* = dx_{j_1} \cdots dx_{j_{n-k}}.$$

(2) If $\varphi = A_1 dx_{I_1} + \cdots + A_m dx_{I_m}$ is any k-form, then

$$\varphi^* = A_1 dx_{I_1}^* + \cdots + A_m dx_{I_m}^*.$$ ◇

EXAMPLE 1.1.10. We give the Hodge operator on $\mathbb{R}^n$ explicitly for $n = 1, 2, 3$. In these formulas, 1 denotes the empty string.

(1) On $\mathbb{R}^1$:
$$1^* = dx, \quad dx^* = 1.$$

(2) On $\mathbb{R}^2$:
$$1^* = dx\, dy, \quad dx^* = dy, \quad dy^* = -dx, \quad (dx\, dy)^* = 1.$$

(3) On $\mathbb{R}^3$:
$$1^* = dx\, dy\, dz, \quad dx^* = dy\, dz, \quad dy^* = dz\, dx,$$
$$dz^* = dx\, dy,$$
$$(dx\, dy)^* = dz, \quad (dy\, dz)^* = dx, \quad (dz\, dx)^* = dy,$$
$$(dx\, dy\, dz)^* = 1.$$ ◇

LEMMA 1.1.11. Let φ be a k-form on $\mathbb{R}^n$. Then

$$(\varphi^*)^* = (-1)^{k(n-k)} \varphi.$$

In particular, if n is odd, then $(\varphi^*)^* = \varphi^*$ for any k. If n is even, then $(\varphi^*)^* = \varphi^*$ for k even and $(\varphi^*)^* = -\varphi^*$ for k odd.

Proof. This is a direct consequence of the definition of $*$ and of Lemma 1.1.4. $\qquad\qquad\qquad\qquad\qquad\qquad\qquad\qquad\qquad\qquad\qquad\quad$ $\square$

1.2 Exterior differentiation

We now introduce the operation of exterior differentiation.

DEFINITION 1.2.1. Let A be a 0-form, i.e., a function on an open set $\mathcal{R}$ in $\mathbb{R}^n$. Its *exterior derivative* dA is the 1-form on $\mathcal{R}$

$$dA = A_{x_1}dx_1 + A_{x_2}dx_2 + \cdots + A_{x_n}dx_n$$

where the subscripts denote partial derivatives.

If $\varphi = A_1 dx_{I_1} + A_2 dx_{I_2} + \cdots + A_m dx_{I_m}$ is a k-form on $\mathcal{R}$, then its *exterior derivative* $d\varphi$ is the $(k+1)$-form

$$d\varphi = dA_1 dx_{I_1} + dA_2 dx_{I_2} + \cdots + dA_m dx_{I_m}. \qquad \diamond$$

REMARK 1.2.2. We have introduced the symbol dx_i, and $\varphi = dx_i$ is certainly a 1-form. We should observe that this 1-form really *is* the exterior derivative of the 0-form (i.e., function) x_i.

To be precise, if $A(x_1, \ldots, x_n) = x_i$, then

$$\begin{aligned} \varphi = dA &= A_{x_1}dx_1 + \cdots + A_{x_i}dx_i + \cdots + A_{x_n}dx_n \\ &= 0dx_1 + \cdots + 1dx_i + \cdots + 0dx_n = dx_i. \end{aligned}$$

This certainly doesn't explain what dx_i means (as we said, we're regarding differential forms as formal objects in this chapter, saving their meaning for later) but it does show that our notation is at least consistent, i.e., that "d" of the function "x_i" is indeed "dx_i." $\qquad \diamond$

EXAMPLE 1.2.3.

(0) Let $\rho = (x^2 + y^3)/z$. Then

$$d\rho = (2x/z)dx + (3y^2/z)dy - \left((x^2 + y^3)/(z^2)\right)dz.$$

(1) Let $\varphi = (x^2 + y^3z)dx + (y^2 - 2xz)dy + (x^4 + y^3 - z^2)dz$. Then

$$
\begin{aligned}
d\varphi &= d(x^2 + y^3z)dx + d(y^2 - 2xz)dy \\
&\quad + d(x^4 + y^3 - z^2)dx \\
&= (2x\,dx + 3y^2z\,dy + y^3\,dz)dx \\
&\quad + (-2z\,dx + 2y\,dy - 2x\,dz)dy \\
&\quad + (4x^3\,dx + 3y^2\,dy - 2z\,dz)dz \\
&= 2x\,dx\,dx + 3y^2z\,dy\,dx + y^3\,dz\,dx \\
&\quad - 2z\,dx\,dy + 2y\,dy\,dy - 2x\,dz\,dy \\
&\quad + 4x^3\,dx\,dz + 3y^2\,dy\,dz - 2z\,dz\,dz \\
&= 0 - 3y^2z\,dx\,dy + y^3\,dz\,dx \\
&\quad - 2z\,dx\,dy + 0 + 2x\,dy\,dz \\
&\quad - 4x^3\,dz\,dx + 3y^2\,dy\,dz + 0 \\
&= (3y^2 + 2x)dy\,dz + (y^3 - 4x^3)dz\,dx \\
&\quad + (-3y^2z - 2z)dx\,dy.
\end{aligned}
$$

(2) Let $\psi = (x^2 + y^3 + z^4)dy\,dz + x^2y^3z^4\,dz\,dx + (x + 2y + 3z + 1)dx\,dy$. Then

$$
\begin{aligned}
d\psi &= d(x^2 + y^3 + z^4)dy\,dz + d(x^2y^3z^4)dz\,dx \\
&\quad + d(x + 2y + 3z + 1)dx\,dy \\
&= (2x\,dx + 3y^2\,dy + 4z^3\,dz)dy\,dz \\
&\quad + (2xy^3z^4\,dx + 3x^2y^2z^4\,dy + 4x^2y^3z^3\,dz)dz\,dx \\
&\quad + (dx + 2dy + 3dz)dx\,dy \\
&= 2x\,dx\,dy\,dx + 0 + 0 \\
&\quad + 0 + 3x^2y^2z^4\,dy\,dz\,dx + 0 \\
&\quad + 0 + 0 + 3\,dz\,dx\,dy \\
&= (2x + 2x^2y^2z^4 + 3)dx\,dy\,dz. \qquad\qquad \diamond
\end{aligned}
$$

We now give some formulas for exterior derivatives in $\mathbb{R}^3$. We will shortly see the significance of these formulas, but we will remark that it is probably easier to use the method of the above example to compute exterior derivatives than it is to memorize these formulas.

PROPOSITION 1.2.4.

(0) If $\varphi = A$, a 0-form, then $d\varphi = A_x\,dx + A_y\,dy + A_z\,dz$.

(1) If $\varphi = A\,dx + B\,dy + C\,dz$, a 1-form, then $d\varphi = (C_y - B_z)dy\,dz + (A_z - C_x)dz\,dx + (B_x - A_y)dx\,dy$.

(2) If $\varphi = A\,dy\,dz + B\,dz\,dx + C\,dx\,dy$, a 2-form, then $d\varphi = (A_x + B_y + C_z)dx\,dy\,dz$.

(3) If $\varphi = A\,dx\,dy\,dz$, a 3-form, then $d\varphi = 0$.

Proof.

(0) This is just the definition of $d\varphi$.

(1) Here

$$
\begin{aligned}
d\varphi &= (dA)dx + (dB)dy + (dC)dz \\
&= \left(A_x\,dx + A_y\,dy + A_z\,dz\right)dx \\
&\quad + \left(B_x\,dx + B_y\,dy + B_z\,dz\right)dy \\
&\quad + \left(C_x\,dx + C_y\,dy + C_z\,dz\right)dz \\
&= 0 - A_y\,dx\,dy + A_z\,dz\,dx \\
&\quad + B_x\,dx\,dy + 0 - B_z\,dy\,dz \\
&\quad - C_x\,dz\,dx + C_y\,dy\,dz + 0 \\
&= \left(C_y - B_z\right)dy\,dz + \left(A_z - C_x\right)dz\,dx \\
&\quad + \left(B_x - A_y\right)dx\,dy.
\end{aligned}
$$

(2) Here

$$
\begin{aligned}
d\varphi &= (dA)dy\,dz + (dB)dz\,dx + (dC)dx\,dy \\
&= \left(A_x\,dx + A_y\,dy + A_z\,dz\right)dy\,dz \\
&\quad + \left(B_x\,dx + B_y\,dy + B_z\,dz\right)dz\,dx \\
&\quad + \left(C_x\,dx + C_y\,dy + C_z\,dz\right)dx\,dy \\
&= A_x\,dx\,dy\,dz + 0 + 0 \\
&\quad + 0 + B_y\,dy\,dz\,dx + 0
\end{aligned}
$$

$$+ 0 + 0 + C_z \, dz \, dx \, dy$$
$$= \left(A_x + B_y + C_z \right) dx \, dy \, dz.$$

(3) $d\varphi$ is a 4-form on a region in $\mathbb{R}^3$, and so must be zero. $\qquad\square$

Exterior differentiation obeys rules similar to those of ordinary differentiation, but the Leibniz rule (or product rule) is a little different, due to the question of sign.

THEOREM 1.2.5.

(1) (Linearity) Let c_1 and c_2 be constants and let φ_1 and φ_2 be k-forms. Then

$$d(c_1\varphi_1 + c_2\varphi_2) = c_1 d\varphi_1 + c_2 d\varphi_2.$$

(2) (Leibniz rule) Let φ be a k-form and let ψ be an ℓ-form. Then

$$d(\varphi\psi) = (d\varphi)\psi + (-1)^k \varphi(d\psi).$$

Proof.

(1) This is immediate from the linearity of differentiation of functions.

(2) Given (1), it suffices to prove this when φ and ψ each consist of a single term. Thus let $\varphi = A dx_I$ and $\psi = B dx_J$, so $\varphi\psi = \left(A dx_I \right)\left(B dx_J \right) = AB \, dx_I dx_J$. Then by the definition of the exterior derivative and by the Leibniz rule for functions

$$d(\varphi\psi) = d \left(AB \, dx_I dx_J \right) = d(AB) dx_I dx_J$$
$$= \left((dA)B + A(dB) \right) dx_I dx_J$$
$$= (dA)B \, dx_I dx_J + A(dB) dx_I dx_J.$$

But now note that

$$(dB) dx_I = (-1)^{1 \cdot k} dx_I dB = (-1)^k dx_I dB$$

by Lemma 1.1.4, since dB is a 1-form and φ is a k-form. Thus, substituting, we find

$$
\begin{aligned}
d(\varphi\psi) &= (dA)B dx_I dx_J + (-1)^k A dx_I (dB) dx_J \\
&= (dA) dx_I B dx_J + (-1)^k A dx_I (dB) dx_J \\
&= (d\varphi)\psi + (-1)^k \varphi(d\psi). \qquad \qquad \square
\end{aligned}
$$

EXAMPLE 1.2.6. Let $\varphi - x^2 \, dx - z^2 \, dy$ and $\psi = y \, dx - x \, dz$. Then Theorem 1.2.5 predicts that $d(\varphi\psi) = (d\varphi)\psi - \varphi(d\psi)$, as φ is a 1-form. Computation shows

$$
\varphi\psi = xz^2 \, dy \, dz + x^3 \, dz \, dx + yz^2 \, dx \, dy,
$$

so

$$
d(\varphi\psi) = (z^2 + 2yz) dx \, dy \, dz,
$$

while

$$
d\varphi = 2z \, dy \, dz
$$

so

$$
(d\varphi)\psi = 2yz \, dx \, dy \, dz,
$$

and

$$
d\psi = -dx \, dy + dz \, dx
$$

so

$$
\varphi(d\psi) = -z^2 \, dx \, dy \, dz,
$$

and this checks.   ◇

One of the most important properties of exterior differentiation is the following (which is often abbreviated as "$d^2 = 0$"):

THEOREM 1.2.7 (*Poincaré's Lemma*). *Let φ be any differential form. Then*

$$
d(d\varphi) = 0.
$$

Proof. By linearity it suffices to prove this when φ consists of a single term, so let $\varphi = A dx_I$. To prove this we simply compute $d(d\varphi)$.

First,

$$d\varphi = (A_{x_1}dx_1 + A_{x_2}dx_2 + \cdots + A_{x_n}dx_n)dx_I$$
$$= A_{x_1}dx_1dx_I + A_{x_2}dx_2dx_I + \cdots + A_{x_n}dx_ndx_I,$$

and then

$$d(d\varphi) = d(A_{x_1})dx_1dx_I + d(A_{x_2})dx_2dx_I + \cdots + d(A_{x_n})dx_ndx_I$$
$$= (A_{x_1x_1}dx_1 + A_{x_1x_2}dx_2 + \cdots + A_{x_1x_n}dx_n)dx_1dx_I$$
$$+ (A_{x_2x_1}dx_1 + A_{x_2x_2}dx_2 + \cdots + A_{x_2x_n}dx_n)dx_2dx_I$$
$$+ \cdots$$
$$+ (A_{x_nx_1}dx_1 + A_{x_nx_2}dx_2 + \cdots + A_{x_nx_n}dx_n)dx_ndx_I.$$

Using the relations $dx_i\,dx_i = 0$ and $dx_idx_j = -dx_jdx_i$ we then find

$$d(d\varphi) = (A_{x_2x_1} - A_{x_1x_2})dx_1dx_2 + (A_{x_3x_1} - A_{x_1x_3})dx_1dx_3$$
$$+ \cdots + (A_{x_nx_{n-1}} - A_{x_{n-1}x_n})dx_{n-1}dx_n.$$

But A is a smooth function, and so we have equality of mixed partials (Clairaut's theorem): $A_{x_ix_j} = A_{x_jx_i}$ for all i, j. Hence $d(d\varphi) = 0$, as claimed. □

We now come to a pair of very important definitions.

DEFINITION 1.2.8. If φ is a differential form with the property that $d\varphi = 0$, then φ is *closed*. ◇

DEFINITION 1.2.9. If φ is a differential form with the property that $\varphi = d\psi$ for some form ψ, or $\varphi = 0$, then φ is *exact*. If $\varphi = d\psi$, then ψ is a *primitive* for φ. ◇

REMARK 1.2.10. If φ is the k-form $\varphi = 0$ for some $k > 0$, then $\varphi = d\psi$ for ψ the $(k-1)$-form $\psi = 0$, so is exact. We also want the 0-form $\varphi = 0$ to be exact. But there are no (-1)-forms ψ, so we cannot have $\varphi = d\psi$ in this case. Thus we must add in the alternative $\varphi = 0$ in Definition 1.2.9 to handle this case. ◇

LEMMA 1.2.11.

(1) Let φ_1 be any form and ψ any closed form. Then $d(\varphi_1 + \psi) = d\varphi_1$.

(2) Let φ_1 and φ_2 be any two forms with $d\varphi_1 = d\varphi_2$. Then $\varphi_2 = \varphi_1 + \psi$ for some closed form ψ.

Proof.

(1) We calculate $d(\varphi_1 + \psi) = d\varphi_1 + d\psi = d\varphi_1 + 0 = d\varphi_1$.

(2) Certainly $\varphi_2 = \varphi_1 + (\varphi_2 - \varphi_1)$, so let $\psi = \varphi_2 - \varphi_1$. Then $d\psi = d(\varphi_2 - \varphi_1) = d\varphi_2 - d\varphi_1 = 0$, so ψ is closed. $\qquad\square$

Let us begin by examining the case of closed 0-forms.
First we prove a preliminary lemma.

LEMMA 1.2.12. *Let $\mathcal{R}$ be a connected region in $\mathbb{R}^n$. If f is a function on $\mathcal{R}$ all of whose first partial derivatives are zero at every point of $\mathcal{R}$, then f is constant on $\mathcal{R}$.*

Proof. For simplicity, we will prove this for a region in $\mathbb{R}^2$.

Note that saying f is constant is the same thing as saying it has the same value at any two given points. Thus, pick a point (x_0, y_0) in $\mathcal{R}$ and consider any other point (x_1, y_1) in $\mathcal{R}$.

Suppose first that $y_1 = y_0$, i.e., that the line joining (x_0, y_0) to $(x_1, y_1) = (x_1, y_0)$ is horizontal, and furthermore that this line lies entirely in $\mathcal{R}$.

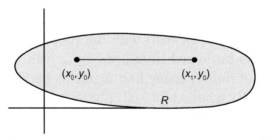

Then we may apply the mean value theorem of ordinary calculus to conclude that $f(x_1, y_0) - f(x_0, y_0) = f_x(x, y_0)(x_1 - x_0)$ for some value of x between x_0 and x_1. (To be precise, we should consider the function $g(x) = f(x, y_0)$ and apply the mean value theorem to it.)

But, by assumption, $f_x(x, y_0) = 0$, so $f(x_1, y_0) - f(x_0, y_0) = 0$; i.e., $f(x_1, y_0) = f(x_0, y_0)$.

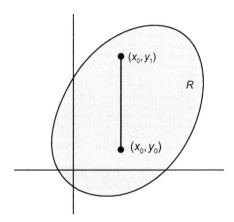

If, on the other hand, $x_1 = x_0$ so that the line joining (x_0, y_0) to $(x_1, y_1) = (x_0, y_1)$ is vertical, and if, furthermore, this line lies entirely in $\mathcal{R}$, we may similarly apply the mean value theorem to conclude $f(x_0, y_1) - f(x_0, y_0) = f_y(x_0, y)(y_1 - y_0)$ for some value of y between y_0 and y_1. Again, by assumption, $f_y(x_0, y) = 0$, and we conclude $f(x_0, y_1) = f(x_0, y_0)$.

Now, of course, for a general point (x_1, y_1) of $\mathcal{R}$, the line joining this point to our starting point (x_0, y_0) will be neither horizontal nor vertical and, furthermore, need not lie in $\mathcal{R}$. However, we are assuming that $\mathcal{R}$ is connected. Furthermore, every point (x_0, y_0) in $\mathbb{R}^2$ has a neighborhood N (an open ball) with the property that every point (x_1, y_1) in N can be joined to (x_0, y_0) by a path lying entirely in N that consists of finitely many line segments parallel to one of the coordinate axes. By a basic topological principle, it then follows that any two points in $\mathcal{R}$ can be joined by a path lying entirely in $\mathcal{R}$ that consists of finitely many line segments parallel to one of the coordinate axes. Then we can successively apply the above arguments to get the values of f at all of the corner points of such a path, and at the endpoint, to equal the value at the starting point, i.e., $f(x_1, y_1) = f(x_0, y_0)$, as required.

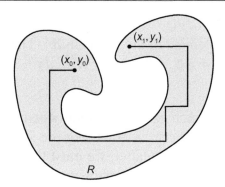

$\square$

COROLLARY 1.2.13. *Let f be a 0-form, i.e., a function, on a connected region $\mathcal{R}$ in $\mathbb{R}^n$. Then f is a closed 0-form on $\mathcal{R}$ if and only if f is constant on $\mathcal{R}$.*

Proof. On the one hand, if f is constant on $\mathcal{R}$, then the partial derivative $f_{x_i} = 0$ for each i at every point of $\mathcal{R}$, so $df = f_{x_1} dx_1 + \cdots + f_{x_n} dx_n = 0$ at every point of $\mathcal{R}$. Thus $f = 0$ at every point of $\mathcal{R}$, and f is closed.

On the other hand, if $df = f_{x_1} dx_1 + \cdots + f_{x_n} dx_n = 0$ at every point of $\mathcal{R}$, then $f_{x_i} = 0$ for each i at every point of $\mathcal{R}$, so, by Lemma 1.2.12, f is constant on $\mathcal{R}$. $\square$

There is a close relationship between the notions of closed and exact forms, which we begin to investigate.

COROLLARY 1.2.14. *Every exact differential form is closed.*

Proof. The corollary claims that if φ is exact, i.e., $\varphi = d\psi$ for some form ψ, then φ is closed, i.e., $d\varphi = 0$. But in this case we have:

$$d\varphi = d(d\psi) = 0$$

by Theorem 1.2.7. $\square$

EXAMPLE 1.2.15.

(1) Let $\varphi = 5x^4 y^2 z^3 \, dx + 2x^5 yz^3 \, dy + 3x^5 y^2 z^2 \, dz$. Note that φ is defined on $\mathbb{R}^3$. Then $d\varphi = 0$, so φ is closed. However, $\varphi = d\psi$, where $\psi = x^5 y^3 z^2$, so in fact φ is exact.

(2) Let $\varphi = (x/z - 2z)dy\,dz + (x^2 e^z - y/z)dz\,dx$. Note that φ is defined on the complement of the z-axis in $\mathbb{R}^3$. Then $d\psi = 0$, so φ is closed. However, $\varphi = d\psi$, where $\psi = x^2 e^z\,dx + z^2\,dy + (xy/z)dz$, so in fact φ is exact. $\diamond$

Observe that it is much easier to decide whether a form is closed than it is to decide whether it is exact. For the former, we need only compute $d\varphi$, whereas for the latter, we must find a suitable form ψ. (Of course, an equivalent way of phrasing Corollary 1.2.14 is that if φ is not closed, then φ is not exact. Thus if $d\varphi \neq 0$, there is no point in even looking for ψ.) We have just seen two examples of closed forms that are exact, so it is reasonable to ask if the converse of Corollary 1.2.14 is true; i.e., Is every closed form exact? The answer to this question is *Sometimes*.

Here is an example of closed but not exact forms.

EXAMPLE 1.2.16. Let φ be a closed 0-form on a connected region $\mathcal{R}$ in $\mathbb{R}^n$. By Corollary 1.2.13, φ is given by $\varphi = r$ for some real number r. If $r \neq 0$, then φ is not exact, as it would have to be the exterior derivative of a (-1)-form, and there aren't any of these. (If $r = 0$, then φ is exact by definition; see Remark 1.2.10.) $\diamond$

Now closed 0-forms are a very simple, and very special, case. Much more interesting is the case of closed k-forms for $k > 0$. As we will see below (Theorem 1.4.1), if $k > 0$ and φ is a closed k-form defined on $\mathbb{R}^n$ (we emphasize–defined on *all of* $\mathbb{R}^n$), then φ is exact. But if φ is a closed k-form that is only defined on a subset of $\mathbb{R}^n$, it may or may not be exact.

EXAMPLE 1.2.17. Let

$$\varphi^1 = \frac{-y}{x^2 + y^2}dx + \frac{x}{x^2 + y^2}dy.$$

We may regard φ^1 as defined on $\mathbb{R}^2$ minus the origin $(0, 0)$, or on $\mathbb{R}^3$ minus the z-axis ($x = y = 0$). Then

$$d\varphi^1 = \frac{y^2 - x^2}{(x^2 + y^2)^2}dy\,dx + \frac{y^2 - x^2}{(x^2 + y^2)^2}dx\,dy = 0,$$

so φ^1 is closed. In fact, φ^1 is *not* exact. $\diamond$

EXAMPLE 1.2.18. Let

$$\varphi^2 = \frac{x\,dy\,dz}{(x^2 + y^2 + z^2)^{3/2}} + \frac{y\,dz\,dx}{(x^2 + y^2 + z^3)^{3/2}} + \frac{z\,dx\,dy}{(x^2 + y^2 + z^2)^{3/2}}.$$

Then φ^2 is defined on $\mathbb{R}^3$ minus the origin $(0, 0, 0)$. You may check that $d\varphi^2 = 0$, so φ^2 is closed, but φ^2 is *not* exact.   ◇

The fact that the forms φ^1 and φ^2 are not exact is a consequence of the Generalized Stokes's Theorem, Theorem 7.1.1, which is one of our main concerns here, and which will take us a while to formulate and prove. But we can give an elementary proof that the form φ^1 is not exact, and we do so now.

PROPOSITION 1.2.19. *The form*

$$\varphi^1 = \frac{-y}{x^2 + y^2}dx + \frac{x}{x^2 + y^2}dy,$$

a closed 1-*form on* $\mathcal{R} = \mathbb{R}^2 - (0, 0)$, *is not exact on* $\mathcal{R}$.

Proof. We shall suppose that φ^1 *is* exact, i.e., $\varphi_1 = df_0$ for some function f_0, and arrive at a contradiction.

Note that Lemma 1.2.11 and Corollary 1.2.13 tell us that if $\varphi^1 = df_1$ for some other function f_1, then $f_1 = f_0 + c$ for some constant c. (For by Lemma 1.2.11, $f_1 = f_0 + f$ where f is closed, and by Corollary 1.2.13 this means f is constant.)

Now we can write down a candidate for the function f_0. (Indeed, this will show us how the mysterious φ^1 came about.) Namely, our candidate is $f_0 = \theta$, where (r, θ) are the polar coordinates of the point (x, y). Note that in order to define this, we must exclude the origin, as θ is meaningless there. However, we shall see that this candidate doesn't work, because θ is not well defined—for a given point (x, y), it may vary by any multiple of 2π.

Recall that θ is given by $\tan\theta = y/x$, valid for $x \neq 0$, or $\cot\theta = x/y$, valid for $y \neq 0$. In fact we want to consider connected regions, so we divide $\mathcal{R}$ into four regions R_1, R_2, R_3, and R_4, where R_1 is the union of the first and second quadrants, R_2 the second and third, R_3 the third and fourth, and R_4 the fourth and first. Note that the equation

$\tan \theta = y/x$ makes sense in each of R_2 and R_4, and $\cot \theta = x/y$ makes sense in each of R_1 and R_3.

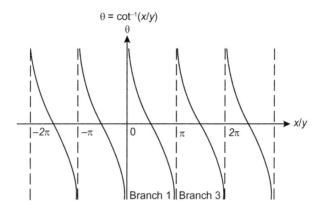

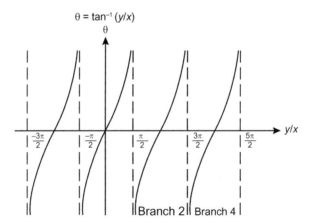

Now in R_2 and R_4 we use the chain rule and the identity $\sec^2 \theta = 1 + \tan^2 \theta$:

$$\tan \theta = y/x$$
$$(\sec^2 \theta) d\theta = (-y/x^2)dx + (1/x)dy$$
$$\left(1 + (y/x)^2\right) d\theta = (-y/x^2)dx + (1/x)dy$$
$$d\theta = \left(-y/(x^2 + y^2)\right) dx$$
$$+ \left(x/(x^2 + y^2)\right) dy = \varphi^1. \qquad (*)$$

Similarly, in R_1 and R_3 we use the chain rule and the identity $\csc^2\theta = 1 + \cot^2\theta$:

$$\cot\theta = x/y$$
$$(-\csc^2\theta)d\theta = (1/y)dx - (x/y^2)dy$$
$$-\left(1 + (y/x)^2\right)d\theta = (1/y)dx - (-y/x^2)dy$$
$$d\theta = \left(-y/(x^2 + y^2)\right)dx$$
$$+ \left(x/(x^2 + y^2)\right)dy = \varphi^1. \qquad (**)$$

With these calculations out of the way we may finish the proof.

Let us begin with the point $(1, 1) \in \mathcal{R}$. We will arrive at our contradiction by showing that f_0, the function we have claimed to satisfy $df_0 = \varphi^1$, would have to have two different values there, which is of course impossible.

By Lemma 1.2.11 we may add any constant to f_0 without changing the property that $df_0 = \varphi^1$, so let us assume $f_0(1, 1) = \pi/4$.

First consider f_0 on the region R_1. On R_1, $\cot\theta = x/y$ with $0 < \theta < \pi$; i.e., θ is given by θ_1, the branch of $\theta = \cot^{-1}(x/y)$ labeled 1 in the figure. We have assumed that f_0 satisfies $df_0 = d\varphi^1$, and we have calculated $d\theta_1 = \varphi^1$ in $(**)$, so $df_0 = d\theta_1$ and, by Lemma 1.2.11 and Corollary 1.2.13, f_0 and θ_1 must differ by a constant. But $f_0(1, 1) = \pi/4 = \theta_1(1, 1)$, so this constant is 0. Hence f_0 agrees with θ_1, i.e., with branch 1 of $\theta = \cot^{-1}(x/y)$, on R_1, and in particular, $f_0(-1, 1) = \theta_1(-1, 1) = 3\pi/4$.

Now consider f_0 on R_2. On R_2, $\tan\theta = y/x$ with $\pi/2 < \theta < 3\pi/2$; i.e., θ is given by θ_2, the branch of $\theta = \tan^{-1}(y/x)$ labeled 2 in the figure. We have assumed that f_0 satisfies $df_0 = \varphi^1$, and we have calculated $d\theta_2 = \varphi^1$ in $(*)$, so $df_0 = d\theta_2$ and, by Lemma 1.2.11 and Corollary 1.2.13, f_0 and θ_2 must differ by a constant. But $f_0(-1, 1) = 3\pi/4 = \theta_2(-1, 1)$, so this constant is 0. Hence f_0 agrees with θ_2, i.e., with branch 2 of $\theta = \tan^{-1}(y/x)$, on R_2, and in particular, $f_0(-1, -1) = \theta_2(-1 - 1) = 5\pi/4$.

You've probably noticed that the argument for f_0 on R_2 was exactly the same as the argument for f_0 on R_1. We may make exactly the same argument on R_3 to conclude that f_0 agrees with θ_3 on R_3, where θ_3 is the branch labeled 3 of $\theta = \cot^{-1}(x/y)$, and hence $f_0(1, -1) = \theta_3(1, -1) = 7\pi/4$. We then make exactly the same argument one more time to conclude that f_0 agrees with θ_4 on R_4, where θ_4 is the branch labeled 4 of $\theta = \tan^{-1}(y/x)$. We thus conclude that $f_0(1, 1) = \theta_4(1, 1) = 9\pi/4$. But this contradicts our original assumption that $f_0(1, 1) = \pi/4$, so no such function can exist. That is, there is *no* function with $df_0 = \varphi^1$ on $\mathcal{R}$, and hence φ^1 is *not* exact on $\mathcal{R}$.   □

REMARK 1.2.20. Note that the proof of Proposition 1.2.19 shows that φ^1 *is* exact on $R_1 \cup R_2 \cup R_3 = \mathbb{R}^2 -$ the nonnegative x-axis, as we can define the function θ there, and then $\varphi^1 = d\theta$. Indeed, by almost the identical argument, φ^1 is exact on any region $\mathcal{R}$ of the form $\mathbb{R}^2 -$ a ray from the origin.   ◇

EXAMPLE 1.2.21.

(1) The form

$$\omega = \frac{x}{x^2 + y^2}dx + \frac{y}{x^2 + y^2}dy,$$

a closed 1-form on $\mathcal{R} = \mathbb{R}^2 - (0, 0)$, is exact on $\mathcal{R}$, as $\omega = dA$, where

$$A(x, y) = \frac{1}{2}\ln(x^2 + y^2).$$

(2) The form

$$\omega^1 = a\left[\frac{x}{(x^2 + y^2 + z^2)^{3/2}}dx + \frac{y}{(x^2 + y^2 + z^2)^{3/2}}dy\right.$$
$$\left. + \frac{z}{(x^2 + y^2 + z^2)^{3/2}}dz\right],$$

a closed 1-form on $\mathcal{R} = \mathbb{R}^3 - (0, 0, 0)$, is exact on $\mathcal{R}$, as $\omega^1 = dA$, where

$$A(x, y, z) = \frac{-a}{(x^2 + y^2 + z^2)^{1/2}}.$$ ◇

We have mentioned that, for all $k > 0$, a closed k-form φ on $\mathbb{R}^n$ is exact. In other words, if φ is a k-form defined on $\mathbb{R}^n$ (for $k > 0$) with $d\varphi = 0$, then φ has a primitive ψ on $\mathbb{R}^n$, i.e., there is a $(k-1)$-form ψ defined on $\mathbb{R}^n$ with $\varphi = d\psi$. Let us see how to actually find ψ, at least for certain values of k and n.

First we consider the case of 1-forms on $\mathbb{R}^n$.

It is convenient to begin with the following general lemma.

LEMMA 1.2.22. *Let* $\varphi = f^1(x_1, \ldots, x_n)dx_1 + f^2(x_1, \ldots, x_n)dx_2 + \cdots + f^n(x_1, \ldots, x_n)dx_n$ *be a closed 1-form defined on a connected region* $\mathcal{R}$ *in* $\mathbb{R}^n$. *Suppose that, for some* i, $f^i(x_1, \ldots, x_n)$ *is a function of* x_i *alone, i.e., that* $f^i(x_1, \ldots, x_n) = g(x_i)$ *for some function* g. *(In particular, this is the case if* $f^i = 0$, *i.e., if the* dx_i *term is missing.) Then for every* $j \neq i$, $f^j(x_1, \ldots, x_n)$ *is a function of the remaining variables* $x_1, \ldots, x_{i-1}, x_{i+1}, \ldots, x_n$, *i.e.,* $f^j(x_1, \ldots, x_n) = h^j(x_1, \ldots, x_{i-1}, x_{i+1}, \ldots, x_n)$ *for some function* h^j.

Proof. Write $\varphi = \psi + \rho$ where $\psi = f^1(x_1, \ldots, x_n)dx_1 + \cdots + f^{i-1}(x_1, \ldots, x_n)dx_{i-1} + f^{i+1}(x_1, \ldots, x_n)dx_{i+1} + \cdots + f^n(x_1, \ldots, x_n)dx_n$ and $\rho = f^i(x_1, \ldots, x_n)dx_i = g(x_i)dx_i$. Then $0 = d\varphi = d(\psi + \rho) = d\psi + d\rho = d\psi + 0 = d\psi$.

Now $d\psi$ is a 2-form. Examining the coefficient of $dx_i dx_j$ in $d\psi$ for each $j \neq i$, we see that it is equal to the partial derivative $f^j_{x_i}$. Hence $f^j_{x_i} = 0$ and so, by the argument in the proof of Lemma 1.2.12, the value of $f^j(x_1, \ldots, x_n)$ is independent of x_i, i.e., $f^j(x_1, \ldots, x_n) = h^j(x_1, \ldots, x_{i-1}, x_{i+1}, \ldots, x_n)$ for some function h^j. $\square$

METHOD 1.2.23. Let $\varphi = f^1(x_1, \ldots, x_n)dx_1 + f^2(x_1, \ldots, x_n)dx_2 + \cdots + f^n(x_1, \ldots, x_n)dx_n$ be a closed 1-form defined on $\mathbb{R}^n$. We find a primitive ψ for φ.

To begin with, let $F^1(x_1, \ldots, x_n)$ be an antiderivative of $f^1(x_1, \ldots, x_n)$ with respect to the variable x_1, so that the partial derivative $F^1_{x_1} = f_{x_1}$. Set $\psi_1 = F^1$. Then

$$d\psi_1 = F^1_{x_1}dx_1 + F^1_{x_2}dx_2 + \cdots + F^1_{x_n}dx_n$$
$$= f_1 dx_1 + F^1_{x_2}dx_2 + \cdots + F^1_{x_n}dx_n$$

so that

$$\varphi - d\psi_1 = (f^2 - F^1_{x_2})dx_2 + \cdots + (f^n - F^1_{x_n})dx_n.$$

Now $d(\varphi - d\psi_1) = d\varphi - d(d\psi_1) = 0 - 0 = 0$ so $\varphi - d\psi_1$ is closed. Hence we may apply Lemma 1.2.22 to conclude that

$f^2 - F_{x_2}^1 = g^2(x_2, \ldots, x_n), \ldots, f^n - F_{x_n}^1 = g^n(x_2, \ldots, x_n)$ for some functions $g^2, \ldots, g^n$. Then let $G^2(x_2, \ldots, x_n)$ be an antiderivative of $g^2(x_2, \ldots, x_n)$ with respect to the variable x_2, so that the partial derivative $G_{x_2}^2 = g_2$. Set $\psi_2 = G^2$. As in the first step, we have

$$d\psi_2 = G_{x_1}^2 dx_1 + G_{x_2}^2 dx_2 + \cdots + G_{x_n}^2 dx_n = g_2 dx_2 + \cdots + G_{x_n}^2 dx_n$$

so that

$$\varphi - d(\psi_1 + \psi_2) = (g^3 - G_{x_3}^2) dx_3 + \cdots + (g^n - G_{x_n}^2) dx_n.$$

Proceeding in this fashion we eventually obtain $\psi_1, \psi_2, \ldots, \psi_n$ with $\varphi - d(\psi_1 + \psi_2 + \cdots + \psi_n) = 0$, i.e.,

$$\varphi = d\psi \text{ for } \psi = \psi_1 + \psi_2 + \cdots + \psi_n. \qquad \diamond$$

EXAMPLE 1.2.24.

(1) Consider

$$\varphi(x, y) = (2xy^3 + 4x^3)dx + (3x^2y^2 + 2y)dy,$$

a closed 1-form on $\mathbb{R}^2$.

Set $\varphi = dA = A_x \, dx + A_y \, dy$. Then $A_x = 2xy^3 + 4x^3$, so

$$A = \int A_x dx = \int (2xy^3 + 4x^3)dx = x^2y^3 + x^4 + c(y).$$

Note that we expect a constant of integration, but to be precise it is constant only with respect to x; so it may depend on y and hence is some function $c(y)$.

Then we have

$$A_y = \frac{\partial}{\partial y}\left(x^2y^3 + 4x^3 + c(y)\right) = 3x^2y^2 + c'(y).$$

But we are given $A_y = 3x^2y^2 + 2y$, so we have $c'(y) = 2y$, and then $c(y) = \int 2y \, dy = y^2 + c$ (c really is a constant this time). Hence $\varphi = dA$, where

$$A = x^2y^3 + x^4 + y^2 + c$$

for any constant c.

(We solved this problem by integrating with respect to x first, and then with respect to y. You should check that we would have obtained the same answer had we integrated with respect to y first, and then with respect to x.)

(2) Consider

$$\varphi(x, y, z) = (2xyz^3 + y^2 + 4z + 2)dx$$
$$+(x^2z^3 + 2xy + 2z^3 - 1)dy$$
$$+(3x^2yz^2 + 6yz^2 + 4x - 4z)dz,$$

a closed 1-form on $\mathbb{R}^3$.

Set $\varphi = dA = A_x\, dx + A_y\, dy + A_z\, dz$. Then $A_x = 2xyz^3 + y^2 + 4z + 2$, so

$$A = \int A_x\, dx = \int (2xyz^3 + y^2 + 4z + 2)dx$$
$$= x^2yz^3 + xy^2 + 4xz + 2x + c_1(y, z),$$

where, by the same logic as before, $c_1(y, z)$ is a function of y and z. Then $x^2z^2 + 2xy + 2z^3 - 1 = A_y = \frac{\partial}{\partial y}(x^2yz^3 + xy^2 + 4xz + 2x + c_1(y, z)) = x^2z^3 + x^2 + \frac{\partial}{\partial y}(c_1(y, z))$; thus we have $\frac{\partial}{\partial y}c_1(y, z) = 2z^3 - 1$ and

$$c_1(y, z) = \int (2z^3 - 1)dy = 2yz^3 - y + c_2(z),$$

where $c_2(z)$ is a function of z.

Thus $A = x^2yz^3 + xy^2 + 4xz + 2x + 2yz^3 - y + c_2(z)$, and $3x^2yz^2 + 6yz^2 + 4x - 4z = \frac{\partial}{\partial z}(x^2yz^3 + xy^2 + 4xz + 2x + 2yz^3 - y + c_2(z)) = 3x^2yz^2 + 4x + 6yz^2 + c_2'(z)$; thus we have $c_2'(z) = -4z$, and

$$c_z(z) = \int -4z\, dz = -2z^2 + c$$

for any constant c. Then we find that $\varphi = dA$, where

$$A = x^2yz^3 + xy^2 + 4xz + 2x2y^3 - y - 2z^2 + c$$

for any constant c.   ◇

REMARK 1.2.25. In both parts of this example we see the familiar "constant of integration." This accords perfectly with Lemma 1.2.11 and Corollary 1.2.13. For if A_1 is any function (i.e., 0-form) with $dA_1 = \varphi$, then, by Lemma 1.2.11, the most general function A with $dA = \varphi$ is of the form $A = A_1 + A_2$, with A_2 a closed 0-form; i.e., by Corollary 1.2.13, a *constant* function $A_2 = c$, so $A = A_1 + c$ and $dA = \varphi$ for any c. Hence we may simply find a single solution, and all solutions differ by a constant.   ◇

REMARK 1.2.26. Let us emphasize where the condition that φ is closed was used in Example 1.2.24. In part (1), say, we had the equation $c'(y) = 2y$, which we solved for $c(y)$. Note that the right-hand side of this equation, which is $2y$, depended *only* on y, so indeed we could integrate it to find a function $c(y)$. If the right-hand side had also involved x, this would have been impossible, as then c would have been a function of both x and y. This didn't occur, precisely because φ was closed.   ◇

REMARK 1.2.27. Actually, this method applies somewhat more generally, even when the form φ may not be defined on all of $\mathbb{R}^n$ (providing, of course, that it is exact), and can be used to obtain the 0-forms A in Example 1.2.21.   ◇

Let us next consider the case of an n-form defined on $\mathbb{R}^n$. This situation turns out to be easy.

LEMMA 1.2.28. *Let φ be an n-form on $\mathbb{R}^n$. In this case, φ is automatically closed and can be written in the form $\varphi = f(x_1, \ldots, x_n)$ $dx_1 \cdots dx_n$. Let $F(x_1, \ldots, x_n)$ be any antiderivative of $f(x_1, \ldots, x_n)$ with respect to the variable x_1. Let*

$$\psi = F(x_1, \ldots, x_n)dx_2 \cdots dx_n.$$

Then $\varphi = d\psi$.

Proof. We simply compute

$$
\begin{aligned}
d\psi &= dF dx_2 \cdots dx_n \\
&= (F_{x_1}dx_1 + F_{x_2}dx_2 + \cdots + F_{x_n}dx_n)dx_2 \cdots dx_n \\
&= f dx_1 dx_2 \cdots dx_n + 0 + \cdots + 0 = \varphi.
\end{aligned}
$$
                                                                        □

REMARK 1.2.29. Let us carefully examine the method provided by Lemma 1.2.28.

(1) Let $n = 2$. In this case we have a 2-form $\varphi = A(x, y)dx\, dy$. If

$$f(x, y) = \int A(x, y)dx,$$

then

$$\psi = f(x, y)dy$$

is a 1-form with $d\psi = \varphi$.

This is easy to check. By the rules of exterior differentiation,

$$\begin{aligned} d\psi &= \left(f_x(x, y)dx + f_y(x, y)dy\right)dy \\ &= A(x, y)dx\, dy + f_y(x, y)dy\, dy \\ &= A(x, y)dx\, dy = \varphi \end{aligned}$$

as required.

Of course, there was nothing special about integrating with respect to x. We could instead have set

$$g(x, y) = \int A(x, y)dy$$

and

$$\psi = -g(x, y)dx$$

(note the minus sign). Then once again

$$\begin{aligned} d\psi &= -\left(g_x(x, y)dx + g_y(x, y)dy\right)dx \\ &= -g_x(x, y)dx\, dx - A(x, y)dy\, dx \\ &= A(x, y)dx\, dy = \varphi \end{aligned}$$

as required.

(2) The case $n = 3$ is just as easy. Here we have a 3-form $\varphi = A(x, y, z)dx\, dy\, dz$, and, being a 3-form on $\mathbb{R}^3$, it is automatically closed. If

$$f(x, y, z) = \int A(x, y, z)dx,$$

then
$$\psi = f(x, y, z)dy\, dz$$

satisfies $d\psi = \varphi$, and so is our desired 2-form.
(Note also that if

$$g(x, y, z) = \int A(x, y, z)dy,$$

then $\psi = g(x, y, z)dz\, dx$ is another solution of $d\psi = \varphi$; and if

$$h(x, y, z) = \int A(x, y, z)dz,$$

then $\psi = h(x, y, z)dx\, dy$ is yet another solution of $d\psi = \varphi$.)

You may wonder why this method produced very different-looking solutions to our problem. But this is only to be expected, for again, by Lemma 1.2.11, we may add a closed 1-form to any solution and still get a solution. Consider, for example, the $n = 2$ case. If we set $\psi_1 = f(x, y)dy$ and $\psi_2 = -g(x, y)dx$, then

$$\psi_2 = \psi_1 + \left(-g(x, y)dx - f(x, y)dy\right)$$

and, indeed, $\rho = -g(x, y)dx - f(x, y)dy$ is closed. (Check for yourself that $d\rho = 0$.)   ◇

EXAMPLE 1.2.30.

(1) Let $\varphi = (12x^2y^3 + 2y)dx\, dy$. If

$$f(x, y) = \int (12x^2y^3 + 2y)dx$$
$$= 4x^3y^3 + 2xy,$$

then
$$\psi = (4x^3y^3 + 2xy)dy$$

satisfies $d\psi = \varphi$.
(Check: $d\psi = \left((12x^2y^3 + 2y)dx + (12x^3y^2 + 2x)dy\right) dy =$
$(12x^2y^3 + 2y)dx\, dy = \varphi$.)

As a second answer, we let

$$g(x, y) = \int (12x^2y^3 + 2y)dy$$
$$= 3x^2y^4 + y^2$$

and

$$\psi = -(3x^2y^4 + y^2)dx$$

satisfies $d\psi = \varphi$.
(Check: $d\psi = -(6xy^4dx + (12x^2y^3 + 2y)dy)dx = -(12x^2y^3 + 2y)dy\,dx = (12x^2y^3 + 2y)dx\,dy = \varphi$.)

(2) Let $\varphi = (2xy^2z^3 + 6x^2y^3z - 4x^3yz^2)dx\,dy\,dz$.
Set

$$f(x, y, z) = \int (2xy^2z^3 + 6x^2y^3z - 4x^3yz^2)dx.$$

Then $\psi(x, y, z) = (x^2y^2z^3 + 2x^3y^3z - x^4yz)dy\,dz$ satisfies $d\psi = \varphi$. $\diamond$

We next consider the case, $k = 2$ and $n = 3$, i.e., the case of a closed 2-form in $\mathbb{R}^3$. As opposed to the cases $k = 1$ and $k = n$, which we were able to handle in general, our method in this case is particular to $\mathbb{R}^3$. But because of the special importance of $\mathbb{R}^3$, we give it here. We work out an illustrative example instead of a general formula.

EXAMPLE 1.2.31. Let $\varphi = xy^2z\,dy\,dz - y^3z\,dz\,dx + (x^2y + y^2z^2)\,dx\,dy$. (You may check that φ is closed.) We look for a 1-form ψ with $d\psi = \varphi$. To make ψ as simple as possible, we look for a 1-form ψ that does not have a dz term. We set

$$\psi_1 = P(x, y, z)dx + Q(x, y, z)dy,$$

where $P(x, y, z)$ and $Q(x, y, z)$ are functions we must find. (You may reasonably ask why we can expect to find this sort of 1-form. How do we know we won't need a term in dz? Here is where we *use* the fact (from Lemma 1.2.11) that ψ is *not* unique, so that if we first found a ψ involving dz, we could get rid of the dz term. Since we can get rid of it, we look for a ψ not having it in the first place.)

We write ψ_1 instead of ψ, as this turns out not to be the final answer. Then $d\psi_1 = -Q_z\,dy\,dz + P_z\,dz\,dx + (-P_y + Q_x)dx\,dy$. The last term is complicated, so let's not worry about it for the moment, and just try to get the first two right. This gives us the following equations:

$$-Q_z = xy^2z \quad \text{and} \quad P_z = -y^3z$$

so

$$Q = -\int xy^2z\,dz = -xy^2z^2/2 \quad \text{and}$$

$$P = \int -y^3z\,dz = -y^3z^2/2.$$

Let

$$\psi_1 = -\frac{y^3z^2}{2}dx - \frac{xy^2z^2}{2}dy.$$

We compute

$$\begin{aligned}
d\psi_1 &= \left(-(3y^2z^2/2)dy - y^3z\,dz\right)dx \\
&\quad + \left(-(y^2z^2/2)dx - xyz^2\,dy - xy^2z\,dz\right)dy \\
&= xy^2z\,dy\,dz - y^3z\,dz\,dx + y^2z^2\,dx\,dy = \varphi_1.
\end{aligned}$$

As expected, the terms involving $dy\,dz$ and $dz\,dx$ are right, but the term involving $dx\,dy$ isn't. Now we set

$$\varphi_2 = \varphi - \varphi_1 = x^2y\,dx\,dy.$$

We look for a 1-form ψ_2 with $d\psi_2 = \varphi_2$, for then, if $\psi = \psi_1 + \psi_2$, we obtain $d\psi = d(\psi_1 + \psi_2) = d\psi_1 + d\psi_2 = \varphi_1 + \varphi_2 = \varphi_1 + (\varphi - \varphi_1) = \varphi$, as required.

But ψ_2 is easy to find, as this is exactly a situation we've considered before. Namely, φ_2 is a 2-form on $\mathbb{R}^2$. As above, we let

$$f(x, y) = \int x^2y\,dx = x^3y/3$$

and set

$$\psi_2 = (x^3y/3)dy.$$

Then our desired 1-form ψ here is

$$\psi = \psi_1 + \psi_2 = -\frac{y^3 z^2}{2} dx + \left(-\frac{xy^2 z^2}{2} + \frac{x^3 y}{3} \right) dy. \qquad \diamond$$

1.3 The fundamental correspondence

We have been talking about differential forms. Let us now see how to translate between this language and the classical language of vector fields. This translation will be via the fundamental correspondence. *We will see that this works fully only in* $\mathbb{R}^3$. But we will also see that this partially works in $\mathbb{R}^n$ in general, and we will end by considering the general situation.

In this section we will usually drop the word "tangent" and simply refer to vector fields, as is classically done.

Recall we introduced the notion of a tangent vector field $\mathbf{v}$ on $\mathbb{R}^3$ in Section 1.0. We first begin by making this notion more precise.

DEFINITION   1.3.1. Let $f^1(x_1, \ldots, x_n), f^2(x_1, \ldots, x_n), \ldots,$ $f^n(x_1, \ldots, x_n)$ be smooth functions defined on a region $\mathcal{R}$ of $\mathbb{R}^n$. Then

$$\mathbf{v} = f^1(x_1, \ldots, x_n)\mathbf{e}^1 + f^2(x_1, \ldots, x_n)\mathbf{e}^2 + \cdots + f^1(x_1, \ldots, x_n)\mathbf{e}^n$$

is a *smooth tangent vector field* on $\mathcal{R}$. $\qquad \diamond$

EXAMPLE 1.3.2.

(1) $\mathbf{F} = x^2\mathbf{i} + (yz + 1)\mathbf{k}$ is a smooth vector field on $\mathcal{R} = \mathbb{R}^3$.

(2) $\mathbf{G} = (xy/z)\mathbf{i} + xe^y\mathbf{j} + 2\mathbf{k}$ is a smooth vector field on $\mathcal{R} = \mathbb{R}^3 -$ the z-axis. $\qquad \diamond$

There are basic operations connecting functions and tangent vector fields.

DEFINITION 1.3.3.

(1) Let $f = f(x_1, x_2, \ldots, x_n)$ be a smooth function on a region $\mathcal{R}$ in $\mathbb{R}^n$. Its *gradient* **grad**(f) is the tangent vector field on $\mathcal{R}$ given by

$$\mathbf{grad}(f) = f_{x_1}\mathbf{e}^1 + f_{x_2}\mathbf{e}^2 + \cdots + f_{x_n}\mathbf{e}^n.$$

(2) Let $\mathbf{F} = f^1(x_1, \ldots, x_n)\mathbf{e}^1 + f^2(x_1, \ldots, x_n)\mathbf{e}^2 + \cdots + f^n(x_1, \ldots, x_n)\mathbf{e}^n$ be a smooth tangent vector field on $\mathcal{R}$. Its *divergence* div($\mathbf{F}$) is the smooth function on $\mathcal{R}$ given by

$$\text{div}(\mathbf{F}) = f^1_{x_1}(x_1, \ldots, x_n) + f^2_{x_2}(x_1, \ldots, x_n)$$
$$+ \cdots + f^n_{x_n}(x_1, \ldots, x_n). \qquad \diamond$$

Note that the gradient and divergence are defined for a region in $\mathbb{R}^n$ for any n. The next operation we introduce, the curl, is only defined for $n = 3$. In order to establish uniformity of notation, we will redefine the gradient and divergence in this case.

DEFINITION 1.3.4. Let $\mathcal{R}$ be a region in $\mathbb{R}^3$.

(1) Let A be a smooth function on $\mathcal{R}$. Its *gradient* $\mathbf{grad}(A)$ is the vector field on $\mathcal{R}$ given by

$$\mathbf{grad}(A) = A_x\mathbf{i} + A_y\mathbf{j} + A_z\mathbf{k}.$$

(2) Let $\mathbf{F} = A\mathbf{i} + B\mathbf{j} + C\mathbf{k}$ be a smooth vector field on $\mathcal{R}$. Its *curl* is the smooth vector field on $\mathcal{R}$ given by

$$\mathbf{curl}(\mathbf{F}) = (C_y - B_z)\mathbf{i} + (A_z - C_x)\mathbf{j} + (B_x - A_y)\mathbf{k}.$$

(3) Let $\mathbf{F} = A\mathbf{i} + B\mathbf{j} + C\mathbf{k}$ be a smooth vector field on $\mathcal{R}$. Its *divergence* is the smooth function on $\mathcal{R}$ given by

$$\text{div}(\mathbf{F}) = A_x + B_y + C_z. \qquad \diamond$$

EXAMPLE 1.3.5.

(1) If $A = (x^2 + y^3)/z$,

$$\mathbf{grad}(A) = (2x/z)\mathbf{i} + (3y^2/z)\mathbf{j} - \left((x^2 + y^3)/z^2\right)\mathbf{k}.$$

(2) If $\mathbf{F} = (x^2 + y^3z)\mathbf{i} + (y^2 - 3xz)\mathbf{j} + (x^4 + y^3 - z^2)\mathbf{k}$,

$$\mathbf{curl}(\mathbf{F}) = (3y^2 - 2x)\mathbf{i} + (y^3 - 4x^3)\mathbf{j} + (-3y^2 - 2z)\mathbf{k}.$$

(3) If $\mathbf{G} = (x^2 + y^3 + z^4)\mathbf{i} + (x^2 y^3 z^4)\mathbf{j} + (x + 2y + 3z + 1)\mathbf{k}$,

$$\text{div}(\mathbf{G}) = 2x + 3x^2 y^2 z^4 + 3. \qquad \diamond$$

We now give the Fundamental Correspondence.

DEFINITION 1.3.6 (*The Fundamental Correspondence in* $\mathbb{R}^3$). Let $\mathcal{R}$ be a region in $\mathbb{R}^3$ and assume that all of the functions/vector fields below are smooth on $\mathcal{R}$.

(0) The 0-form $\varphi = A$ corresponds to the function A.

(1) The 1-form $\varphi = A\,dx + B\,dy + C\,dz$ corresponds to the vector field $\mathbf{F} = A\mathbf{i} + B\mathbf{j} + C\mathbf{k}$.

(2) The 2-form $\varphi = A\,dy\,dz + B\,dz\,dx + C\,dx\,dy$ corresponds to the vector field $\mathbf{F} = A\mathbf{i} + B\mathbf{j} + C\mathbf{k}$.

(3) The 3-form $\varphi = A\,dx\,dy\,dz$ corresponds to the function A.

In cases (0) and (3) (resp. (1) and (2)), we will write $\varphi \longleftrightarrow A$ (resp. $\varphi \longleftrightarrow \mathbf{F}$). $\qquad \diamond$

Definition 1.3.6 gives a pretty simple and direct correspondence. Its utility, however, comes from the fact that it also involves exterior differentiation and the operations grad, curl, and div of Definition 1.3.4. We state this in standard mathematical language. Since this language is a bit cryptic, we will then explain what we mean.

THEOREM 1.3.7. *Let* $\longleftrightarrow$ *denote the fundamental correspondence and* d *denote exterior differentiation. Then the following diagram commutes*:

$$
\begin{array}{ccc}
\{0 - forms\} & \longleftrightarrow & \{functions\} \\
d \downarrow & & \downarrow grad \\
\{1 - forms\} & \longleftrightarrow & \{vector\ fields\} \\
d \downarrow & & \downarrow curl \\
\{2 - forms\} & \longleftrightarrow & \{vector\ fields\} \\
d \downarrow & & \downarrow div \\
\{3 - forms\} & \longleftrightarrow & \{functions\}.
\end{array}
$$

Here we assume that all functions, vector fields, and differential forms are smooth on some region $\mathcal{R}$ in $\mathbb{R}^3$.

To say that a diagram commutes is to say that the result of following a path of arrows from any point of the diagram to any other point of the diagram only depends on the starting and ending points and not on the particular path chosen. Thus, the commutativity of the top, middle, and bottom squares says the following:

(0) If the 0-form $\varphi \longleftrightarrow$ the function A, then the 1-form $d\varphi \longleftrightarrow$ the vector field $\mathbf{grad}(A)$.

(1) If the 1-form $\varphi \longleftrightarrow$ the vector field $\mathbf{F}$, then the 2-form $d\varphi \longleftrightarrow$ the vector field $\mathbf{curl}(\mathbf{F})$.

(2) If the 2-form $\varphi \longleftrightarrow$ the vector field $\mathbf{F}$, then the 3-form $d\varphi \longleftrightarrow$ the function $\mathrm{div}(\mathbf{F})$.

Proof.

(0) Let $\varphi = A$, the 0-form corresponding to the function A. From Proposition 1.2.4 we see that $d\varphi = A_x\, dx + A_y\, dy + A_z\, dz$, while from Definition 1.3.4 we see that $\mathbf{grad}(A) = A_x\mathbf{i} + A_y\mathbf{j} + A_z\mathbf{k}$. Then, from Definition 1.3.6, we see that $d\varphi$ and $\mathbf{grad}(A)$ correspond.

(1) Let $\varphi = A\, dx + B\, dy + C\, dz$, the 1-form corresponding to the vector field $\mathbf{F} = A\mathbf{i} + B\mathbf{j} + C\mathbf{k}$. From Proposition 1.2.4 we see that $d\varphi = (C_y - B_z)dy\, dz + (A_z - C_x)dz\, dx + (B_x - A_y)dx\, dy$, while from Definition 1.3.4 we see that $\mathbf{curl}(\mathbf{F}) = (C_y - B_z)\mathbf{i} + (A_z - C_x)\mathbf{j} + (B_x - A_y)\mathbf{k}$. Then, from Definition 1.3.6, we see that $d\varphi$ and $\mathbf{curl}(\mathbf{F})$ correspond.

(2) Let $\varphi = A\, dy\, dz + B\, dz\, dx + C\, dx\, dy$, the 2-form corresponding to the vector field $\mathbf{F} = A\mathbf{i} + B\mathbf{j} + C\mathbf{k}$. From Proposition 1.2.4 we see that $d\varphi = (A_x + B_y + C_z)dx\, dy\, dz$, while from Definition 1.3.4 we see that $\mathrm{div}(\mathbf{F}) = A_x + B_y + C_z$. Then, from Definition 1.3.6, we see that $d\varphi$ and $\mathrm{div}(\mathbf{F})$ correspond. $\square$

EXAMPLE 1.3.8.

(0) Let $\varphi = (x^2+y^3)/z \longleftrightarrow A = (x^2+y^3)/z$. In Example 1.2.3 we computed that $d\varphi = (2x/z)dx + (3y^2/z)dy - \left((x^2+y^3)/z^2\right)dz$, and in Example 1.3.5 we computed that $\mathbf{grad}(A) = (2x/z)\mathbf{i} + (3y^2/z)\mathbf{j} - \left((x^2+y^3)/z^2\right)\mathbf{k}$, and these correspond.

(1) Let $\varphi = (x^2 + y^3z)dx + (y^2 - 2xz)dy + (x^4 + y^3 - z^2)dz$. Then $\varphi \longleftrightarrow \mathbf{F} = (x^2 + y^3z)\mathbf{i} + (y^2 - 2xz)\mathbf{j} + (x^4 + y^3 - z^2)\mathbf{k}$. In Example 1.2.3 we computed that $d\varphi = (3y^2 - 3x)dy\,dz + (y^3 - 4x^3)dz\,dx + (-3y^2 - 2z)dx\,dy$, and in Example 1.3.5 we computed that $\mathbf{curl}(\mathbf{F}) = (3y^2 - 2x)\mathbf{i} + (y^3 - 4x^3)\mathbf{j} + (-3y^2 - 2z)\mathbf{k}$, and these correspond.

(2) Let $\varphi = (x^2 + y^3 + z^4)dydz + (x^2y^3z^4)dz\,dx + (x + 2y + 3z + 1)dx\,dy$. Then $\varphi \longleftrightarrow \mathbf{F} = (x^2 + y^3 + z^4)\mathbf{i} + (x^2y^3z^4)\mathbf{j} + (x + 2y + 3z + 1)\mathbf{k}$. In Example 1.2.3 we computed that $d\varphi = (2x + 3x^2y^2z^4 + 3)dx\,dy\,dz$, and in Example 1.3.5 we computed that $\mathrm{div}(\mathbf{F}) = 2x + 3x^2y^2z^4 + 3$, and these correspond.   ◇

REMARK 1.3.9. We would like to emphasize the importance of Theorem 1.3.7. While its proof is quite easy—just calculation—its implications are important. It unites the disparate operations of grad, curl, and divergence, and, using the fundamental correspondence, shows they are all special cases of exterior differentiation.   ◇

One consequence of this theorem is a pair of well-known formulas.

COROLLARY 1.3.10. *Let $\mathcal{R}$ be a region in $\mathbb{R}^3$.*

(1) For any smooth function A on $\mathcal{R}$, $\mathbf{curl}\left(\mathbf{grad}(A)\right) = 0$.

(2) For any smooth vector field $\mathbf{F}$ on $\mathcal{R}$, $\mathrm{div}\left(\mathbf{curl}(\mathbf{F})\right) = 0$.

Proof. Using the fundamental correspondence, this is just the relation $d^2 = 0$. To be precise:

(1) Let $\varphi \longleftrightarrow A$. Then $d\varphi \longleftrightarrow \mathbf{grad}(A)$, so $d(d\varphi) \longleftrightarrow$ $\mathbf{curl}\big(\mathbf{grad}(A)\big)$. But $d(d\varphi) = 0$ by Theorem 1.2.7, and so $\mathbf{curl}\big(\mathbf{grad}(A)\big) = 0$.

(2) Let $\varphi \longleftrightarrow \mathbf{F}$. Then $d\varphi \longleftrightarrow \mathbf{curl}(\mathbf{F})$, and so $d(d\varphi) \longleftrightarrow$ $\mathrm{div}\big(\mathbf{curl}(\mathbf{F})\big)$. But $d(d\varphi) = 0$ by Theorem 1.2.7, and so $\mathrm{div}\big(\mathbf{curl}(\mathbf{F})\big) = 0$. $\qquad\square$

REMARK 1.3.11. The fundamental correspondence also gives a mathematical translation of another physical term.

A vector field $\mathbf{F}$ on a region $\mathcal{R}$ in $\mathbb{R}^3$ is called *conservative* if $\mathbf{F} = \mathbf{grad}(f)$ for some function f on R. The function f is then called a *potential* for $\mathbf{F}$. By the fundamental correspondence and Theorem 1.3.7, we see that this is equivalent to the following. Let the 1-form $\varphi \longleftrightarrow \mathbf{F}$. Then $\mathbf{F}$ is conservative if $\varphi = df$, i.e., if φ is exact. Now recall from Corollary 1.2.14 that every exact form is closed. Thus here φ is closed, i.e., $d\varphi = 0$. By the fundamental correspondence, $d\varphi \longleftrightarrow \mathbf{curl}(\mathbf{F})$, so here $\mathbf{curl}(\mathbf{F}) = 0$. Putting this all together, we see that if $\mathbf{F}$ is a conservative vector field, then $\mathbf{curl}(\mathbf{F}) = 0$.

On the other hand, suppose that $\mathbf{curl}(\mathbf{F}) = 0$. Then $\mathbf{F}$ is often but not always conservative. Let us suppose that we are in a case where $\mathbf{F}$ is conservative, and consider the problem of finding a potential for $\mathbf{F}$. In fact, this is a problem we have essentially already solved! That is, suppose $\mathbf{F}$ is a conservative vector field. By definition, this means that $\mathbf{F} \longleftrightarrow \varphi$ where the form φ is exact. Then $\varphi = df$ for some function f, and then f is a potential for $\mathbf{F}$. We are thus looking for f with $\varphi = df$, i.e., for a primitive f of φ. (Again we have a translation: A potential for the conservative vector field $\mathbf{F}$ is the same as a primitive for the corresponding exact 1-form φ.) But recall that we developed a method for finding f in the last section (Method 1.2.23). $\qquad\diamond$

Let us do a physically important example here.

EXAMPLE 1.3.12. Let $\mathbf{F}(x, y, z)$ be the conservative vector field

$$\mathbf{F}(x, y, z) = a\left[\frac{x}{(x^2 + y^2 + z^2)^{3/2}}\mathbf{i} + \frac{y}{(x^2 + y^2 + z^2)^{3/2}}\mathbf{j}\right.$$
$$\left. + \frac{z}{(x^2 + y^2 + z^2)^{3/2}}\mathbf{k}\right].$$

This is an inverse square vector field with respect to the origin and represents, for example, the electrical attraction or repulsion exerted on a charged particle at the point (x, y, z) by another charged particle located at the origin. Note we may write

$$\mathbf{F}(x, y, z) = \frac{a}{\|(x, y, z)\|^3} \begin{bmatrix} x \\ y \\ z \end{bmatrix},$$

where $\|(x, y, z)\|$ denotes the distance from the origin $(0, 0, 0)$ to the point (x, y, z).

Thus we see that $\mathbf{F}$ is directed along the radius from $(0, 0, 0)$ to (x, y, z); furthermore $\mathbf{F}\left(t(x, y, z)\right) = \frac{1}{t^2}\mathbf{F}(x, y, z)$ (check this for yourself), so the force falls off with the inverse square of the distance of (x, y, z) from the origin.

We wish to find a potential for $\mathbf{F}$. To do so, we first use the fundamental correspondence to get a 1-form ω^1 with $\omega^1 \longleftrightarrow \mathbf{F}$. This ω is simply the 1-form of Example 1.2.21(2). In that example we found a function A with $\omega^1 = dA$. Thus the function $f = A$ of Example 1.2.21(2), given by

$$f(x, y, z) = \frac{-a}{(x^2 + y^2 + z^2)^{1/2}} = \frac{-a}{\|(x, y, z)\|},$$

is a potential for $\mathbf{F}$.   ◇

Let us now bring the Hodge $*$-operator on $\mathbb{R}^3$ into the picture.

First, using terms of the $*$-operator, we have an alternative way of expressing the fundamental correspondence. (You can easily check for yourself that this definition agrees with our previous definition.)

DEFINITION 1.3.13 (*The Fundamental Correspondence in $\mathbb{R}^3$*).

(0) The 0-form $\varphi = A$ corresponds to the function A.

(1) The 1-form $\varphi = A\,dx + b\,dy + c\,dz$ corresponds to the vector field $\mathbf{F} = A\mathbf{i} + B\mathbf{j} + C\mathbf{k}$.

(2) The 2-form $\varphi = A\,dy\,dz + B\,dz\,dx + C\,dx\,dy$ corresponds to the vector field $\mathbf{F}$ that corresponds to the 1-form φ^* as in (1).

(3) The 3-form $\varphi = A\,dx\,dy\,dz$ corresponds to the function A that corresponds to the 0-form φ^* as in (0). ◇

Second, we can express the usual vector operations in $\mathbb{R}^3$ in terms of the $*$-operator and the fundamental correspondence.

LEMMA 1.3.14. *Let the 1-form α correspond to the constant vector field $\mathbf{v}$ on $\mathbb{R}^3$, and let the 1-form β correspond to the constant vector field $\mathbf{w}$ on $\mathbb{R}^3$.*

(0) The constant 0-form $\left(\alpha\beta^*\right)^*$ corresponds to the constant function that is the dot product $\mathbf{v} \cdot \mathbf{w}$ at every point of $\mathbb{R}^3$.

(1) The constant 1-form $\left(\alpha\beta\right)^*$ corresponds to the constant vector field that is the cross product $\mathbf{v} \times \mathbf{w}$ at every point of $\mathbb{R}^3$.

Proof. We leave this proof as an exercise. □

While the full fundamental correspondence only works in $\mathbb{R}^3$, it is easy to generalize part of it to $\mathbb{R}^n$.

DEFINITION 1.3.15 (*The Fundamental Correspondence in $\mathbb{R}^n$*).
Let $\mathcal{R}$ be a region in $\mathbb{R}^n$ and assume that all of the functions/vector fields below are smooth on $\mathcal{R}$.

(0) The 0-form $\varphi = A$ corresponds to the function A.

(1) The 1-form $\varphi = \sum_{i=1}^{n} A^i dx_i$ corresponds to the vector field $\mathbf{F} = \sum_{i=1}^{n} A^i \mathbf{e}^i$.

(2) The $(n-1)$-form $\varphi = \sum_{i=1}^{n} (-1)^{i-1} A^i dx_1 \cdots \widehat{dx_i} \cdots dx_n$ corresponds to the vector field $\mathbf{F} = \sum_{i=1}^{n} A^i \mathbf{e}^i$. (Here $\widehat{dx_i}$ means that dx_i is omitted.)

(3) The n-form $\varphi = A dx_1 \cdots dx_n$ corresponds to the function A.

In cases (0) and (3) (resp. (1) and (2)), we will write $\varphi \longleftrightarrow A$ (resp. $\varphi \longleftrightarrow \mathbf{F}$). ◇

We can also define the fundamental correspondence in $\mathbb{R}^n$ by using the Hodge $*$-operator.

DEFINITION 1.3.16 (*The fundamental correspondence in* $\mathbb{R}^n$).

(0) The 0-form $\varphi = A$ corresponds to the function A.

(1) The 1-form $\varphi = \sum_{i=1}^n A^i \, dx_i$ corresponds to the vector field $\mathbf{F} = \sum_{i=1}^n A^i \mathbf{e}^i$.

(2) The $(n-1)$-form φ corresponds to the vector field $\mathbf{F}$ that corresponds to the 1-form φ^* as in (1).

(3) The n-form φ corresponds to the function A that corresponds to the 0-form φ^* as in (0).   ◇

REMARK 1.3.17. Note the alternation of signs in Definition 1.3.15(2). We avoided this in the case of $\mathbb{R}^3$ by writing the second term of a 2-form as $B \, dz \, dx$ (rather than $-B \, dx \, dz$).

We note that there is no apparent alternation of signs in Definition 1.3.16. But there is a hidden alternation, as if φ is the 1-form $\varphi = \sum_{i=1}^n A^i \, dx_i$, then φ^* is the $(n-1)$-form $\varphi^* = \sum_{i=1}^n (-1)^{i-1} A^i \, dx_1 \cdots \widehat{dx_i} \cdots dx_n$.   ◇

THEOREM 1.3.18. *Let* $\longleftrightarrow$ *denote the fundamental correspondence and d denote exterior differentiation. Then each of the following diagrams commutes:*

$$\{0 - forms\} \longleftrightarrow \{functions\}$$
$$d \downarrow \qquad\qquad\qquad \downarrow grad$$
$$\{1 - forms\} \longleftrightarrow \{vector\ fields\}$$

and

$$\{(n-1) - forms\} \longleftrightarrow \{vector\ fields\}$$
$$d \downarrow \qquad\qquad\qquad \downarrow div$$
$$\{n - forms\} \qquad \longleftrightarrow \{functions\}.$$

Here we assume that all functions, vector fields, and differential forms are smooth on some region $\mathcal{R}$ in $\mathbb{R}^3$.

Proof. Exactly the same as the proof of Theorem 1.3.7.   □

REMARK 1.3.19. $\mathbb{R}^3$ has two peculiarities: Vector calculus *only* works in $\mathbb{R}^3$, not in $\mathbb{R}^n$ for $n \neq 3$, and the cross product exists *only* for vectors in $\mathbb{R}^3$, not for vectors in $\mathbb{R}^n$ for any $n \neq 3$. These peculiarities of $\mathbb{R}^3$ are *explained* by Definition 1.3.13 and Lemma 1.3.14.

Suppose we take any $n \neq 3$ and try to mimic these constructions. If we begin with a vector field in $\mathbb{R}^n$, which corresponds to a 1-form, and try to define something analogous to its curl, we would obtain something that, under the correspondence of Definition 1.3.13, would correspond to an $(n-2)$-form, not to another 1-form corresponding to another vector field. Similarly, if we begin with a pair of vector fields in $\mathbb{R}^n$ and try to define something analogous to their cross product as in Lemma 1.3.14, we again obtain an $(n-2)$-form, rather than a 1-form corresponding to a vector field.

On the other hand, as we have seen, the differential algebra of differential forms, i.e., differential forms, their product, and their exterior derivative, works equally well in $\mathbb{R}^n$ for any value of n. ◇

1.4 The Converse of Poincaré's Lemma, I

Recall Poincare's Lemma "$d^2 = 0$" (Theorem 1.2.7), which had as an immediate consequence that every exact differential form is closed (Corollary 1.2.14). We have already begun to investigate the question of when a closed differential form must be exact. We saw in Proposition 1.2.19 that this is not always the case. However, it *is* the case in many important situations, and that is what we begin to prove now.

To be precise, we will state and prove a sequence of partial converses to Poincaré's Lemma, each of which is more general than the preceding one. We do this because the proofs involve different ideas, and because each proof requires less background than the following one. These converses are Theorem 1.4.1 in this section, Theorem 2.4.3 in Chapter 2, Corollary 3.5.11 in Chapter 3, and Corollary 4.7.6 in Chapter 4.

We wish to alert the reader to the fact that although the proof of Theorem 1.4.1 does not require much background, it is rather subtle.

THEOREM 1.4.1 (*Converse of Poincaré's Lemma*). *Let φ be a closed k-form defined on $\mathbb{R}^n$, $k > 0$. Then φ is exact.*

Proof. We prove this by induction, where the inductive hypothesis is cleverly (and carefully) chosen.

Let us denote by d_p exterior differentiation with respect to the first p variables $x_1, \ldots, x_p$ (so we regard $x_{p+1}, \ldots, x_n$ as constants). Thus, for example, for a function f,

$$d_p(f) = f_{x_1}dx_1 + \cdots + f_{x_p}dx_p.$$

Our inductive hypothesis is as follows:

Let φ be a k-form defined on $\mathbb{R}^n$ which involves only $dx_1, \ldots, dx_p$ (though it may involve all the variables $x_1, \ldots, x_n$). Suppose that $d_p(\varphi) = 0$. Then there is a $(k-1)$-form ψ with $\varphi = d_p(\psi)$. Furthermore, ψ involves only $dx_1, \ldots, dx_{p-1}$. Also, if for some $i > p$, φ does not involve the variable x_i, neither does ψ.

Our language here requires a bit of explanation. By φ not involving dx_j for some j we mean that when φ is written as in Definition 1.1.2, dx_j does not appear in any term $f dx_{i_1} \cdots dx_{i_k}$. By φ not involving x_j for some j we mean that when φ is written in this form, every function f is constant as a function of x_j, or, equivalently, that every f is a function of $x_1, \ldots, x_{j-1}, x_{j+1}, \ldots, x_n$.

Of course, we are only interested in the case $p = n$, for then $d_p = d_n = d$, and this gives the conclusion of the lemma. However, the intermediate steps are necessary for the proof, and the strengthened conclusions about ψ are necessary for it to work. (Also, this proof provides us with the method of finding ψ that we used in Example 1.2.31. Thus we may regard this proof as a vast elaboration of that method.)

We begin the induction with the case $p = 1$, which is easy. Then φ involves only dx_1, and so, in fact,

$$\varphi = A(x_1, \ldots, x_n)dx_1.$$

Define the function $B(x_1, \ldots, x_n)$ by

$$B(x_1, \ldots, x_n) = \int_0^{x_1} A(t, x_2, \ldots, x_n)dt.$$

Then, by the fundamental theorem of calculus, $\frac{\partial B}{\partial x_1} = A$. Thus, setting $\psi = B$, we see that $d_1\psi = \varphi$. (Note also that ψ does not involve any dx_i. Moreover, if A does not involve some x_i for $i > 1$, the same is true of B.)

Now for the inductive step. Assume our hypothesis is true for $p = m - 1$.

First let us establish some notation. Let d^m be exterior differentiation with respect to the variable x_m alone; e.g., for a function f, $d^m(f) = f_{x_m} dx_m$. Then

$$d_m = d_{m-1} + d^m,$$

by which we mean that for any form α, $d_m(\alpha) = d_{m-1}(\alpha) + d^m(\alpha)$.

Now we need to make a couple of observations for future use:

(1) For any m, $d_m(d_m(\alpha)) = 0$.
This is analogous to the basic property $d^2 = 0$ of exterior differentiation and the proof is the same. (Just forget about the variables $x_{m+1}, \ldots, x_n$.)

(2) If α is a form with $d_m(\alpha) = 0$ and α involves only $dx_1, \ldots, dx_{m-1}$, then also $d_{m-1}(\alpha) = 0$ and $d^m(\alpha) = 0$.

To see this, consider $d_m(\alpha) = d_{m-1}(\alpha) + d^m(\alpha)$. For a form α involving only $dx_1, \ldots, dx_{m-1}$, the form $d_{m-1}(\alpha)$ also involves only $dx_1, \ldots, dx_{m-1}$, while every term in $d^m(\alpha)$ involves dx_m. Thus, if $d_m(\alpha) = 0$, each of these summands must be zero (as there cannot be any cancellation between them).

With these preliminaries out of the way we get down to business. Suppose φ involves only $dx_1, \ldots, dx_m$ and $d_m(\varphi) = 0$. Let us write

$$\varphi = \varphi_{m-1} + \varphi^m$$

where φ_{m-1} consists of all terms in φ that *do not* involve dx_m (i.e., that involve only $dx_1, \ldots, dx_{m-1}$) and φ^m consists of all terms in φ that *do* involve dx_n.

By our definition of φ^m, we may write φ^m as

$$\varphi^m = dx_m(A^1 dx_{I_1} + A^2 dx_{I_2} + \cdots).$$

Set

$$B^i = \int_0^{x_m} A^i(x_1, \ldots, x_{m-1}, t, x_{m+1}, \ldots, x_n)dt.$$

Then, by the fundamental theorem of calculus again, $\frac{\partial B^i}{\partial x_m} = A^i$.

Next set

$$\psi^m = B^1 dx_{I_1} + B^2 dx_{I_2} + \cdots .$$

We see that

$$d^m(\psi^m) = \varphi^m$$

and that ψ^m involves only $dx_1, \ldots, dx_{m-1}$; we also see that if φ^m does not involve some variable x_i for $i > m$, neither does ψ^m.

Now consider

$$\tilde{\varphi} = \varphi - d_m(\psi^m).$$

We claim that $\tilde{\varphi}$ does not involve dx_m. Indeed, we chose ψ^m for that purpose. This follows immediately from our construction, but we will prove it explicitly:

$$\begin{aligned}
\tilde{\varphi} = \varphi - d_m(\psi^m) &= \varphi - (d_{m-1} + d^m)(\psi^m) \\
&= \varphi_{m-1} + \varphi^m - (d_{m-1} + d^m)(\psi^m) \\
&= \varphi_{m-1} + \varphi^m - d_{m-1}(\psi^m) - d^m(\psi^m) \\
&= \varphi_{m-1} - d_{m-1}(\psi^m),
\end{aligned}$$

and neither of the terms on the right-hand side involves dx_m.

Now

$$d_m(\tilde{\varphi}) = d_m(\varphi - d_m(\psi^m)) = d_m(\varphi) - d_m(d_m(\psi)) = 0,$$

as $d_m(\varphi) = 0$ by assumption and $d_m(d_m(\psi)) = 0$ by observation (1).

Since $d_m(\tilde{\varphi}) = 0$ and $\tilde{\varphi}$ involves only $dx_1, \ldots, dx_{m-1}$, then also

$$d_{m-1}(\tilde{\varphi}) = 0$$

by observation (2).

Also, by observation (2)

$$d^m(\tilde{\varphi}) = 0.$$

Let us see what this means. We have that $\tilde{\varphi}$ involves only $dx_1, \ldots,$ dx_{m-1}, so we may write it as

$$\tilde{\varphi} = \tilde{A}^1 dx_{I_1} + \tilde{A}^2 dx_{I_2} + \cdots$$

with $dx_{I_1}, dx_{I_2}, \ldots$ not involving dx_m. Then

$$0 = d^m(\tilde{\varphi}) = \tilde{A}^1_{x_m} dx_m dx_{I_1} + \tilde{A}^2_{x_m} dx_m dx_{I_2} + \cdots .$$

The thing to note here is that each of the strings $dx_{I_1}, dx_{I_2}, \ldots$ is distinct, and hence each of the strings $dx_m dx_{I_1}, dx_m dx_{I_1}, \ldots$ is also distinct. Thus there can be no cancellation between different terms, so each term must be zero; i.e.,

$$\tilde{A}^1_{x_m} = \tilde{A}^2_{x_m} = \cdots = 0.$$

In other words, $\tilde{\varphi}$ does not involve x_m.

We may now apply our inductive hypothesis to conclude that there is a form $\tilde{\psi}$ with

$$\tilde{\varphi} = d_{m-1}(\tilde{\psi})$$

and that, furthermore, $\tilde{\psi}$ involves only $dx_1, \ldots, dx_{m-2}$; in particular, $\tilde{\psi}$ only involves $dx_1, \ldots, dx_{m-1}$ and does not involve x_m, as $i = m > p = m - 1$.

But if $\tilde{\psi}$ does not involve x_m, then certainly

$$d^m(\tilde{\psi}) = 0,$$

and thus

$$d_m(\tilde{\psi}) = (d_{m-1} + d^m)(\tilde{\psi}) = d_{m-1}(\tilde{\psi}) + d^m(\tilde{\psi}) = \tilde{\varphi} + 0 = \tilde{\varphi}.$$

Now set

$$\psi = \tilde{\psi} + \psi^m.$$

We claim that ψ is our desired form. Clearly, ψ involves only $dx_1, \ldots,$ dx_{m-1}; and if φ does not involve some x_i for $i > m$, neither does ψ. What we need to check is the crucial equality $d_m(\psi) = \varphi$, which we do by direct computation:

$$
\begin{aligned}
d_m(\psi) &= d_m(\tilde{\psi} + \psi^m) \\
&= d_m(\tilde{\psi}) + d_m(\psi^m) \\
&= \tilde{\varphi} + d_m(\psi^m) \\
&= \tilde{\varphi}
\end{aligned}
$$

as required.

This completes the proof of the inductive step, and so the theorem follows.   □

REMARK 1.4.2. As a matter of terminology, there is disagreement among modern authors as to what should be called Poincaré's Lemma. Some call Theorem 1.2.7 Poincaré's Lemma, as we do here. Others call one of the partial converses Poincaré's Lemma. Our attribution here follows that of Elie Cartan, the developer of the theory of differential forms—he ought to know.   ◇

1.5   Exercises

In Exercises 1–9,

$$
\begin{aligned}
f_1 &= x^2 y^3 z - 2xyz^2, \\
f_2 &= x^2 + y^2 - 3z^4, \\
f_3 &= x^2 y + y^2 z - z^2 x, \\
f_4 &= 2x + 3y - 4z + 5, \\
\varphi_1 &= 2x^2\, dx + (x + y)dy, \\
\varphi_2 &= -x\, dx + (x - 2y)dy, \\
\varphi_3 &= x^3\, dx + yz\, dy - (x^2 + y^2 + z^2)dz, \\
\varphi_4 &= y^2 z\, dx - xz\, dy + (2x + 1)dz, \\
\psi_1 &= (x^2 - y^2)dx\, dy, \\
\psi_2 &= (x - y)dx\, dy, \\
\psi_3 &= (x^2 + y^2)dy\, dz + (x - y^2)dz\, dx + 3x\, dx\, dy, \\
\psi_4 &= (x^2 - y^2)dy\, dz + (x + y - z^2)dz\, dx - 6xy\, dx\, dy.
\end{aligned}
$$

1. Find: (a) $2\varphi_1+\varphi_2$, (b) $\varphi_1-x\varphi_2$, (c) $3\varphi_1-4\varphi_2$, (d) $x\varphi_1+y\varphi_2$.

2. Find: (a) $\varphi_1\varphi_2$, (b) $\varphi_1\varphi_3$, (c) $\varphi_1\varphi_4$, (d) $\varphi_2\varphi_3$, (e) $\varphi_2\varphi_4$, (f) $\varphi_3\varphi_4$.

3. Find: (a) $y\psi_1 + x^2\psi_2$, (b) $-\psi_1 + (x + y)\psi_2$.

4. Find: (a) $\varphi_1\psi_3$, (b) $\varphi_1\psi_4$, (c) $\varphi_2\psi_3$, (d) $\varphi_2\psi_4$, (e) $\varphi_3\psi_3$, (f) $\varphi_3\psi_4$, (g) $\varphi_4\psi_3$, (h) $\varphi_4\psi_4$.

5. Find: (a) df_1, (b) df_2, (c) df_3, (d) df_4.

6. Find: (a) $d\varphi_1$, (b) $d\varphi_2$, (c) $d\varphi_3$, (d) $d\varphi_4$.

7. Find: (a) $d\psi_3$, (b) $d\psi_4$.

8. Find: (a) $d(f_1\varphi_1)$, (b) $d(f_1\varphi_2)$, (c) $d(f_2\varphi_1)$, (d) $d(f_2\varphi_2)$, (e) $d(f_1\psi_1)$, (f) $d(f_1\psi_2)$, (g) $d(f_1\psi_3)$, (h) $d(f_1\psi_4)$, (i) $d(f_2\psi_1)$, (j) $d(f_2\psi_2)$, (k) $d(f_2\psi_3)$, (l) $d(f_2\psi_4)$, (m) $d(\varphi_1\varphi_2)$, (n) $d(\varphi_1\varphi_3)$, (o) $d(\varphi_1\varphi_4)$, (p) $d(\varphi_2\varphi_3)$, (q) $d(\varphi_2\varphi_4)$, (r) $d(\varphi_3\varphi_4)$, and verify that Theorem 1.2.5(2) is true here.

9. Find: (a) $d^2(f_1)$, (b) $d^2(f_2)$, (c) $d^2(f_3)$, (d) $d^2(f_4)$, (e) $d^2(\varphi_1)$, (f) $d^2(\varphi_2)$, (g) $d^2(\varphi_3)$, (h) $d^2(\varphi_4)$, and verify that Theorem 1.2.7 is true here.

10. Each of the following 1-forms φ is exact. Find a function A with $\varphi = dA$:

   (a) $\varphi = y^2dx + 2xy\,dy$, (b) $\varphi = (3x^2 + y)dx + (x + 2y)dy$,

   (c) $\varphi = 2x\,dx + 3y^2\,dy + 4z^3\,dz$, (d) $\varphi = 2xy\,dx + (x^2 + 2y + z)dy + (y - 3z^2)dz$.

11. Each of the following 2-forms φ is exact. Find a 1-form ψ with $\varphi = d\psi$:

   (a) $\varphi = 24x^3y^2\,dx\,dy$, (b) $\varphi = (6x^2y - 3xy^2)dx\,dy$,

   (c) $\varphi = 2yz\,dy\,dz - 2xz\,dz\,dx - 4xy\,dx\,dy$,

   (d) $\varphi = (3y^2 - 2z)dy\,dz + (1 - 3x^2)dz\,dx + (2x - 1)dx\,dy$.

12. Each of the following 3-forms φ is exact. Find a 2-form ψ with $\varphi = d\psi$:

    (a) $\varphi = 12x^2y^3z^4dx\,dy\,dz$,   (b) $\varphi = (x^3 - 4y + z^2)dx\,dy\,dz$,

    (c) $\varphi = (2xz + 3yz^2)dx\,dy\,dz$,   (d) $\varphi = 2(x + y + z)dx\,dy\,dz$.

13. The Laplacian of a C^2-function f on $\mathbb{R}^n$ is given by

$$\nabla^2 f = \frac{\partial^2 f}{\partial x_1^2} + \frac{\partial^2 f}{\partial x_2^2} + \cdots + \frac{\partial^2 f}{\partial x_n^2}.$$

    (a) Show that $\nabla^2 f = (*d)^2 f$; i.e., $\nabla^2 f = (*d*d)(f)$.

    (b) Show that $\nabla^2(fg) = f\nabla^2 g + g\nabla^2 f + 2\left((df)((dg)^*)\right)^*$.

14. A function f with $\nabla^2 f = 0$ is called *harmonic*. Show that each of the following functions is harmonic:

    (a) $f = ax + by + cz + d$ for any fixed a, b, c, d (i.e., any linear function),

    (b) $f = ax^2 + by^2 + cz^2$ for $a + b + c = 0$,

    (c) $f = x^3 - 3xy^2$,   (d) $f = (x^2 + y^2 + z^2)^{-1/2}$.

15. Coordinatize $\mathbb{R}^3$ by (u, v, w). Let β be the 1-form

$$\beta = \frac{-2uw^2}{(u^2 + v^2 - 4)^2 + w^4}du + \frac{-2vw^2}{(u^2 + v^2 - 4)^2 + w^4}dv$$
$$+ \frac{2w(u^2 + v^2 - 4)}{(u^2 + v^2 - 4)^2 + w^4}dw.$$

Note that β is defined on the complement of the circle $\{(u, v, 0)\mid u^2 + v^2 = 4\}$ in $\mathbb{R}^3$. Show that β is closed. (We will see later that β is not exact.)

16. Let φ^n be the n-form

$$\varphi^n = \sum_{i=1}^{n+1} \frac{(-1)^{i-1} x_i \, dx_1 \cdots \widehat{dx_i} \cdots dx_{n+1}}{\left(\sum_{i=1}^{n+1} x_i^2 \right)^{(n+1)/2}}.$$

Note that φ^n is defined on $\mathbb{R}^{n+1} - \{(0, \ldots, 0)\}$. Show that φ^n is closed. (We will see later that φ^n is not exact.)

2 Differential Forms in $\mathbb{R}^n$, II

In the last chapter we introduced differential forms in $\mathbb{R}^n$ as formal objects. In this chapter we see their true meaning. Again our discussion is general, but most of our examples will be drawn from $\mathbb{R}^1$, $\mathbb{R}^2$, or $\mathbb{R}^3$.

2.1   1-Forms

We begin by redefining 1-forms.

DEFINITION 2.1.1. Let $\mathcal{R}$ be a region in $\mathbb{R}^n$. A 1-*form* φ on $\mathcal{R}$ is a smooth linear function on the tangent spaces $T_p\mathbb{R}^n$ at each point p of $\mathcal{R}$. That is:

(1) (Linearity) For any fixed point $p \in \mathcal{R}$, $\varphi : T_p\mathbb{R}^n \longrightarrow \mathbb{R}$ is a linear function, i.e., $\varphi(c\mathbf{v}_p + d\mathbf{w}_p) = c\varphi(\mathbf{v}_p) + d\varphi(\mathbf{w}_p)$.

(2) (Smoothness) For any smooth vector field $\mathbf{F}$ on $\mathcal{R}$, the function $e : \mathcal{R} \longrightarrow \mathbb{R}$ given by $e(p) = \varphi(\mathbf{F}(p))$ is a smooth function.   ◇

Our first goal is to show that Definition 2.1.1 agrees with our previous definition of 1-forms. We now head toward that goal.

DEFINITION 2.1.2. The 1-form dx_i on $\mathbb{R}^n$ is defined as follows. For any point p in $\mathbb{R}^n$, dx_i is the linear function on $T_p\mathbb{R}^n$ given by:

$$dx_i(\mathbf{e}_p^i) = 1 \quad \text{and} \quad dx_i(\mathbf{e}_p^j) = 0 \text{ for } j \neq i.$$ ◇

REMARK 2.1.3. By linearity, this suffices to define dx_i, as a linear function on a vector space is determined by its values on a basis, and, for each p, $\{\mathbf{e}_p^1, \ldots, \mathbf{e}_p^n\}$ is a basis of $T_p\mathbb{R}^n$.

Differential Forms, Second Edition. http://dx.doi.org/10.1016/B978-0-12-394403-0.00002-5

Explicitly,

$$dx_i \left(\begin{bmatrix} a_1 \\ \vdots \\ a_n \end{bmatrix}_p \right) = a_i,$$

so dx_i "picks out" the ith entry of a tangent vector at every point. ◇

REMARK 2.1.4. In our notation in $\mathbb{R}^3$, for every point p in $\mathbb{R}^3$:

$$dx(\mathbf{i}_p) = 1, \ dx(\mathbf{j}_p) = 0, \ dx(\mathbf{k}_p) = 0;$$
$$dy(\mathbf{i}_p) = 0, \ dy(\mathbf{j}_p) = 1, \ dy(\mathbf{k}_p) = 0;$$
$$dz(\mathbf{i}_p) = 0, \ dz(\mathbf{j}_p) = 0, \ dz(\mathbf{k}_p) = 1.$$ ◇

THEOREM 2.1.5. *Let $\mathcal{R}$ be a region in $\mathbb{R}^n$.*
Let $A^1(x_1, \ldots, x_n), \ldots, A^1(x_1, \ldots, x_n)$ be smooth functions on $\mathcal{R}$.
Then

$$\varphi = A^1 \, dx_1 + \cdots + A^n \, dx_n$$

is a 1-form on $\mathcal{R}$. Conversely, every 1-form φ on $\mathcal{R}$ can be written in this way, for unique smooth functions $A^1, \ldots, A^n$ on $\mathcal{R}$.

Proof. This formula certainly defines a function on tangent vectors to points p in $\mathcal{R}$:

$$\varphi(\mathbf{v}_p) = A^1(p)dx_1(\mathbf{v}_p) + \cdots + A^n(p)dx_n(\mathbf{v}_p).$$

We must show linearity and smoothness.
   Linearity: Each function dx_i is linear on $T_p\mathbb{R}^n$, by definition. Then φ is linear on $T_p\mathbb{R}^n$, as it is linear combination of linear functions.
   Smoothness: Let $\mathbf{F}$ be the smooth vector field on $\mathcal{R}$ given by

$$\mathbf{F} = f^1\mathbf{e}^1 + \cdots + f^n\mathbf{e}^n,$$

i.e., by

$$\mathbf{F}(p) = f^1(p)\mathbf{e}_p^1 + \cdots + f^n(p)\mathbf{e}_p^n$$

for every point p.
   Then $dx_i(\mathbf{F}(p)) = f^i(p)$, so

$$e(p) = \varphi(\mathbf{F}(p)) = A^1(p)f^1(p) + \cdots + A^n(p)f^n(p).$$

Since each function A^i and each function f^i is smooth on $\mathcal{R}$, the function e is also smooth on $\mathcal{R}$.

Thus φ defined in this way is a 1-form.

Conversely, let φ be a 1-form on $\mathcal{R}$. For each i, define the function A^i by

$$A^i(p) = \varphi(e^i_p).$$

Now e^i is a constant vector field, so it is certainly a smooth vector field. Then, by part (2) of Definition 2.1.1, each A^i is smooth. The functions $A^1, \ldots, A^n$ are unique as each is determined by φ. Let

$$\psi = A^1\, dx_1 + \cdots + A^n\, dx_n.$$

As we observed, ψ is a linear function on each tangent space $T_p\mathbb{R}^n$. Furthermore, for any i,

$$\psi(e^i_p) = (A^1(p)dx_1 + \cdots + A^i(p)dx_i + \cdots + A^n(p)dx_n)(e^i_p)$$
$$= A^i(p) = \varphi(e^i_p).$$

Thus φ and ψ are linear functions that agree on the basis $\{e^1_p, \ldots, e^n_p\}$ of $T_p\mathbb{R}^n$, so $\varphi = \psi$ on $T_p\mathbb{R}^n$. Since this is true for every $p \in \mathcal{R}$, $\varphi = \psi$, as claimed. $\qquad\square$

We record part of the computation we have just done separately for future reference.

COROLLARY 2.1.6. *Let $\mathcal{R}$ be a region in $\mathbb{R}^n$. Let*

$$\varphi = A^1\, dx_1 + \cdots + A^n\, dx_n$$

be a 1-form on $\mathcal{R}$ and let

$$\mathbf{F} = f^1 \mathbf{e}^1 + \cdots + f^n \mathbf{e}^n$$

be a vector field on $\mathcal{R}$.

Then for any point $p \in \mathcal{R}$,

$$\varphi(\mathbf{F}(p)) = A^1(p)f^1(p) + \cdots + A^n(p)f^n(p).$$

Proof. Directly from the proof of Theorem 2.1.5. $\qquad\square$

EXAMPLE 2.1.7.

(1) Let $\mathcal{R}$ be $\mathbb{R}^3$. Let φ be the constant 1-form $\varphi = dx + 2\,dy - 3\,dz$, and let $\mathbf{v}$ be the constant vector field $\begin{bmatrix} 2 \\ 4 \\ 3 \end{bmatrix}$. Then at any point p of $\mathcal{R}$,

$$\begin{aligned}
\varphi(\mathbf{v}_p) &= (dx + 2\,dy - dz)(2\mathbf{i}_p + 4\mathbf{j}_p + 3\mathbf{k}_p) \\
&= 2\,dx(\mathbf{i}_p) + 4\,dx(\mathbf{j}_p) + 3\,dx(\mathbf{k}_p) \\
&\quad + 4\,dy(\mathbf{i}_p) + 8\,dy(\mathbf{j}_p) + 6\,dy(\mathbf{k}_p) \\
&\quad - 6\,dz(\mathbf{i}_p) - 12\,dz(\mathbf{j}_p) - 9\,dz(\mathbf{k}_p) \\
&= 2 \cdot 1 + 4 \cdot 0 + 3 \cdot 0 \\
&\quad + 4 \cdot 0 + 8 \cdot 1 + 6 \cdot 0 \\
&\quad - 6 \cdot 0 - 12 \cdot 0 - 9 \cdot 1 \\
&= 2 + 8 - 9 = 1.
\end{aligned}$$

(2) Let $\mathcal{R}$ be $\mathbb{R}^2 - (0,0)$. Let $\mathbf{v}$ be the vector field given by $\mathbf{v}_p = \begin{bmatrix} -y \\ x \end{bmatrix}_p$ for any point $p \in \mathcal{R}$. Let φ^1 be the 1-form of Example 1.2.17,

$$\varphi^1 = \frac{-y}{(x^2 + y^2)}dx + \frac{x}{(x^2 + y^2)}dy.$$

Then $\varphi^1(\mathbf{v}_p) = 1$ (at every point p).

On the other hand, if ω is the 1-form

$$\omega = \frac{x}{(x^2 + y^2)}dx + \frac{y}{(x^2 + y^2)}dy,$$

then $\omega(\mathbf{v}_p) = 0$ (at every point p). $\diamond$

Now that we have seen the geometric meaning of 1-forms, we can derive the geometric meaning of the exterior derivative of a function (the result of which is a 1-form).

THEOREM 2.1.8. *Let $\mathcal{R}$ be a region in $\mathbb{R}^n$ and let g be a smooth function on $\mathcal{R}$. Let p be any point of $\mathcal{R}$ and let $\mathbf{v}_p \in T_p\mathbb{R}^n$ be any* *tangent vector at p. Let* $r : (-\delta, \delta) \longrightarrow \mathcal{R}$ *be any smooth function with* $r(0) = p$ *and* $r'(0) = \mathbf{v}_p$, *where* δ *is any positive real number. Define the function* $e : (-\delta, \delta) \longrightarrow \mathbb{R}$ *by* $e(t) = g(r(t))$. *Then*

$$dg(\mathbf{v}_p) = e'(0).$$

Proof.   The proof is nothing other than an application of the (multivariable) chain rule, which we write out explicitly. Let $\mathbf{v}_p = \begin{bmatrix} a_1 \\ \vdots \\ a_n \end{bmatrix}_p$.

Then, by the chain rule, Corollary 2.1.6, and the definition of dg (Definition 1.2.1),

$$e'(0) = g'(r(0))r'(0) = g'(p)r'(0)$$

$$= \begin{bmatrix} g_{x_1}(p) \cdots g_{x_n}(p) \end{bmatrix} \begin{bmatrix} a_1 \\ \vdots \\ a_n \end{bmatrix}_p$$

$$= g_{x_1}(p)a_1 + \cdots + g_{x_n}(p)a_n$$

$$= (g_{x_1}(p)dx_1 + \cdots + g_{x_n}(p)dx_n)(\mathbf{v}_p)$$

$$= dg(p)(\mathbf{v}_p)$$

as claimed.

(Note that this theorem implicitly claims that $dg(\mathbf{v}_p)$ is independent of the choice of the function r (provided, of course, that $r(0) = p$ and $r'(0) = \mathbf{v}_p$), and we have shown that in the course of the proof.) $\square$

In fact Theorem 2.1.8 shows that the exterior derivative is closely related to an elementary construction.

DEFINITION 2.1.9. Let $\mathcal{R}$ be a region in $\mathbb{R}^n$ and let g be a smooth function on $\mathcal{R}$. Let p be any point of $\mathcal{R}$ and let $\mathbf{v}_p \in T_p\mathbb{R}^n$ be any tangent vector to $\mathbb{R}^n$ at p. Let $\mathbf{v}$ be the constant vector field on $\mathcal{R}$ whose value at p is $\mathbf{v}_p$. Then $D_{\mathbf{v}_p}(g)$, the *derivative of g along the vector* $\mathbf{v}_p$, is

$$D_{\mathbf{v}_p}(g) = lim_{t \to 0} \frac{g(p + t\mathbf{v}) - g(p)}{t}. \qquad \diamond$$

REMARK 2.1.10. In case $\mathbf{v}_p$ is a unit vector, this is just the directional derivative of g at p in the direction of $\mathbf{v}_p$. But this formula makes sense for any tangent vector $\mathbf{v}_p$, not just a unit vector.   ◇

COROLLARY 2.1.11. *Let $\mathcal{R}$ be a region in $\mathbb{R}^n$ and let g be a smooth function on $\mathcal{R}$. Let p be any point of $\mathcal{R}$ and let $\mathbf{v}_p \in T_p\mathbb{R}^n$ be any tangent vector at p. Then*

$$dg(\mathbf{v}_p) = D_{\mathbf{v}_p}(g).$$

Proof. By Theorem 2.1.8, we may use any smooth function r with $r(0) = p$ and $r'(0) = \mathbf{v}_p$ to compute $dg(\mathbf{v}_p)$. Choose $r(t) = p + t\mathbf{v}$, where $\mathbf{v}$ is the constant vector field on $\mathcal{R}$ whose value at p is $\mathbf{v}_p$ (so that the image of r is a straight line segment). We then have that

$$dg(\mathbf{v}_p) = e'(0),$$

where $e(t) = g(f(t))$. But in this case, by the definition of the derivative,

$$
\begin{aligned}
e'(0) &= lim_{t \to 0} \frac{e(t) - e(0)}{t} \\
&= lim_{t \to 0} \frac{g(p + t\mathbf{v}) - g(p)}{t} \\
&= D_{\mathbf{v}_p}(g).
\end{aligned}
$$
$\square$

REMARK 2.1.12. Note from Definition 1.2.1 that Corollary 2.1.11 gives us the meaning of the exterior derivative in general (as "all the action" in taking the exterior derivative of a form occurs in taking the exterior derivatives of the functions that appear in its expression.) This will also be reflected in Theorem 2.2.22, where we give a formula for the exterior derivative of a k-form.

We thus see that, from this point of view, we can regard the process of exterior differentiation as a vast generalization of that of taking directional derivatives.   ◇

REMARK 2.1.13. A linear function on tangent vectors is often called a *cotangent vector*. Thus in this language a 1-form on $\mathcal{R}$ is a smooth cotangent vector field on $\mathcal{R}$.   ◇

2.2 *k*-Forms

The definition of a k-form is quite analogous to the definition of a
1-form, though there is one new wrinkle.

DEFINITION 2.2.1. A k-form on a region $\mathcal{R}$ in $\mathbb{R}^n$ is a smooth multilinear alternating function on k-tuples of tangent vectors to $\mathbb{R}^n$, all
of which are based at the same point of $\mathcal{R}$.
    Thus, if φ is a k-form on M, and $\mathbf{v}_p^1, \ldots, \mathbf{v}_p^k$ are vectors all of which
are tangent to $\mathbb{R}^n$ at some point $p \in \mathcal{R}$, then $\varphi(\mathbf{v}_p^1, \ldots, \mathbf{v}_p^k)$ is defined.
Furthermore,

(1) (Multilinearity) If $\mathbf{v}_p^i = c\mathbf{u}_p + d\mathbf{w}_p$ then $\varphi(\mathbf{v}_p^1, \ldots, \mathbf{v}_p^i, \ldots,$
$\mathbf{v}_p^k) = c\varphi(\mathbf{v}_p^1, \ldots, \mathbf{u}_p, \ldots, \mathbf{v}_p^k) + d\varphi(\mathbf{v}_p^1, \ldots, \mathbf{w}_p, \ldots, \mathbf{v}_p^k)$.
(Thus multilinearity is linearity in each argument separately.)

(2) (Alternation) If $\mathbf{v}_p^i = \mathbf{v}_p^j$ for some $i \neq j$, then $\varphi(\mathbf{v}_p^1, \ldots, \mathbf{v}_p^i, \ldots,$
$\mathbf{v}_p^j, \ldots, \mathbf{v}_p^k) = 0$.

(3) (Smoothness) For any k smooth vector fields $\mathbf{F}_1, \ldots, \mathbf{F}^k$ on $\mathcal{R}$,
the function $e : \mathcal{R} \longrightarrow \mathbb{R}$ given by $e(p) = \varphi(\mathbf{F}_1, \ldots, \mathbf{F}^k)$ is a smooth
function.   ◇

    The alternation condition may seem odd at first glance, but there
is an important geometric reason for imposing it, which we will see
below. Note that for a 1-form, this condition is vacuous (and hence we
did not include it in our definition) as there are no distinct arguments
to be equal.
    Again our first goal is to show that Definition 2.2.1 agrees with our
previous definition of k-forms, and again we head toward that goal.
    Note that the following lemma shows why property (2) in Definition 2.2.1 is called alternation.

LEMMA 2.2.2.

*(1) Let α be a multilinear function on $T_p\mathbb{R}^n$. The following are
equivalent:*

(a) If $\mathbf{v}_p^i = \mathbf{v}_p^j$ for some $i \neq j$, then $\alpha(\mathbf{v}_p^1, \ldots, \mathbf{v}_p^k) = 0$.

(b) For any $i \neq j, \alpha(\mathbf{v}_p^1, \ldots, \mathbf{v}_p^i, \ldots, \mathbf{v}_p^j, \ldots, \mathbf{v}_p^k) = -\alpha(\mathbf{v}_p^1,$
$\ldots, \mathbf{v}_p^j, \ldots, \mathbf{v}_p^i, \ldots, \mathbf{v}_p^k)$. *(In other words, if any two arguments to* φ *are transposed, i.e., interchanged, the value of* φ *changes sign.)*

(2) Suppose that α *is multilinear and alternating. Let* σ *be any permutation (i.e., a reordering) of* $\{1, \ldots, k\}$. *Then* $\alpha(\mathbf{v}_p^{\sigma(1)}, \ldots,$
$\mathbf{v}_p^{\sigma(k)}) = \text{sign}(\sigma)\alpha(\mathbf{v}_p^1, \ldots, \mathbf{v}_p^k)$.

Proof.

(1) On the one hand, suppose that (a) is true. Set $\mathbf{w}_p = \mathbf{v}_p^i + \mathbf{v}_p^j$. Then by multilinearity,

$$0 = \alpha(\mathbf{v}_p^1, \ldots, \mathbf{w}_p, \ldots, \mathbf{w}_p, \ldots, \mathbf{v}_p^k)$$
$$0 = \alpha(\mathbf{v}_p^1, \ldots, \mathbf{v}_p^i + \mathbf{v}_p^j, \ldots, \mathbf{v}_p^i + \mathbf{v}_p^j, \ldots, \mathbf{v}_p^k)$$
$$0 = \alpha(\mathbf{v}_p^1, \ldots, \mathbf{v}_p^i, \ldots, \mathbf{v}_p^i, \ldots, \mathbf{v}_p^k)$$
$$+ \alpha(\mathbf{v}_p^1, \ldots, \mathbf{v}_p^i, \ldots, \mathbf{v}_p^j, \ldots, \mathbf{v}_p^k)$$
$$+ \alpha(\mathbf{v}_p^1, \ldots, \mathbf{v}_p^j, \ldots, \mathbf{v}_p^i, \ldots, \mathbf{v}_p^k)$$
$$+ \alpha(\mathbf{v}_p^1, \ldots, \mathbf{v}_p^j, \ldots, \mathbf{v}_p^j, \ldots, \mathbf{v}_p^k)$$
$$= 0 + \alpha(\mathbf{v}_p^1, \ldots, \mathbf{v}_p^i, \ldots, \mathbf{v}_p^j, \ldots, \mathbf{v}_p^k)$$
$$+ \alpha(\mathbf{v}_p^1, \ldots, \mathbf{v}_p^j, \ldots, \mathbf{v}_p^i, \ldots, \mathbf{v}_p^k) + 0$$

and so $\alpha(\mathbf{v}_p^1, \ldots, \mathbf{v}_p^i, \ldots, \mathbf{v}_p^j, \ldots, \mathbf{v}_p^k) = - \alpha(\mathbf{v}_p^1, \ldots, \mathbf{v}_p^j, \ldots,$
$\mathbf{v}_p^i, \ldots, \mathbf{v}_p^k)$.
On the other hand, assume that (b) is true. Suppose that $\mathbf{v}_p^i = \mathbf{v}_p^j$ for some $i \neq j$. Call this common value $\mathbf{u}_p$. Then

$$\alpha(\mathbf{v}_p^1, \ldots, \mathbf{u}_p, \ldots, \mathbf{u}_p, \ldots, \mathbf{v}_p^k)$$
$$= \alpha(\mathbf{v}_p^1, \ldots, \mathbf{v}_p^i, \ldots, \mathbf{v}_p^j, \ldots, \mathbf{v}_p^k)$$
$$= -\alpha(\mathbf{v}_p^1, \ldots, \mathbf{v}_p^j, \ldots, \mathbf{v}_p^i, \ldots, \mathbf{v}_p^k)$$
$$= -\alpha(\mathbf{v}_p^1, \ldots, \mathbf{u}_p, \ldots, \mathbf{u}_p, \ldots, \mathbf{v}_p^k).$$

Hence $0 = 2\alpha(\mathbf{v}_p^1, \ldots, \mathbf{u}_p, \ldots, \mathbf{u}_p, \ldots, \mathbf{v}_p^k)$ so

$$0 = \alpha(\mathbf{v}_p^1, \ldots, \mathbf{u}_p, \ldots, \mathbf{u}_p, \ldots, \mathbf{v}_p^k)$$
$$= \alpha(\mathbf{v}_p^1, \ldots, \mathbf{v}_p^i, \ldots, \mathbf{v}_p^j, \ldots, \mathbf{v}_p^k).$$

(2) This follows directly from the definition of the sign of a permutation. If σ is written as the product of t transpositions, then $\text{sign}(\sigma) = (-1)^t$. (Note that t is not well-defined but the parity of t is well-defined and hence $(-1)^t$ is well-defined.) But by the alternation property, this is exactly the factor by which the value of φ is multiplied when the arguments are permuted by σ. □

DEFINITION 2.2.3. Let $\mathcal{R} = \mathbb{R}^n$ and let p be any point of $\mathcal{R}$.

(1) Let $I = \{i_1, \ldots, i_k\}$ be any ordered k-element subset of $\{1, \ldots, n\}$. (This implicitly implies all the elements of I are distinct.) Then $dx_I = dx_{i_1} \ldots dx_{i_k}$ is the k-form on $\mathcal{R}$ defined as follows. Let p be any point of $\mathcal{R}$. Then:

(a) $dx_I(\mathbf{e}_p^{i_1}, \ldots, \mathbf{e}_p^{i_k}) = 1$;

(b) $dx_I(\mathbf{e}_p^{j_1}, \ldots, \mathbf{e}_p^{j_k}) = 0$ for J any ordered subset of $\{1, \ldots, n\}$

that is not a permutation of I.

(2) If $I = \{i_1, \ldots, i_k\}$ with not all elements distinct, then $dx_I = dx_{i_1} \ldots dx_{i_k}$ is identically 0. ◇

COROLLARY 2.2.4. *Let σ_0 be any permutation of $I = \{i_1, \ldots, i_k\}$ and let $J = \sigma_0(I)$, i.e., $J = \{j_1, \ldots, j_k\}$ with $j_1 = \sigma_0(i_1), \ldots, j_k = \sigma_0(i_k)$. Then*

$$dx_I(\mathbf{e}^{j_1}, \ldots, \mathbf{e}^{j_k}) = \text{sign}(\sigma_0).$$

Proof. This is just a special case of Definition 2.2.3(1)(a) followed by Lemma 2.2.2(2). □

REMARK 2.2.5. Note that this defines $dx_I(\mathbf{e}_p^{j_1}, \ldots, \mathbf{e}_p^{j_k})$ in general. For in addition to the cases given in Corollary 2.2.4, we only have the cases when the elements of J are not distinct, or when they are distinct but J is not a permutation of I. In both of these cases $dx_I(\mathbf{e}_p^{j_1}, \ldots, \mathbf{e}_p^{j_k}) = 0$, by a special case of Lemma 2.2.2(1) and by Definition 2.2.3(2) respectively. Then, once we have all these covered, the value of $dx_I(\mathbf{v}_p^1, \ldots, \mathbf{v}_p^k)$ for any k-tuple $\{\mathbf{v}_p^1, \ldots, \mathbf{v}_p^k\}$ is determined by multilinearity.   ◇

REMARK 2.2.6. In our notation in $\mathbb{R}^3$, for every point p in $\mathbb{R}^3$:

$$dx\, dy(\mathbf{i}_p, \mathbf{j}_p) = 1, \; dx\, dy(\mathbf{j}_p, \mathbf{k}_p) = 0, \; dx\, dy(\mathbf{k}_p, \mathbf{i}_p) = 0;$$
$$dy\, dz(\mathbf{i}_p, \mathbf{j}_p) = 0, \; dy\, dz(\mathbf{j}_p, \mathbf{k}_p) = 1, \; dy\, dz(\mathbf{k}_p, \mathbf{i}_p) = 0;$$
$$dz\, dx(\mathbf{i}_p, \mathbf{j}_p) = 0, \; dz\, dx(\mathbf{j}_p, \mathbf{k}_p) = 0, \; dz\, dx(\mathbf{k}_p, \mathbf{i}_p) = 1;$$

and

$$dx\, dy\, dz(\mathbf{i}_p, \mathbf{j}_p, \mathbf{k}_p) = 1.$$ ◇

LEMMA 2.2.7. *Let σ be any permutation of $I = \{i_1, \ldots, i_k\}$. Then*

$$dx_{\sigma(I)} = \text{sign}(\sigma) dx_I.$$

Proof. For $J = \{j_1, \ldots, j_k\}$, let us write $\mathbf{e}_p^J$ for the k-tuple $(\mathbf{e}_p^{j_1}, \ldots, \mathbf{e}_p^{j_k})$. With this notation we need to show that for any J,

$$dx_{\sigma(I)}(\mathbf{e}_p^J) = \text{sign}(\sigma) dx_I(\mathbf{e}_p^J).$$

First the uninteresting cases: If the elements of J are not distinct, or if they are distinct but J is not a permutation of I (and hence not a permutation of $\sigma(I)$ either), both sides are 0 and we certainly have equality.

Now for the interesting case. Suppose $J = \sigma_0(I)$ for some permutation σ_0. Then, by Corollary 2.2.4,

$$dx_I(\mathbf{e}_p^J) = \text{sign}(\sigma_0).$$

Let $I' = \sigma(I)$. Then

$$J = \sigma_0(I) = \sigma_0(\sigma^{-1}\sigma(I)) = (\sigma_0\sigma^{-1})\sigma(I) = \sigma_1(I') \quad \text{where}$$
$$\sigma_1 = \sigma_0\sigma^{-1}.$$

Then, by, Corollary 2.2.4 applied to I',

$$d_{I'}(\mathbf{e}_p^J) = \text{sign}(\sigma_1).$$

Now $\text{sign}(\sigma_1) = \text{sign}(\sigma_0 \sigma^{-1}) = \text{sign}(\sigma_0)\text{sign}(\sigma^{-1})$. Thus we see

$$dx_{\sigma(I)}(\mathbf{e}_p^J) = d_{I'}(\mathbf{e}_p^J) = \text{sign}(\sigma_0)\text{sign}(\sigma^{-1})$$
$$= \text{sign}(\sigma^{-1})dx_I(\mathbf{e}_p^J).$$

But for any permutation σ,

$$\text{sign}(\sigma^{-1}) = \text{sign}(\sigma),$$

completing the proof. $\qquad\qquad\qquad\qquad\qquad\qquad\qquad\qquad$ □

COROLLARY 2.2.8. *For any monomial* $dx_{i_1} \cdots dx_{i_k}$,

$$dx_{i_1} \cdots dx_{i_{j_2}} \cdots dx_{i_{j_1}} \cdots dx_{i_k} = -dx_{i_1} \cdots dx_{i_{j_1}} \cdots dx_{i_{j_2}} \cdots dx_{i_k}.$$

Proof. This is just the special case of Lemma 2.2.7 where σ consists of the single transposition that interchanges $dx_{i_{j_1}}$ and $dx_{i_{j_2}}$. □

REMARK 2.2.9. We have stated this corollary explicitly in order to observe that the relation we obtain here gives the relation $dx_i\, dx_j = -dx_j\, dx_i$ in our formal definition of k-forms in Chapter 1. ◇

THEOREM 2.2.10. *Let $\mathcal{R}$ be a region in $\mathbb{R}^n$.*
Let $\mathcal{I} = \{I_1, I_2, \ldots\}$ where each I_j is an ordered k-element subset of $\{1, \ldots, n\}$, and for each $I \in \mathcal{I}$, let A^I be a smooth function on $\mathcal{R}$. Then

$$\varphi = \sum_{I \in \mathcal{I}} A^I\, dx_I$$

is a k-form on $\mathcal{R}$.
  Conversely, consider all k-element subsets of $\{1, \ldots, n\}$ and choose an order on each of these; let the resulting ordered sets be $I_1, I_2, \ldots$ Let $\mathcal{I} = \{I_1, I_2, \ldots\}$. Then every k-form φ on $\mathcal{R}$ can be written as

$$\varphi = \sum_{I \in \mathcal{I}} A^I\, dx_I$$

for unique smooth functions $A^{I_1}, A^{I_2}, \ldots$ on $\mathcal{R}$.

Proof. The proof is very analogous to the proof of Theorem 2.1.5, and so we merely sketch it.

In one direction, since each dx_I is an alternating multilinear function, so is any linear combination of terms of this form, and since each coefficient function is smooth, φ is a k-form.

In the other direction, let φ be any k-form on $\mathcal{R}$. For each $I \in \mathcal{I}$, define the function A^I by

$$A^I(p) = \varphi(\mathbf{e}_p^I).$$

Then each A^I is a smooth function and it is easy to check that φ is as claimed. $\qquad\qquad\qquad\qquad\qquad\qquad\qquad\qquad\qquad\qquad\qquad\qquad\qquad\quad\square$

With this in hand, we now turn to multiplication of differential forms from the geometric point of view.

Before considering the issue of how to multiply forms in general, let us consider the issue of how to multiply a pair of 1-forms. This leads us to two questions:

First, in general, how do we get from a pair of 1-forms φ_1 and φ_2 to their exterior product, the 2-form $\psi = \varphi_1 \varphi_2$?

Second, in particular, how should we define the product so that, for example, the product of the 1-forms dx_i and dx_j is the 2-form $dx_i \, dx_j$, to be consistent with our work above?

We consider the first question first.

Now a 1-form on a region $\mathcal{R}$ in $\mathbb{R}^n$ is a function of a single tangent vector at each point of $\mathcal{R}$, and a 2-form is a function of a pair of tangent vectors at each point of $\mathcal{R}$, so there is an obvious thing to try—we can try setting $\psi(\mathbf{v}_p, \mathbf{w}_p) = \varphi_1(\mathbf{v}_p)\varphi_2(\mathbf{w}_p)$, where p is any point of $\mathcal{R}$. However, this obvious choice is doomed to fail because we need ψ to be alternating, and there is no reason to expect $\varphi_1(\mathbf{v}_p)\varphi_2(\mathbf{w}_p)$ to have any relationship whatsoever with $\varphi_1(\mathbf{w}_p)\varphi_2(\mathbf{v}_p)$. (For example, in $\mathbb{R}^3$, if $\varphi_1 = dx, \varphi_2 = dy, \mathbf{v} = \mathbf{i}$, and $\mathbf{w} = \mathbf{j}$, then $\varphi_1(\mathbf{v}_p)\varphi_2(\mathbf{w}_p) = dx(\mathbf{i}_p)dy(\mathbf{j}_p) = 1 \cdot 1 = 1$ but $\varphi_1(\mathbf{w}_p)\varphi_2(\mathbf{v}_p) = dx(\mathbf{j}_p)dy(\mathbf{i}_p) = 0 \cdot 0 = 0$.)

The solution to this problem is simple—we simply modify our choice slightly in a way that *forces* ψ to be alternating. Instead of defining $\psi(\mathbf{v}_p, \mathbf{w}_p) = \varphi_1(\mathbf{v}_p)\varphi_2(\mathbf{w}_p)$, we define

$$\psi(\mathbf{v}_p, \mathbf{w}_p) = \varphi_1(\mathbf{v}_p)\varphi_2(\mathbf{w}_p) - \varphi_1(\mathbf{w}_p)\varphi_2(\mathbf{v}_p).$$

It is then easy to check either of the equivalent definitions of alternation: In case $\mathbf{w}_p = \mathbf{v}_p$, then $\psi(\mathbf{v}_p, \mathbf{w}_p) = \varphi_1(\mathbf{v}_p)\varphi_2(\mathbf{v}_p) - \varphi_1(\mathbf{v}_p)\varphi_2(\mathbf{v}_p) = 0$, or for any two tangent vectors $\mathbf{v}_p$ and $\mathbf{w}_p$, $\psi(\mathbf{w}_p, \mathbf{v}_p) = \varphi_1(\mathbf{w}_p)\varphi_2(\mathbf{v}_p) - \varphi_1(\mathbf{v}_p)\varphi_2(\mathbf{w}_p) = -\psi(\mathbf{v}_p, \mathbf{w}_p)$.

Second, we observe that this solution to our first problem also provides a solution to our second problem. For example:

$$dx_1\, dx_2(\mathbf{i}_p, \mathbf{j}_p) = dx_1(\mathbf{i}_p)dx_2(\mathbf{j}_p) - dx_1(\mathbf{j}_p)dx_2(\mathbf{i}_p)$$
$$= 1 \cdot 1 - 0 \cdot 0 = 1$$

while

$$dx_1\, dx_2(\mathbf{i}_p, \mathbf{k}_p) = dx_1(\mathbf{i}_p)dx_2(\mathbf{k}_p) - dx_1(\mathbf{k}_p)dx_2(\mathbf{i}_p)$$
$$= 1 \cdot 0 - 0 \cdot 0 = 0$$

and

$$dx_1\, dx_2(\mathbf{j}_p, \mathbf{k}_p) = dx_1(\mathbf{j}_p)dx_2(\mathbf{k}_p) - dx_1(\mathbf{k}_p)dx_2(\mathbf{j}_p)$$
$$= 0 \cdot 0 - 0 \cdot 1 = 0.$$

With this motivation in hand, we proceed more generally. We will first define the product of k 1-forms, and then we will define the product of a k-form and a ℓ-form. Finally, we will assemble these definitions to define the product of any number of forms of any degree.

We establish a bit of notation. We let S_k denote the *symmetric group* on k symbols, which we regard as the group of functions from a k-element set (the set $\{1, 2, \ldots, k\}$ unless otherwise specified) to itself, with the group law being given by composition of functions, i.e., if σ and τ are elements of S_k, then $\rho = \sigma\tau$ is the element of S_k given by $\rho(i) = \sigma(\tau(i))$ for any $i \in \{1, 2, \ldots, k\}$.

DEFINITION 2.2.11. Let $\varphi_1, \varphi_2, \ldots, \varphi_k$ be 1-forms on a region $\mathcal{R}$ of $\mathbb{R}^n$. Then $\psi = \varphi_1\varphi_2 \cdots \varphi_k$ is the k-form on $\mathcal{R}$ defined by

$$\psi(\mathbf{v}_p^1, \mathbf{v}_p^2, \ldots, \mathbf{v}_p^k) = \sum_{\sigma \in S_k} \text{sign}(\sigma)\varphi_1(\mathbf{v}_p^{\sigma(1)})\varphi_2(\mathbf{v}_p^{\sigma(2)}) \cdots \varphi_k(\mathbf{v}_p^{\sigma(k)})$$

where p is any point of $\mathcal{R}$ and $(\mathbf{v}_p^1, \mathbf{v}_p^2, \ldots, \mathbf{v}_p^k)$ is any k-tuple of tangent vectors to $\mathbb{R}^n$ at p.   ◇

LEMMA 2.2.12.

(a) ψ *as given by Definition 2.2.11 is a k-form.*

(b) If $\varphi_1 = dx_{i_1}, \varphi_2 = dx_{i_2}, \ldots, \varphi_k = dx_{i_k},$ *then* $\psi = dx_{i_1} dx_{i_2} \cdots$ $dx_{i_k}.$

Proof.

(a) We need to show that ψ is multilinear and alternating. Multilinearity is almost immediate. Since each φ_i is linear, each term in ψ is multilinear, and hence ψ itself is multilinear.
To prove alternation, suppose $\mathbf{v}_p^i = \mathbf{v}_p^j$. Let τ be the transposition in S_n that interchanges i and j, i.e., $\tau(i) = j, \tau(j) = i$, and $\tau(m) = m$ if $m \neq i, j$. Group the elements of S_n in pairs (σ, σ') where $\sigma' = \sigma\tau$, and let T be the set of such pairs. Note that $\mathrm{sign}(\sigma') = -\mathrm{sign}(\sigma)$. Then

$$\psi(\mathbf{v}_p^1, \mathbf{v}_2^p, \ldots, \mathbf{v}_p^k)$$
$$= \sum_{\sigma \in T} \mathrm{sign}(\sigma)\varphi_1(\mathbf{v}_p^{\sigma(1)})\varphi_2(\mathbf{v}_p^{\sigma(2)}) \cdots \varphi_k(\mathbf{v}_p^{\sigma(k)})$$
$$+ \mathrm{sign}(\sigma')\varphi_1(\mathbf{v}_p^{\sigma'(1)})\varphi_2(\mathbf{v}_p^{\sigma'(2)}) \cdots \varphi_k(\mathbf{v}_p^{\sigma'(k)})$$
$$= \sum_{\sigma \in T} \mathrm{sign}(\sigma)\varphi_1(\mathbf{v}_p^{\sigma(1)})\varphi_2(\mathbf{v}_p^{\sigma(2)}) \cdots \varphi_k(\mathbf{v}_p^{\sigma(k)})$$
$$- \mathrm{sign}(\sigma)\varphi_1(\mathbf{v}_p^{\sigma(1)})\varphi_2(\mathbf{v}_p^{\sigma(2)}) \cdots \varphi_k(\mathbf{v}_p^{\sigma(k)})$$
$$= \sum_{\sigma \in T} 0 = 0$$

as required.

(b) We need to verify that in this case the value of ψ agrees with the value of the k-form $dx_{i_1} dx_{i_2} \cdots dx_{i_k}$ as defined in Definition 2.2.3 on any k-tuple of tangent vectors at any point p.

Let $I = (i_1, \ldots, i_k)$ be a k-tuple of distinct elements. If $J = I$ then in the definition of ψ, the term given by the identity permutation in the definition of ψ gives the value

$$dx_{i_1}(\mathbf{e}_p^{i_1})dx_{i_2}(\mathbf{e}_p^{i_2}) \cdots dx_{i_k}(\mathbf{e}_p^{i_k}) = 1 \cdot 1 \cdots 1 = 1$$

and every other term gives 0, as for any permutation σ that is not the identity, there is some i with $\sigma(i) \neq i$, and then the corresponding term in the definition of ψ has a factor $dx_i(\mathbf{e}_p^{\sigma(i)}) = 0$, so the value of ψ on this k-tuple is equal to 1.

If J is not a permutation of I then for every term in the definition of ψ, there is some $j \in J$ with $\sigma(j) \neq i$ for some $i \in I$, so every term has a factor $dx_i(\mathbf{e}_p^{\sigma(j)}) = 0$, and the value of ψ on this k-tuple is equal to 0.

Now let $I = (i_1, \ldots, i_k)$ with the entries not distinct. Then there are $j_1 \neq j_2$ with $i_{j_1} = i_{j_2}$. Just as in the proof of alternation in part (a), we see that the terms in the summation for value of ψ on any k-tuple of tangent vectors cancel in pairs, so $\psi = 0$ in this case.   □

Before we can define the product of forms of higher degrees we need to recall a bit more group theory. Let G be a group and H a subgroup. A set $C = \{g_1, g_2, \ldots\}$ of elements of G is a set of *left coset representatives* of H in G if

$$G = \cup_{g \in C} gH \quad \text{and} \quad g_i H \cap g_j H = \emptyset \text{ for } g_i \neq g_j \in C.$$

Here $gH = \{gh \mid h \in H\}$.

In particular, we will want to apply this notion where $m = k + \ell$, $G = S_m$, and $H = S_k \times S_\ell$ where we regard $S_k \times S_\ell$ as the subgroup of S_m consisting of products of permutations $\sigma_1 \times \sigma_2$ where σ_1 permutes the *first* k entries of an m-tuple and σ_2 permutes the *last* ℓ entries of an m-tuple. Note that G has $(k + \ell)!$ elements and H has $k!\ell!$ elements, so C has $\frac{(k+\ell)!}{k!\ell!}$ elements.

DEFINITION 2.2.13. Let φ_1 be a k-form and φ_2 an ℓ-form on a region $\mathcal{R}$ of $\mathbb{R}^n$. Let $m = k + \ell$, and let C be a set of left coset representatives of $S_k \times S_\ell$ in S_m. Then $\psi = \varphi_1 \varphi_2$ is the m-form on $\mathcal{R}$ defined by

$$\psi(\mathbf{v}_p^1, \mathbf{v}_p^2, \ldots, \mathbf{v}_p^m) = \sum_{\sigma \in C} \text{sign}(\sigma) \varphi_1(\mathbf{v}_p^{\sigma(1)}, \ldots, \mathbf{v}_p^{\sigma(k)})$$
$$\times \varphi_2(\mathbf{v}_p^{\sigma(k+1)}, \ldots, \mathbf{v}_p^{\sigma(m)}),$$

where p is any point of $\mathcal{R}$ and $(\mathbf{v}_p^1, \mathbf{v}_p^2, \ldots, \mathbf{v}_p^m)$ is any m-tuple of tangent vectors to $\mathbb{R}^n$ at p.   ◇

LEMMA 2.2.14.

(a) ψ as given by Definition 2.2.13 is an m-form.

(b) If $\varphi_1 = dx_{i_1} \cdots dx_{i_k}$, and $\varphi_2 = dx_{i_{k+1}} \cdots dx_{i_{k+\ell}}$, then $\psi = dx_{i_1} \cdots dx_{i_m}$.

Proof. The proof of this lemma is very similar to the proof of Lemma 2.2.12.

For the proof of part (a), there is nothing new to say.

For the proof of part (b), we simply need to observe that for any $J = \{j_1, \ldots, j_m\}$ there will be exactly one left coset representative σ with $\sigma(j_1) = i_1, \ldots, \sigma(j_m) = i_m$. □

We will need the following special case of Definition 2.2.13, and so we work it out explicitly.

LEMMA 2.2.15. *Let φ_1 be a 1-form and φ_2 an ℓ-form on a region $\mathcal{R}$ of $\mathbb{R}^n$. Then $\psi = \varphi_1 \varphi_2$ is the $m = \ell + 1$-form on $\mathcal{R}$ defined by*

$$
\begin{aligned}
\psi(\mathbf{v}_p^1, \mathbf{v}_p^2, \ldots, \mathbf{v}_p^m) = {} & \varphi_1(\mathbf{v}_p^1)\varphi_2(\mathbf{v}_p^2, \mathbf{v}_p^3, \ldots, \mathbf{v}_p^m) \\
& - \varphi_1(\mathbf{v}_p^2)\varphi_2(\mathbf{v}_p^1, \mathbf{v}_p^3, \ldots, \mathbf{v}_p^m) \\
& + \varphi_1(\mathbf{v}_p^3)\varphi_2(\mathbf{v}_p^1, \mathbf{v}_p^2, \ldots, \mathbf{v}_p^m) + \cdots \\
& + (-1)^{m-1}\varphi_1(\mathbf{v}_p^m)\varphi_2(\mathbf{v}_p^1, \mathbf{v}_p^2, \ldots, \mathbf{v}_p^{m-1}),
\end{aligned}
$$

where p is any point of $\mathcal{R}$ and $(\mathbf{v}_p^1, \mathbf{v}_p^2, \ldots, \mathbf{v}_p^m)$ is any m-tuple of tangent vectors to $\mathbb{R}^n$ at p.

Proof. We may take the set of left coset representatives of $S_1 \times S_\ell$ in S_m to be $C = \{\sigma_0, \sigma_1, \ldots, \sigma_{m-1}\}$ where σ_0 is the identity and, for $j = 1, \ldots, m-1, \sigma_j$ is the permutation given by $\sigma_j(i) = i+1$ for $1 \leq i \leq j, \sigma_j(j+1) = 1, \sigma_j(i) = i$ for $j+1 < i \leq m$. Note that $\text{sign}(\sigma_j) = (-1)^j$. □

EXAMPLE 2.2.16. Let φ be a smooth 1-form and ψ a smooth 2-form on a region $\mathcal{R}$ in $\mathbb{R}^n$, and let $\rho = \varphi\psi$.

Here $k = 1$, $\ell = 2$, and $m = 3$. The subgroup $S_1 \times S_2 \subset S_3$ has a set of left coset representatives $\{\sigma_1, \sigma_2, \sigma_3\}$ given by

$$\sigma_1(1) = 1, \ \sigma_1(2) = 2, \ \sigma_1(3) = 3;$$
$$\sigma_2(1) = 2, \ \sigma_2(2) = 3, \ \sigma_2(3) = 1;$$
$$\sigma_3(1) = 3, \ \sigma_3(2) = 1, \ \sigma_1(3) = 2.$$

Note $\operatorname{sign}(\sigma_1) = \operatorname{sign}(\sigma_2) = \operatorname{sign}(\sigma_3) = 1$. (This choice of left coset representatives differs from our choice in the previous lemma.) Then for any triple of vector fields $(\mathbf{u}, \mathbf{v}, \mathbf{w})$ on $\mathcal{R}$, and any point p of $\mathcal{R}$,

$$\rho(\mathbf{u}_p, \mathbf{v}_p, \mathbf{w}_p) = \varphi(\mathbf{u}_p)\psi(\mathbf{v}_p, \mathbf{w}_p) + \varphi(\mathbf{v}_p)\psi(\mathbf{w}_p, \mathbf{u}_p)$$
$$+ \varphi(\mathbf{w}_p)\psi(\mathbf{u}_p, \mathbf{v}_p). \qquad \diamond$$

It is sometimes more convenient to be able to sum over the entire symmetric group rather than having to choose a set of left coset representatives. To this end we have the following alternate formula.

COROLLARY 2.2.17. *Let φ_1 be a k-form and φ_2 an ℓ-form on a region $\mathcal{R}$ of $\mathbb{R}^n$. Let $m = k + \ell$. Then $\psi = \varphi_1\varphi_2$ is the m-form on $\mathcal{R}$ defined by*

$$\psi(\mathbf{v}_p^1, \mathbf{v}_p^2, \dots, \mathbf{v}_p^m) = \frac{1}{k!\ell!} \sum_{\sigma \in S_m} \operatorname{sign}(\sigma)\varphi_1(\mathbf{v}_p^{\sigma(1)}, \dots, \mathbf{v}_p^{\sigma(k)})$$
$$\times \varphi_2(\mathbf{v}_p^{\sigma(k+1)}, \dots, \mathbf{v}_p^{\sigma(m)}),$$

where p is any point of $\mathcal{R}$ and $(\mathbf{v}_p^1, \mathbf{v}_p^2, \dots, \mathbf{v}_p^m)$ is any m-tuple of tangent vectors to $\mathbb{R}^n$ at p.

Proof. We are regarding $S_k \times S_\ell$ as the subgroup of S_m consisting of products of permutations $\sigma_1 \times \sigma_2$ where σ_1 permutes the *first* k entries of an m-tuple and σ_2 permutes the *last* ℓ entries of an m-tuple. Note that $\operatorname{sign}(\sigma_1 \times \sigma_2) = \operatorname{sign}(\sigma_1) + \operatorname{sign}(\sigma_2)$.
Now for any $\sigma_1 \in S_k$,

$$\operatorname{sign}(\sigma_1)\varphi_1(\mathbf{v}_p^{\sigma(1)}, \dots, \mathbf{v}_p^{\sigma(k)}) = \varphi(\mathbf{v}_p^1, \dots, \mathbf{v}_p^k)$$

for any k-tuple of tangent vectors $(\mathbf{v}_p^1, \ldots, \mathbf{v}_p^k)$, so

$$\sum_{\sigma_1 \in S_k} \text{sign}(\sigma_1) \varphi_1(\mathbf{v}_p^{\sigma(1)}, \ldots, \mathbf{v}_p^{\sigma(k)}) = k! \varphi(\mathbf{v}_p^1, \ldots, \mathbf{v}_p^k)$$

and similarly

$$\sum_{\sigma_2 \in S_\ell} \text{sign}(\sigma_2) \varphi_2(\mathbf{v}_p^{\sigma(k+1)}, \ldots, \mathbf{v}_p^{\sigma(k+\ell)}) = \ell! \varphi(\mathbf{v}_p^{k+1}, \ldots, \mathbf{v}_p^{k+\ell}).$$

Thus we see that

$$\frac{1}{k!\ell!} \sum_{\sigma = \sigma_1 \times \sigma_2 \in S_k \times S_\ell} \psi(\mathbf{v}_p^{\sigma(1)}, \ldots, \mathbf{v}_p^{\sigma(m)}) = \psi(\mathbf{v}_p^1, \ldots, \mathbf{v}_p^m).$$

But S_m is the disjoint union of the left cosets of $S_k \times S_\ell$, so summing over elements of S_m is the same as summing over (elements of $S_k \times S_\ell$) $\times$ elements of C. $\qquad\square$

We have a common generalization of these constructions.

COROLLARY 2.2.18. *Let φ_1 be a k_1-form, φ_2 a k_2-form, $\ldots$, and φ_j be a k_j-form on a region $\mathcal{R}$ of $\mathbb{R}^n$. Let $m = k_1 + k_2 + \cdots + k_j$, and let C be a set of left coset representatives of $S_{k_1} \times S_{k_2} \times \cdots \times S_{k_j}$ in S_m. Let $\psi = \varphi_1 \varphi_2 \cdots \varphi_j$ be defined by*

$$\psi(\mathbf{v}_p^1, \mathbf{v}_p^2, \ldots, \mathbf{v}_p^m)$$
$$= \sum_{\sigma \in C} \text{sign}(\sigma) \varphi_1(\mathbf{v}_p^{\sigma(1)}, \ldots, \mathbf{v}_p^{\sigma(k_1)}) \varphi_2(\mathbf{v}_p^{\sigma(k_1+1)}, \ldots, \mathbf{v}_p^{\sigma(k_1+k_2)})$$
$$\cdots \varphi_j(\mathbf{v}_p^{\sigma(m-k_j+1)}, \ldots, \mathbf{v}_p^{\sigma(m)}),$$

where p is any point of $\mathcal{R}$ and $(\mathbf{v}_p^1, \mathbf{v}_p^2, \ldots, \mathbf{v}_p^m)$ is any m-tuple of tangent vectors to $\mathbb{R}^n$ at p.

Then ψ is a m-form on $\mathcal{R}$.
Furthermore,

$$\psi(\mathbf{v}_p^1, \mathbf{v}_p^2, \ldots, \mathbf{v}_p^m)$$
$$= \frac{1}{k_1! k_2! \cdots k_j!} \sum_{\sigma \in S_m} \text{sign}(\sigma) \varphi_1(\mathbf{v}_p^{\sigma(1)}, \ldots, \mathbf{v}_p^{\sigma(k_1)})$$

$$\times \varphi_2(\mathbf{v}_p^{\sigma(k_1+1)}, \ldots, \mathbf{v}_p^{\sigma(k_1+k_2)})$$
$$\cdots \varphi_j(\mathbf{v}_p^{\sigma(m-k_j+1)}, \ldots, \mathbf{v}_p^{\sigma(m)}).$$

Also, if $\varphi_1 = dx_{i_1} \cdots dx_{i_{k_1}}$, $\varphi_2 = dx_{i_{k_1+1}} \cdots dx_{i_{k_1+k_2}}, \ldots$, *and* $\varphi_j = dx_{i_{m-k_j+1}} \cdots dx_{i_m}$, *then* $\psi = dx_{i_1} \cdots dx_{i_m}$.

Proof. Nothing new. $\qquad\qquad\qquad\qquad\qquad\qquad\qquad\qquad\qquad$ $\square$

EXAMPLE 2.2.19. For a pair of 1-forms φ_1 and φ_2, letting $\psi = \varphi_1\varphi_2$, we have the formula

$$\psi(\mathbf{v}, \mathbf{w}) = \varphi_1(\mathbf{v})\varphi_2(\mathbf{w}) - \varphi_1(\mathbf{w})\varphi_2(\mathbf{v}).$$

(a) Let ψ be the 2-form $\psi = (2x + y)dx\, dy$, let $p = (2, 1)$, and let $\mathbf{v}_p = \begin{bmatrix} 2 \\ 3 \end{bmatrix}_p$ and $\mathbf{w}_p = \begin{bmatrix} 1 \\ 5 \end{bmatrix}_p$. Then

$$\psi(\mathbf{v}_p, \mathbf{w}_p) = (2 \cdot 2 + 1)\left(dx\left(\begin{bmatrix} 2 \\ 3 \end{bmatrix}_p\right) dy\left(\begin{bmatrix} 1 \\ 5 \end{bmatrix}_p\right)\right.$$
$$\left. - dx\left(\begin{bmatrix} 1 \\ 5 \end{bmatrix}_p\right) dy\left(\begin{bmatrix} 2 \\ 3 \end{bmatrix}_p\right)\right)$$
$$= 5(2 \cdot 5 - 1 \cdot 3) = 35.$$

(b) Let ψ be the 2-form $\psi = x^2yz\, dy\, dz - xy^2\, dz\, dx - 40dx\, dy$, let $p = (2, 5, 3)$, and let $\mathbf{v}_p = \begin{bmatrix} 2 \\ 3 \\ 1 \end{bmatrix}_p$ and $\mathbf{w}_p = \begin{bmatrix} -1 \\ 0 \\ 4 \end{bmatrix}_p$. Then

$$\psi(\mathbf{v}_p, \mathbf{w}_p) = (2^2 \cdot 5 \cdot -3)\left(dy\left(\begin{bmatrix} 2 \\ 3 \\ 1 \end{bmatrix}_p\right) dz\left(\begin{bmatrix} -1 \\ 0 \\ 4 \end{bmatrix}_p\right)\right.$$
$$\left. - dy\left(\begin{bmatrix} -1 \\ 0 \\ 4 \end{bmatrix}_p\right) dz\left(\begin{bmatrix} 2 \\ 3 \\ 1 \end{bmatrix}_p\right)\right)$$

$$
- (2 \cdot 5^2) \left(dz \left(\begin{bmatrix} 2 \\ 3 \\ 1 \end{bmatrix}_p \right) dx \left(\begin{bmatrix} -1 \\ 0 \\ 4 \end{bmatrix}_p \right) \right.
$$

$$
\left. - dz \left(\begin{bmatrix} -1 \\ 0 \\ 4 \end{bmatrix}_p \right) dx \left(\begin{bmatrix} 2 \\ 3 \\ 1 \end{bmatrix}_p \right) \right)
$$

$$
- (40) \left(dx \left(\begin{bmatrix} 2 \\ 3 \\ 1 \end{bmatrix}_p \right) dy \left(\begin{bmatrix} -1 \\ 0 \\ 4 \end{bmatrix}_p \right) \right.
$$

$$
\left. - dx \left(\begin{bmatrix} -1 \\ 0 \\ 4 \end{bmatrix}_p \right) dy \left(\begin{bmatrix} 2 \\ 3 \\ 1 \end{bmatrix}_p \right) \right)
$$

$$
= -60(3 \cdot 4 - 0 \cdot 1) - 50(1 \cdot (-1) - 4 \cdot 2)
$$
$$
- 40(2 \cdot 0 - (-1) \cdot 3)
$$
$$
= -390.
$$

For a triple of 1-forms φ_1, φ_2, and φ_3, letting $\rho = \varphi_1 \varphi_2 \varphi_3$, we have the formula

$$
\rho(\mathbf{u}, \mathbf{v}, \mathbf{w}) = \varphi_1(\mathbf{u})\varphi_2(\mathbf{v})\varphi_3(\mathbf{w}) + \varphi_1(\mathbf{v})\varphi_2(\mathbf{w})\varphi_3(\mathbf{u})
$$
$$
+ \varphi_1(\mathbf{w})\varphi_2(\mathbf{u})\varphi_3(\mathbf{v}) - \varphi_1(\mathbf{v})\varphi_2(\mathbf{u})\varphi_3(\mathbf{w})
$$
$$
- \varphi_1(\mathbf{u})\varphi_2(\mathbf{w})\varphi_3(\mathbf{v}) - \varphi_1(\mathbf{w})\varphi_2(\mathbf{v})\varphi_3(\mathbf{u}).
$$

(c) Let ρ be the 3-form $\rho = (x + 2y - z)dx\,dy\,dz$, let $p = (1, 2, 3)$, and let $\mathbf{u}_p = \begin{bmatrix} 1 \\ 2 \\ -1 \end{bmatrix}_p$, $\mathbf{v}_p = \begin{bmatrix} 3 \\ 5 \\ -2 \end{bmatrix}_p$, and $\mathbf{w}_p = \begin{bmatrix} 4 \\ -2 \\ 1 \end{bmatrix}_p$. Then

$$
\rho(\mathbf{u}_p, \mathbf{v}_p, \mathbf{w}_p)
$$
$$
= \left(1 + 2 \cdot 2 - 3\right) dx\,dy\,dz(\mathbf{u}_p, \mathbf{v}_p, \mathbf{w}_p)
$$
$$
= 2 \left(dx \left(\begin{bmatrix} 1 \\ 2 \\ -1 \end{bmatrix}_p \right) dy \left(\begin{bmatrix} 3 \\ 5 \\ -2 \end{bmatrix}_p \right) dz \left(\begin{bmatrix} 4 \\ -2 \\ 1 \end{bmatrix}_p \right) \right.
$$

$$+\, dx\left(\begin{bmatrix} 3 \\ 5 \\ -2 \end{bmatrix}_p\right) dy\left(\begin{bmatrix} 4 \\ -2 \\ 1 \end{bmatrix}_p\right) dz\left(\begin{bmatrix} 1 \\ 2 \\ -1 \end{bmatrix}_p\right)$$

$$+\, dx\left(\begin{bmatrix} 4 \\ -2 \\ 1 \end{bmatrix}_p\right) dy\left(\begin{bmatrix} 1 \\ 2 \\ -1 \end{bmatrix}_p\right) dz\left(\begin{bmatrix} 3 \\ 5 \\ -2 \end{bmatrix}_p\right)$$

$$-\, dx\left(\begin{bmatrix} 3 \\ 5 \\ -2 \end{bmatrix}_p\right) dy\left(\begin{bmatrix} 1 \\ 2 \\ -1 \end{bmatrix}_p\right) dz\left(\begin{bmatrix} 4 \\ -2 \\ 1 \end{bmatrix}_p\right)$$

$$-\, dx\left(\begin{bmatrix} 1 \\ 2 \\ -1 \end{bmatrix}_p\right) dy\left(\begin{bmatrix} 4 \\ -2 \\ 1 \end{bmatrix}_p\right) dz\left(\begin{bmatrix} 3 \\ 5 \\ -2 \end{bmatrix}_p\right)$$

$$-\, dx\left(\begin{bmatrix} 4 \\ -2 \\ 1 \end{bmatrix}_p\right) dy\left(\begin{bmatrix} 3 \\ 5 \\ -2 \end{bmatrix}_p\right) dz\left(\begin{bmatrix} 1 \\ 2 \\ -1 \end{bmatrix}_p\right)\right)$$

$$= 2(1 \cdot 5 \cdot 1 + 3 \cdot (-2) \cdot (-1) + 4 \cdot 2 \cdot (-2)$$
$$-\, 3 \cdot 2 \cdot 1 - 1 \cdot (-2) \cdot (-2) - 4 \cdot 5 \cdot (-1))$$
$$= 2(5) = 10.$$

For a 1-form φ and a 2-form ψ, letting $\rho = \varphi\psi$, we have the formula

$$\rho(\mathbf{u}_p, \mathbf{v}_p, \mathbf{w}_p) = \varphi(\mathbf{u}_p)\psi(\mathbf{v}_p, \mathbf{w}_p) + \varphi(\mathbf{v}_p)\psi(\mathbf{w}_p, \mathbf{u}_p)$$
$$+\, \varphi(\mathbf{w}_p)\psi(\mathbf{u}_p, \mathbf{v}_p).$$

(d) Let φ be the 1-form $\varphi = dx + y\,dy - z\,dz$, let ψ be the 2-form $\psi = x\,dy\,dz + 2\,dz\,dx - dx\,dy$, let $p = (1, 2, 3)$, and let $\mathbf{u}_p = \begin{bmatrix} 1 \\ 2 \\ 1 \end{bmatrix}_p$, $\mathbf{v}_p = \begin{bmatrix} 3 \\ 5 \\ -2 \end{bmatrix}_p$, and $\mathbf{w}_p = \begin{bmatrix} 4 \\ 2 \\ -1 \end{bmatrix}_p$. Let $\rho = \varphi\psi$. Then, by computations similar to those we have just done,

$$\varphi(\mathbf{u}_p) = 8, \quad \varphi(\mathbf{v}_p) = 19, \quad \varphi(\mathbf{w}_p) = -3$$

and

$$\psi(\mathbf{v}_p, \mathbf{w}_p) = -47, \quad \psi(\mathbf{w}_p, \mathbf{u}_p) = 20, \quad \psi(\mathbf{u}_p, \mathbf{v}_p) = -2.$$

Thus

$$\rho(\mathbf{u}_p, \mathbf{v}_p, \mathbf{w}_p) = 8 \cdot (-47) + 19 \cdot 20 + (-3) \cdot (-2) = 10.$$

However, as you may also check, $\rho = \varphi\psi = (x + 2y - z)dx\,dy\,dz$ is just the 3-form of part (c) of this example, so this computation checks.   ◇

The reader may have noticed a similarity between some of our formulas above and some formulas that arise in evaluating determinants. This is no coincidence, as we shall now see.

Let us introduce some notation. Given k vectors $\mathbf{v}_p^1, \ldots, \mathbf{v}_p^k$ in $T_p\mathbb{R}^n$, we let $M = \left[\mathbf{v}_p^1 \Big| \cdots \Big| \mathbf{v}_p^k\right]$ be the n-by-k matrix with columns $\mathbf{v}_p^1, \ldots, \mathbf{v}_p^k$. Also, given any k-tuple $I = (i_1, \ldots, i_k)$ of integers between 1 and n, we let $P_I(M)$ be the k-by-k matrix whose rows are the rows $i_1 \ldots, i_k$ (in that order) of M.

THEOREM 2.2.20. *Let* $\mathbf{v}_p^1, \ldots, \mathbf{v}_p^k$ *be* k *vectors in* $T_p\mathbb{R}^n$, *for some point* p, *and let* $M = \left[\mathbf{v}_p^1 \Big| \cdots \Big| \mathbf{v}_p^k\right]$. *Let* $I = (i_1, \ldots, i_k)$ *be any* k-*tuple of integers between* 1 *and* n. *Then*

$$dx_I(\mathbf{v}_p^1, \ldots, \mathbf{v}_p^k) = \det(P_I(M)).$$

In particular,

$$dx_1 \cdots dx_n(\mathbf{v}_p^1, \ldots, \mathbf{v}_p^n) = \det\left(\left[\mathbf{v}_p^1 \Big| \cdots \Big| \mathbf{v}_p^n\right]\right).$$

Proof. First consider the case $I = (1, \ldots, n)$, where $P_I(M) = M$. It is a basic property of the determinant of a matrix M that it is multilinear and alternating in the columns of M. It is easy to check that for arbitrary I, $\det(P_I(M))$ is also multilinear and alternating in the columns of M, i.e., in the vectors $\mathbf{v}_p^1, \ldots, \mathbf{v}_p^k$. It thus suffices to check the claimed equality when $(\mathbf{v}_p^1, \ldots, \mathbf{v}_p^k) = (\mathbf{e}_p^{j_1}, \ldots, \mathbf{e}_p^{j_k})$ for arbitrary $J = (j_1, \ldots, j_k)$.

First suppose that the entries of I are not distinct. Then for any k-tuple $(\mathbf{v}_p^1, \ldots, \mathbf{v}_p^k)$, the matrix $P_I(M)$ has two equal rows, so its determinant is 0.

Now suppose the entries of I are distinct. If $J \neq I$ as unordered sets, then some row of the matrix $P_I(M)$ is 0, so its determinant is 0. Now suppose $J = I$ as unordered sets. Then, by alternation, it suffices to consider the case where $J = I$ as ordered sets. But it this case $P_I(M)$ is the identity matrix, so its determinant is 1.

Thus, comparing Definition 2.2.3, we see that the claimed formula holds in every case, so we are done. $\qquad\square$

Let us redo parts of Example 2.2.19 from this point of view.

EXAMPLE 2.2.21.

(a) Let ψ be the 2-form $\psi = (2x + y)dxdy$, let $p = (2, 1)$, and let $\mathbf{v}_p = \begin{bmatrix} 2 \\ 3 \end{bmatrix}_p$ and $\mathbf{w}_p = \begin{bmatrix} 1 \\ 5 \end{bmatrix}_p$. Then

$$\psi(\mathbf{v}_p, \mathbf{w}_p) = (2 \cdot 2 + 1) \det \left(\begin{bmatrix} 2 & 1 \\ 3 & 5 \end{bmatrix} \right)$$
$$= 5(7) = 35.$$

(b) Let ψ be the 2-form $\psi = x^2 yz\, dy\, dz - xy^2\, dz\, dx - 40\, dx\, dy$, let $p = (2, 5, 3)$, and let $\mathbf{v}_p = \begin{bmatrix} 2 \\ 3 \\ 1 \end{bmatrix}_p$ and $\mathbf{w}_p = \begin{bmatrix} -1 \\ 0 \\ 4 \end{bmatrix}_p$. Then

$$\psi(\mathbf{v}_p, \mathbf{w}_p) = (2^2 \cdot 5 \cdot -3) \det \left(\begin{bmatrix} 3 & 0 \\ 1 & 4 \end{bmatrix} \right)$$
$$- (2 \cdot 5^2) \det \left(\begin{bmatrix} 1 & 4 \\ 2 & -1 \end{bmatrix} \right)$$
$$- (40) \det \left(\begin{bmatrix} 2 & -1 \\ 3 & 0 \end{bmatrix} \right)$$
$$= -60(12) - 50(-9) - 40(3) = -390.$$

(c) Let ρ be the 3-form $\rho = (x + 2y - z)dx\, dy\, dz$, let $p = (1, 2, 3)$, and let $\mathbf{u}_p = \begin{bmatrix} 1 \\ 2 \\ -1 \end{bmatrix}_p$, $\mathbf{v}_p = \begin{bmatrix} 3 \\ 5 \\ -2 \end{bmatrix}_p$, and $\mathbf{w}_p = \begin{bmatrix} 4 \\ -2 \\ 1 \end{bmatrix}_p$.

Then

$$\rho(\mathbf{u}_p, \mathbf{v}_p, \mathbf{w}_p) = \left(1 + 2 \cdot 2 - 3\right) \det \left(\begin{bmatrix} 1 & 3 & 4 \\ 2 & 5 & -2 \\ -1 & -2 & 1 \end{bmatrix} \right)$$

$$= 2(5) = 10. \qquad \qquad \diamond$$

It is of course purely a matter of taste whether one uses the method of Example 2.2.19 or of 2.2.21.

Our next step is to investigate exterior differentiation from the geometric point of view. We already did so for 1-forms in the last section, and now we consider k-forms in general.

THEOREM 2.2.22. *Let φ be a k-form defined on a region $\mathcal{R}$ of $\mathbb{R}^n$. Then $d\varphi$ is the $(k+1)$-form defined by the following. Let $\mathbf{v}^1, \ldots, \mathbf{v}^{k+1}$ be a $(k+1)$-tuple of constant vector fields on $\mathcal{R}$. Then*

$$d\varphi(\mathbf{v}_p^1, \ldots, \mathbf{v}_p^{k+1}) =$$

$$\sum_{m=1}^{k+1} (-1)^{m-1} D_{\mathbf{v}_p^m} \varphi(\mathbf{v}_p^1, \ldots, \mathbf{v}_p^{m-1}, \mathbf{v}_p^{m+1}, \ldots, \mathbf{v}_p^{k+1}).$$

Proof. Let ψ_m denote the mth term on the right-hand side, $m = 1, \ldots, k+1$, and let $\psi = \psi_1 + \cdots + \psi_m$. First we check that ψ is indeed a k-form, i.e., that it is multilinear, alternating, and smooth.

Since φ is smooth, it is clear that each ψ_i is smooth and hence that ψ is smooth.

Since φ is multilinear, it is clear that, for any value of m, each ψ_i is linear in $\mathbf{v}_p^m$, except possibly ψ_m. But by Corollary 2.1.11 (i.e., the case of 1-forms), $\psi_m(\mathbf{v}_p^1, \ldots, \mathbf{v}_p^{k+1}) = (-1)^{m-1} dg(\mathbf{v}_p^m)$ where g is the function $g(p) = \varphi(\mathbf{v}_p^1, \ldots, \mathbf{v}_p^{m-1}, \mathbf{v}_p^{m+1}, \ldots, \mathbf{v}_p^{k+1})$, and dg is a linear function of its argument $\mathbf{v}_p^m$.

As for alternation, suppose that $\mathbf{v}_p^i = \mathbf{v}_p^j$ for some $i < j$. Call this common value $\mathbf{u}_p$. Then for any $m \neq i$ or j, $\psi_m(\mathbf{v}_p^1, \ldots, \mathbf{v}_p^{k+1}) = (-1)^{m-1} \varphi(\mathbf{v}_p^1, \ldots, \mathbf{v}_p^{m-1}, \mathbf{v}_p^{m+1}, \ldots, \mathbf{v}_p^{k+1}) = 0$ as two of the arguments of φ are equal to $\mathbf{u}_p$. Hence

$$\psi(\mathbf{v}_p^1, \ldots, \mathbf{v}_p^{k+1})$$

$$= \psi_i(\mathbf{v}_p^1, \ldots, \mathbf{v}_p^{k+1}) + \psi_j(\mathbf{v}_p^1, \ldots, \mathbf{v}_p^{k+1})$$

$$= (-1)^{i-1} D_{\mathbf{u}_p} \varphi(\mathbf{v}_p^1, \ldots, \mathbf{v}_p^{i-1}, \mathbf{v}_p^{i+1}, \ldots, \mathbf{v}_p^{k+1})$$
$$+ (-1)^{j-1} D_{\mathbf{u}_p} \varphi(\mathbf{v}_p^1, \ldots, \mathbf{v}_p^{j-1}, \mathbf{v}_p^{j+1}, \ldots, \mathbf{v}_p^{k+1})$$
$$= (-1)^{i-1} D_{\mathbf{u}_p} \varphi(\mathbf{v}_p^1, \ldots, \mathbf{v}_p^{i-1}, \mathbf{v}_p^{i+1}, \ldots, \mathbf{v}_p^{j-1},$$
$$\mathbf{u}_p, \mathbf{v}_p^{j+1}, \ldots, \mathbf{v}_p^{k+1}) + (-1)^{j-1} D_{\mathbf{u}_p} \varphi(\mathbf{v}_p^1, \ldots, \mathbf{v}_p^{i-1}, \mathbf{u}_p,$$
$$\mathbf{v}_p^{i+1}, \ldots, \mathbf{v}_p^{j-1}, \mathbf{v}_p^{j+1}, \ldots, \mathbf{v}_p^{k+1}).$$

Note that these two terms have the same arguments, but in a different order. To get from the second order to the first we must first transpose $\mathbf{u}_p$ with $\mathbf{v}_p^{i+1}$, then transpose $\mathbf{u}_p$ with $\mathbf{v}_p^{i+2}$, etc., and finally transpose $\mathbf{u}_p$ with $\mathbf{v}_p^{j-1}$. This is a total of $j - i - 1$ transpositions, so is a permutation with sign $(-1)^{j-i-1}$. Hence

$$\psi(\mathbf{v}_p^1, \ldots, \mathbf{v}_p^{k+1}) = s D_{\mathbf{u}_p} \varphi(\mathbf{v}_p^1, \ldots, \mathbf{v}_p^{i-1}, \mathbf{v}_p^{i+1}, \ldots, \mathbf{v}_p^{j-1},$$
$$\mathbf{u}_p, \mathbf{v}_p^{j+1}, \ldots, \mathbf{v}_p^{k+1}),$$

where $s = \left((-1)^{i-1} + (-1)^{j-i-1}(-1)^{j-1}\right)$. But $s = ((-1)^{i-1} + (-1)^{j-i-1}(-1)^{j-1}) = (-1)^{i-1} + (-1)^{2j-i-2} = (-1)^{i-1}(1 + (-1)^{2j-2i-1}) = (-1)^{i-1}(1 + (-1)^{-1}) = (-1)^{i-1}(1 - 1) = 0$ as required.

Thus ψ is indeed a k-form, and we are claiming that $d\varphi = \psi$.

Given Theorem 2.2.10, and linearity, we need only verify that this equality holds for $\varphi = f \, dx_{i_1} \cdots dx_{i_k} = f \, dx_I$ where $I = \{i_1, \ldots, i_k\}$ (and the elements of I are distinct). Then, by linearity again, we need only check that this holds when $(\mathbf{v}^1, \ldots, \mathbf{v}^{k+1}) = (\mathbf{e}^{j_1}, \ldots, \mathbf{e}^{j_{k+1}}) = \mathbf{e}^J$ where $J = \{j_1, \ldots, j_{k+1}\}$. For ease of notation, set $J_m = \{j_1, \ldots, j_{m-1}, j_{m+1}, \ldots, j_{k+1}\}$. Thus, in this notation, we need to check the equality

$$d(f \, dx_I)(\mathbf{e}_p^J) = \sum_{m=1}^{k+1} (-1)^{m-1} D_{\mathbf{e}_p^{j_m}} \left(f \, dx_I(\mathbf{e}_p^{J_m}) \right)$$

for every I and every J.

By alternation, if the elements of J are not distinct, then both sides are 0 and we are done. Hence we may assume the elements of J are distinct.

Thus we are reduced to checking two cases:

(a) $I \subset J$ as unordered sets;

(b) $I \not\subset J$ as unordered sets.

Also, by alternation, if we verify this for some ordering of J, it is true for every ordering of J.

Let us explicitly write out $d(f\, dx_I)$. By definition

$$d(f\, dx_I) = \sum_{\ell=1}^{n} f_{x_\ell}\, dx_\ell\, dx_I$$

$$= \sum_{\ell \notin I} f_{x_\ell}\, dx_\ell\, dx_I$$

as if $\ell \in I$, i.e., if $\ell = i_m$ for some m, then $dx_\ell\, dx_I = 0$.

Case (b) is almost immediate. J has $k+1$ elements and I has k elements, so in this case every term on each side of the equation is 0. Each term on the left-hand side is 0 as J is never equal to $\{\ell\} \cup I$, and each term on the right-hand side is 0 as no J_m is equal to I (both as unordered sets).

Now for case (a). In this case we may assume (choosing a particular order) that $I = \{i_1, \ldots, i_k\}$ and $J = \{\ell\} \cup I = \{\ell, i_1, \ldots, i_k\}$. There is a single nonzero term on the left-hand side, the term

$$f_{x_\ell}(p)dx_\ell\, dx_I(\mathbf{e}_p^J) = f_{x_\ell}(p).$$

There is a single nonzero term on the right-hand side, the term with $m = 1$ (as we delete the first entry in J to obtain I). In other words, $J_1 = I$, as ordered sets. That term is equal to

$$(-1)^{1-1} D_{\mathbf{e}_p^\ell}\left(f\, dx_I(\mathbf{e}_p^{J_1})\right) = D_{\mathbf{e}_p^\ell}(f) = f_{x_\ell}(p)$$

as $dx_I(\mathbf{e}_p^{J_1}) = 1$.

Thus we have equality in this case as well, completing the proof. □

At this point we have given geometric meaning to all of our formal constructions in Chapter 1. Before we leave k-forms, we introduce another operation, contraction, that takes a k-form and a vector field to a $(k-1)$-form. This operation is quite natural, as it is simply (partial) evaluation. We should alert the reader that there is no standard notation for the contraction, so we simply choose one.

DEFINITION 2.2.23. Let $k \geq 1$. Let φ be a k-form on a region $\mathcal{R}$ in $\mathbb{R}^n$, and let $\mathbf{v}$ be a vector field on $\mathcal{R}$. The *contraction* of φ and $\mathbf{v}$ is the $(k-1)$-form $\varphi_\mathbf{v}$ on $\mathcal{R}$ given as follows:
If $k = 1$, then
$$\varphi_\mathbf{v} = \varphi(\mathbf{v}),$$

a function on $\mathcal{R}$.
If $k > 1$, then

$$\varphi_\mathbf{v}(\mathbf{v}^1, \ldots, \mathbf{v}^{k-1}) = \varphi(\mathbf{v}, \mathbf{v}^1, \ldots, \mathbf{v}^{k-1})$$

for any $(k-1)$-tuple of vector fields $(\mathbf{v}^1, \ldots, \mathbf{v}^{k-1})$ on $\mathcal{R}$.   ◇

LEMMA 2.2.24.

(a) In the situation of Definition 2.2.23, $\varphi_\mathbf{v}$ *is a $(k-1)$-form on $\mathcal{R}$.*

(b) Contraction is linear with respect to smooth functions (not just constants) in each of its arguments, i.e.,

$$\varphi_{(f\mathbf{v}+g\mathbf{w})} = f\varphi_\mathbf{v} + g\varphi_\mathbf{w}$$

and

$$(f\varphi + g\psi)_\mathbf{v} = f\varphi_\mathbf{v} + g\psi_\mathbf{v}$$

for any smooth functions f and g on $\mathcal{R}$.

(c) For any k-form φ and any ℓ-form ψ,

$$(\varphi\psi)_\mathbf{v} = \varphi_\mathbf{v}\psi + (-1)^k\varphi\psi_\mathbf{v}.$$

(d) For any smooth function f on $\mathcal{R}$,

$$(df)_\mathbf{v} = D_\mathbf{v}f.$$

Proof.   Straightforward.   □

REMARK 2.2.25. The exceptionally observant reader may have been bothered by one incongruity in our development. A k-form for $k > 0$ is a particular kind of function on k-tuples of tangent vectors, but a 0-form is simply a function on points, with no reference to tangent vectors whatsoever. But in fact they are there. For we may define a 0-form as a function on 0-tuples of tangent vectors. But a 0-tuple is the empty set $\emptyset$. However, there is not just one empty set but rather an empty set at each point p of the region $\mathcal{R}$ on which the form is defined, namely $\emptyset_p \subset T_p\mathbb{R}^n$. Just as the condition of alternation only appears for $k \geq 2$, the condition of linearity only appears for $k \geq 1$. Thus a 0-form is a smooth function $f(\emptyset_p)$ for $p \in \mathcal{R}$. But it is a bit ridiculous to write $f(\emptyset_p)$ rather than just $f(p)$ (logically they are the same thing, as in each case the value of the function only depends on the point p), so neither we nor anybody else does it.   ◇

2.3   Orientation and signed volume

We begin this section by reviewing the notion of an orientation of a finite-dimensional real vector space. Using this notion, we develop the notion of signed volume. Then, at the end, we relate this to differential forms.

As we will be discussing general vector spaces, we change our notation slightly. We denote a vector space by V and vectors in V by u, v, w, etc.

Recall that $\mathbf{R}^n$ denotes the space of column vectors,

$$\mathbf{R}^n = \left\{ \begin{bmatrix} a_1 \\ \vdots \\ a_n \end{bmatrix} \right\}$$

with each a_i a real number, and we let $e_i \in \mathbf{R}^n$ be the vector whose ith entry is 1 and all of whose other entries are 0.

To begin with we recall some familiar notions.

DEFINITION 2.3.1. The *standard basis* of $\mathbf{R}^n$ is

$$\mathcal{E} = \{e_1, \ldots, e_n\}.$$ ◇

We let V be a real vector space of dimension $n > 0$.

DEFINITION 2.3.2. Let $\mathcal{B} = \{v_1, \ldots, v_n\}$ be a basis of V. Let $u \in V$ be any vector. Then

$$u = c_1 v_1 + \cdots + c_n v_n$$

for unique real numbers $c_1, \ldots, c_n$. In this case the *coordinate vector* of u with respect to the basis $\mathcal{B}$ is the vector

$$[u]_{\mathcal{B}} = \begin{bmatrix} c_1 \\ \vdots \\ c_n \end{bmatrix}. \qquad \diamond$$

REMARK 2.3.3. In the special case $V = \mathbf{R}^n$ with basis $\mathcal{E}$, if

$$v = \begin{bmatrix} a_1 \\ \vdots \\ a_n \end{bmatrix}$$

then

$$[v]_{\mathcal{E}} = \begin{bmatrix} a_1 \\ \vdots \\ a_n \end{bmatrix}. \qquad \diamond$$

DEFINITION 2.3.4. Let $\mathcal{B} = \{v_1, \ldots, v_n\}$ and $\mathcal{C} = \{w_1, \ldots, w_n\}$ be two bases of V. The *change of basis matrix* from $\mathcal{B}$ to $\mathcal{C}$ is the matrix

$$P_{\mathcal{C} \leftarrow \mathcal{B}} = \begin{bmatrix} [v_1]_{\mathcal{C}} | \ldots | [v_n]_{\mathcal{C}} \end{bmatrix}. \qquad \diamond$$

The following is a standard result from linear algebra.

THEOREM 2.3.5.

(a) *For any vector* $v \in V$,

$$[v]_{\mathcal{C}} = P_{\mathcal{C} \leftarrow \mathcal{B}} [v]_{\mathcal{B}}.$$

(b) *For any bases* $\mathcal{B}, \mathcal{C}$, *and* $\mathcal{D}$ *of* V:

(i) $P_{\mathcal{B} \leftarrow \mathcal{B}} = I$ *(the identity matrix)*;

(ii) $P_{\mathcal{B} \leftarrow \mathcal{C}} = \left(P_{\mathcal{C} \leftarrow \mathcal{B}}\right)^{-1}$;

(iii) $P_{\mathcal{D} \leftarrow \mathcal{B}} = P_{\mathcal{D} \leftarrow \mathcal{C}} P_{\mathcal{C} \leftarrow \mathcal{B}}$.

Now we come to the definition of orientation.

DEFINITION 2.3.6. Two bases $\mathcal{B}$ and $\mathcal{C}$ of V define the *same orientation* of V if $\det\left(P_{\mathcal{C} \leftarrow \mathcal{B}}\right) > 0$. They define *opposite orientations* of V if $\det\left(P_{\mathcal{C} \leftarrow \mathcal{B}}\right) < 0$. ◇

LEMMA 2.3.7. *"Defining the same orientation" on V is an equivalence relation on bases of V.*

Proof.   Immediate from Theorem 2.3.5(b).   □

DEFINITION 2.3.8. An *orientation* of V is an equivalence class of bases of V under the equivalence relation of defining the same orientation of V. If a basis $\mathcal{B}$ of V is in that equivalence class, it *gives* that orientation of V. ◇

LEMMA 2.3.9. *A vector space V has exactly two orientations.*

Proof.   If $\mathcal{B}$ is any basis of V, the two orientations are: the set of bases $\mathcal{C}$ with $\det\left(P_{\mathcal{C} \leftarrow \mathcal{B}}\right) > 0$, and the set of bases $\mathcal{C}$ with $\det\left(P_{\mathcal{C} \leftarrow \mathcal{B}}\right) < 0$.   □

It is useful to note the following.

LEMMA 2.3.10. *Let $\mathcal{B} = \{v_1, \ldots, v_n\}$ be any basis of V. Let σ be a permutation of $\{1, \ldots, n\}$, and let $\mathcal{C}$ be the basis $\mathcal{C} = \{v_{\sigma(1)}, \ldots, v_{\sigma(n)}\}$. Then $\mathcal{C}$ gives the same orientation of V as $\mathcal{B}$ does if $\mathrm{sign}(\sigma) = 1$ and $\mathcal{C}$ gives the opposite orientation of V as $\mathcal{B}$ does if $\mathrm{sign}(\sigma) = -1$.*

Proof.   This follows immediately from the fact the determinant of a matrix is an alternating function of its columns.   □

We have defined orientation for a vector space V of dimension $n > 0$. It is convenient to extend the definition to $n = 0$. But the only basis for the vector space $V = \{0\}$ is the empty set $\mathcal{B} = \{\}$, so this does not fit into the general pattern. Instead, a special definition is needed.

DEFINITION 2.3.11. An *orientation* for a 0-dimensional vector space V is a choice of sign ± 1.   ◇

In general there is no preferred orientation for a vector space V. But there is a preferred orientation for $\mathbf{R}^n$.

DEFINITION 2.3.12. The *standard orientation* of $\mathbf{R}^0$ is $+1$ and the *nonstandard orientation* of $\mathbf{R}^0$ is -1.

For $n > 0$, the *standard orientation* of $\mathbf{R}^n$ is the orientation given by the standard basis $\mathcal{E}$ of $\mathbf{R}^n$. The *nonstandard orientation* of $\mathbf{R}^n$ is the orientation opposite to the standard orientation.   ◇

LEMMA 2.3.13. *Let $n > 0$. Let $\mathcal{B} = \{v_1, \ldots, v_n\}$ be a basis of $\mathbf{R}^n$. Then $\mathcal{B}$ gives the standard basis of $\mathbf{R}^n$ if*

$$\det\left(\left[v_1|\ldots|v_n\right]\right) > 0;$$

otherwise it gives the nonstandard orientation.

Proof. Immediate from Remark 2.3.3 and Definition 2.3.6.   □

We have given a precise and purely algebraic definition of the standard orientation of $\mathbf{R}^n$. But how let us see what this means geometrically in the case of $\mathbf{R}^1$, $\mathbf{R}^2$, and $\mathbf{R}^3$.

REMARK 2.3.14. Consider $\mathbf{R}^1$. Then the standard basis $\mathcal{E} = \{e_1\}$ of $\mathbf{R}^1$ is as pictured below:

Note that e_1 points in the direction of increasing x. A basis $\mathcal{B} = \{v_1\}$ gives the same orientation of $\mathbf{R}^1$ if v_1 points in the same direction, and the opposite orientation if it points in the opposite direction.

Thus we see that the standard orientation of $\mathbf{R}^1$ is "increasing," and the nonstandard orientation of $\mathbf{R}^1$ is "decreasing."   ◇

REMARK 2.3.15. Consider $\mathbf{R}^2$. Then the standard basis $\mathcal{E} = \{e_1, e_2\}$ of $\mathbf{R}^2$ is as pictured below:

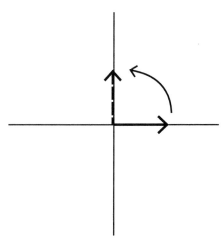

Here we have drawn the first vector in the basis as a solid line and the second as a dashed line. We have also drawn the direction of rotation from the first vector to the second vector. Note we get from the first vector to the second vector by a 90° rotation counterclockwise. It is this direction of rotation that gives the orientation.

Consider a basis $\mathcal{B} = \{v_1, v_2\}$ of $\mathbf{R}^2$. The vectors v_1 and v_2 need not be perpendicular to each other. But they cannot point in the same or opposite directions as, if they did, they would be linearly dependent and $\mathcal{B}$ would not be a basis. This enables us to decide whether we go counterclockwise or clockwise from v_1 to v_2 as we rotate the direction of v_1 to the direction of v_2 through an angle of less than 180°.

The standard orientation of $\mathbf{R}^2$ is "counterclockwise," and the nonstandard orientation of $\mathbf{R}^2$ is "clockwise." (It requires some work to check that if this direction of rotation is counterclockwise, then indeed $\det\left(\left[v_1 \mid v_2\right]\right) > 0$; we will present a geometric argument in Lemma 2.3.21.)

We can then see that, for example, the basis $\{v_2, -v_1\}$ also gives the standard orientation of $\mathbf{R}^2$ (again with rotation 90° counterclockwise) while, for example, the bases $\{v_1, -v_2\}$ and $\{v_2, v_1\}$ both give the nonstandard orientation of $\mathbf{R}^2$ (with, in each case, a rotation of 90° clockwise). ◇

REMARK 2.3.16. Consider $\mathbf{R}^3$. Then the standard basis $\mathcal{E} = \{e_1, e_2, e_3\}$ of $\mathbb{R}^3$ has its first vector pointing in the "x" direction, its second vector in the "y" direction, and its third vector pointed in the "z" direction. This is a right-handed orientation, as pictured below:

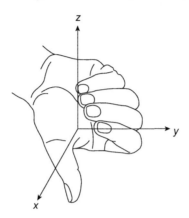

That is, this orientation is given by the right-hand rule: If you point the thumb of your right hand in the direction of the first vector, and the body of your hand in the direction of the second vector, then your fingers curl in the direction of the third vector. (The reader who has seen the right-hand rule before will note that this is not how it is usually phrased. But this phraseology is equivalent and fits in much better with our development, as will become apparent later.)

Again for an arbitrary basis $\mathcal{B} = \{v_1, v_2, v_3\}$ of $\mathbf{R}^3$ this will always make sense, even though the vectors v_1, v_2, and v_3 need not be perpendicular. Again it requires some work, which we shall not give, to check that $\det\left(\begin{bmatrix} v_1 \mid v_2 \mid v_3 \end{bmatrix}\right) > 0$ precisely when these this basis satisfies the right-hand rule.

Thus, for $\mathbf{R}^3$, the standard orientation is "right-handed" and the nonstandard orientation is "left-handed". $\diamond$

Now in $\mathbf{R}^n$, we have the usual notions of the length of a vector and the n-dimensional volume of an n-parallelogram (i.e., of a line segment in case $n = 1$, of a parallelogram in case $n = 2$, of a parallelopiped in case $n = 3$, etc.). But in fact we have a more precise notion, that of signed volume, which we now develop.

DEFINITION 2.3.17.

(a) Let $n > 0$. Let $\mathcal{B} = \{v_1, \ldots, v_n\}$ be a basis of $\mathbf{R}^n$. Then the *oriented n-parallelogram determined by* $\mathcal{B}$ is the n-parallelogram P whose 2^n vertices are the points

$$\{0 + \varepsilon_1 v_1 + \cdots + \varepsilon_n v_n | \varepsilon_i = 0 \text{ or } 1\}$$

with the orientation given by $\mathcal{B}$.

(b) For $n = 0$, there is a unique 0-parallelogram $P = \{0\}$ (with $2^0 = 1$ vertex) and with orientation ± 1.   ◇

REMARK 2.3.18. Let $n > 0$. Note that if we obtain the basis $\mathcal{B}'$ by changing the order of the elements of $\mathcal{B}$, and P' is the corresponding oriented n-parallelogram, then P and P' are identical as n-parallelograms, as we see from Definition 2.3.17, but may or may not have the same orientation, as we see from Lemma 2.3.10.   ◇

DEFINITION 2.3.19. Let $\mathbf{R}^n$ have its standard orientation.

(a) For $n > 0$, let $\mathcal{B} = \{v_1, \ldots, v_n\}$ be a basis of $\mathbf{R}^n$ and let P be the oriented n-parallelogram determined by $\mathcal{B}$. Then the *signed volume* of P is $\varepsilon \cdot$ the volume of P where $\varepsilon = 1$ if $\mathcal{B}$ gives the standard orientation of $\mathbf{R}^n$ and $\varepsilon = -1$ if $\mathcal{B}$ gives the nonstandard orientation of $\mathbf{R}^n$.

(b) For $n = 0$, the unique 0-parallelogram $P = \{0\}$ has volume 1 and *signed volume* $\varepsilon \cdot 1 = \varepsilon$ where $\varepsilon = 1$ if P has the orientation $+1$, agreeing with the standard orientation of $\mathbf{R}^0$, and $\varepsilon = -1$ if P has the orientation -1, agreeing with the nonstandard orientation of $\mathbf{R}^0$.   ◇

We will now give a geometric way of determining signed volume in case $n = 1$ or $n = 2$ before giving an algebraic way in general. The case $n = 1$ is almost trivial but we start with it to establish the pattern.

LEMMA 2.3.20. *Let* $\mathcal{B} = \{v_1 = [x]\}$ *be a basis of* $\mathbf{R}^1$ *and let* P *be the oriented 1-parallelogram determined by* $\mathcal{B}$. *Then the signed volume of* P *is equal to* $x = \det([v_1])$.

Proof. In this case P is simply the line segment between the origin 0 and the point x, so P has length (i.e., 1-dimensional volume) $|x|$, and signed volume $+|x|$ if $x > 0$ and signed volume $-|x|$ if $x < 0$. But in either case this is just equal to x. □

LEMMA 2.3.21. *Let* $\mathcal{B} = \left\{ v_1 = \begin{bmatrix} x_1 \\ y_1 \end{bmatrix}, v_2 = \begin{bmatrix} x_2 \\ y_2 \end{bmatrix} \right\}$ *be a basis of* $\mathbb{R}^2$ *and let* P *be the oriented 2-parallelogram determined by* $\mathcal{B}$. *Then the signed volume of* P *is equal to* $x_1 y_2 - x_2 y_1 = \det \left(\begin{bmatrix} v_1 | v_2 \end{bmatrix} \right)$.

Proof. It suffices to prove this in case $\mathcal{B}$ gives the standard orientation, in which case we will see that the signed volume is positive. For if the lemma is true in that case, then on the one hand switching the order of the vectors in $\mathcal{B}$ changes the orientation and hence the sign of the orientated volume, and on the other hand it changes the sign of the determinant, so it is true in case $\mathcal{B}$ gives the nonstandard orientation as well.

Actually, for simplicity we will restrict our attention to the case when v_1 and v_2 are in the first quadrant, so x_1, y_1, x_2, and y_2 are all positive. We then have the following diagram, with lengths as labeled:

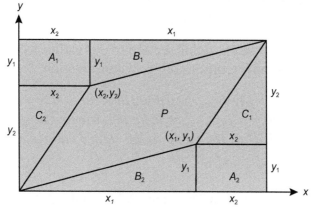

Note that rectangles A_1 and A_2 are congruent, as are triangles B_1 and B_2, and triangles C_1 and C_2. Let R be the big rectangle. Then

$$
\begin{aligned}
Area(P) &= Area(R) - 2Area(A_1) - 2Area(B_1) - 2Area(C_1) \\
&= (x_1 + x_2)(y_1 + y_2) - 2x_2 y_1 - x_1 y_1 - x_2 y_2 \\
&= x_1 y_2 - x_2 y_1
\end{aligned}
$$

as claimed.   □

Now for the general result. Henceforth we assume $n > 0$ except in Definition 2.3.25(b).

THEOREM 2.3.22. *Let* $S = \{v_1, \ldots, v_n\}$ *be a set of vectors in* $\mathbf{R}^n$ *and let* P *be the oriented n-parallelogram determined by* S. *Then the signed volume of* P *is equal to* $\det\left(\left[v_1| \ldots |v_n\right]\right)$.

Proof.   Signed volume is multilinear and alternating as a function of the vectors in S, and the signed volume of the unit cube with the standard orientation (which is the oriented n-parallelogram given by the standard basis $\mathcal{E}$) is 1.

But it is a result of linear algebra that there is a unique function of a matrix that is multilinear and alternating as a function of its columns, and that takes the value 1 on the identity matrix, namely the determinant.

Hence, by this uniqueness, the signed volume is just given by the determinant.   □

REMARK 2.3.23. Note that if S is not a basis of $\mathbf{R}^n$, Theorem 2.3.22 is true with both sides of the equality being equal to 0.   ◇

Let us relate our work in this section to our earlier work on differential forms. We now return to our previous notation.

COROLLARY 2.3.24. *Let* p *be a point in* $\mathbb{R}^n$ *and let* $S = \{v_p^1, \ldots, v_p^n\}$ *be n vectors in* $\mathbf{R}_p^n = T_p\mathbb{R}^n$. *Let* P *be the oriented n-parallelogram in* $\mathbf{R}_p^n$ *given by* S. *Then the value of the signed volume of* P *is equal to* $dx_1 \cdots dx_n(v_p^1, \ldots, v_p^n)$.

Proof.   Immediate from Theorems 2.2.20 and 2.3.22.   □

This corollary justifies the following definition.

DEFINITION 2.3.25.

(a) For $n > 0$, the n-form $dx_1 \cdots dx_n$ is the *volume form* in the standard orientation of $\mathbf{R}^n$.

(b) For $n = 0$, the constant function 1 is the *volume form* in the standard orientation of $\mathbf{R}^n$.   ◇

We have thus given a geometric meaning to an n-form in $\mathbf{R}^n$. We can also give a geometric meaning to a k-form in $\mathbf{R}^n$.

COROLLARY 2.3.26. *Let* $I = (i_1, \ldots, i_k)$ *be a k-tuple of distinct integers between 1 and n.*

Let p be a point in $\mathbb{R}^n$ and let $S = \{\mathbf{v}_p^1, \ldots, \mathbf{v}_p^k\}$ be k vectors in $T_p\mathbb{R}^n$. Let P be the oriented k-parallelogram in $\mathbf{R}_p^n = T_p\mathbb{R}^n$ given by S. Let Q be the image of P under the map $(x_1, \ldots, x_n) \mapsto (x_{i_1}, \ldots, x_{i_k})$. Then the value of the signed volume of Q is equal to $dx_{i_1} \cdots dx_{i_k}(\mathbf{v}_p^1, \ldots, \mathbf{v}_p^k)$.

Proof. Similar to the proof of Theorem 2.3.22.   □

REMARK 2.3.27. Geometrically speaking, P is a k-parallelogram in $(x_1, \ldots, x_n)$-space, and we may think of Q is its "shadow" in $(x_{i_1}, \ldots, x_{i_k})$-space.   ◇

2.4   The converse of Poincaré's Lemma, II

Recall Poincaré's Lemma "$d^2 = 0$" (Theorem 1.2.7), which had as an immediate consequence that every exact differential form is closed (Corollary 1.2.14).

Recall also that we proved a partial converse to Poincaré's Lemma as Theorem 1.4.1. In this section we prove a more general partial converse.

DEFINITION 2.4.1. A region $\mathcal{R}$ in $\mathbb{R}^n$ is *star-shaped* if there is a point p in $\mathcal{R}$ such that for any point q in $\mathcal{R}$, the straight line segment joining p and q lies entirely in $\mathcal{R}$. The point p (which need not be unique) is called a *center* of $\mathcal{R}$.   ◇

EXAMPLE 2.4.2.

(a) $\mathbb{R}^n$ is star-shaped with any point as center.

(b) The open ball of any positive radius around any point p of $\mathbb{R}^n$ is star-shaped, with p as a center.

(c) $\mathbb{R}^2 - \{(x, 0) | x \leq 0\}$ is star-shaped with $(1, 0)$ as a center.

(d) The regions in the figure below are star-shaped with center p.

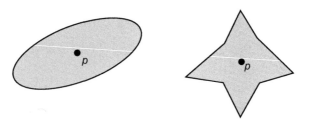

$\diamond$

THEOREM 2.4.3 (*Converse of Poincaré's Lemma*). *Let φ be a closed k-form defined on a star-shaped region $\mathcal{R}$ in $\mathbb{R}^n$, $k > 0$. Then φ is exact.*

Proof. For any k-form φ defined on $\mathcal{R}$, closed or not, we will construct a $(k - 1)$-form $\xi(\varphi)$ defined on $\mathcal{R}$ with the property that $\xi(0) = 0$ and that

$$(*_1) \qquad \varphi = d(\xi(\varphi)) + \xi(d\varphi).$$

Then, if φ is closed (so $d\varphi = 0$), we obtain

$$\varphi = d(\xi(\varphi)) + \xi(d\varphi) = d(\xi(\varphi)) + \xi(0) = d(\xi(\varphi)) + 0 = d(\xi(\varphi)).$$

Thus if $\psi = \xi(\varphi)$, then $\varphi = d\psi$, and so φ is exact.

For convenience, we let $\mathcal{R}$ have center c the origin, $c = (0, \ldots, 0)$, which we abbreviate to $c = 0$.

For a point $p = (x_1, \ldots, x_n)$ in $\mathcal{R}$, we let $\mathbf{p}$ be the constant vector field given by $\mathbf{p}_q = \begin{bmatrix} x_1 \\ \vdots \\ x_n \end{bmatrix}_q$ for any point q. We observe that the line segment joining c to p consists of all points $tp = 0 + t\mathbf{p} = (tx_1, \ldots, tx_n)$ for $0 \leq t \leq 1$.

For an ℓ-form θ and any $(\ell - 1)$ constant vector fields $\mathbf{v}^1, \ldots, \mathbf{v}^{\ell-1}$ define $\xi(\theta)$ by the formula

$$(*_2) \qquad \xi(\theta)(\mathbf{v}_p^1, \ldots, \mathbf{v}_p^{\ell-1}) = \int_0^1 t^{\ell-1} \theta_{\mathbf{p}}(\mathbf{v}_{tp}^1, \ldots, \mathbf{v}_{tp}^{\ell-1}) dt.$$

Before proceeding, let us examine this formula a little more closely.

We wish to define $\xi(\theta)$ as an $(\ell - 1)$-form on $\mathcal{R}$, and to do so we must specify its value on an arbitrary $(\ell - 1)$-tuple of vectors all based at some point of $\mathcal{R}$. This is exactly the argument to the left-hand side.

As for the right-hand side, θ is an ℓ-form. Recall that $\theta_\mathbf{p}$ denotes the contraction of θ and $\mathbf{p}$ (as defined in Definition 2.2.23) and so $\theta_\mathbf{p}$ is an $(\ell - 1)$-form, so it makes sense to evaluate it on an $(\ell - 1)$-tuple of tangent vectors all based at the same point tp. Then the integrand on the right-hand side is a function of t, and we take its (ordinary) definite integral to get the value of the right-hand side.

Since $\theta_\mathbf{p}$ is an $(\ell - 1)$-form, it is multilinear and alternating, and then by the linearity of the integral, $\xi(\theta)$ is multilinear and alternating as well, so $\xi(\theta)$ is indeed an $(\ell - 1)$-form.

Now we must verify that $\xi(\theta)$ satisfies the properties we have claimed.

First, note that if $\theta = 0$, the integrand on the right-hand side of $(*_2)$ is identically zero (regardless of $p, \mathbf{v}^1, \ldots, \mathbf{v}^{\ell-1}$) and so we see $\xi(0) = 0$. Second, note that ξ is linear; i.e., $\xi(c_1\theta_1 + c_2\theta_2) = c_1\xi(\theta_1) + c_2\xi(\theta_2)$ for any two ℓ-forms θ_1 and θ_2 and any two constants c_1 and c_2. Thus, in order to verify $(*_2)$, it suffices to verify it for φ consisting of a single term.

Thus, suppose

$$(*_3) \qquad \varphi = A(x_1, \ldots, x_n)dx_{i_1} \cdots dx_{i_k} = A(p)dx_I,$$

where p is the point $(x_1, \ldots, x_n)$ and I is the multi-index $(i_1, \ldots, i_k)$.

In order to verify $(*_2)$ it suffices to check that

$$\varphi(\mathbf{e}_p^J) = d(\xi(\varphi))(\mathbf{e}_p^J) + \xi(d\varphi)(\mathbf{e}_p^J)$$

for every multi-index J.

As a first step, let us obtain a formula for $\xi(\varphi)$. We know we can write

$$\xi(\varphi) = \sum Z^H(p)dx_H,$$

where $H = (h_1, \ldots, h_{k-1})$ ranges over all multi-indices of degree $k - 1$. Furthermore, we know that $Z^H(p) = \xi(\varphi)(\mathbf{e}_p^H)$.

We need one more piece of notation. Given a multi-index $I = (i_1, \ldots, i_k)$ of degree k and an integer m between 1 and k, we let $I_m = (i_1, \ldots, i_{m-1}, i_{m+1}, \ldots, i_k)$ be the multi-index of degree $k - 1$ obtained by omitting i_m from I. Observe that for any multi-index H of degree $k - 1$, either $H = I_m$ for some m, or else H has an entry that is not in I.

Now $dx_I = dx_{i_1} dx_{i_2} \cdots dx_{i_k}$ so we may use Definition 2.2.11 to evaluate $dx_I(\mathbf{p}_q, \mathbf{e}_q^H)$ at any point q. We see that $dx_I(\mathbf{p}_q, \mathbf{e}_q^H)$ is a product of terms

$$\pm dx_{i_1}(\mathbf{w}_q^1) dx_{i_2}(\mathbf{w}_q^2) \cdots dx_{i_k}(\mathbf{w}_q^k),$$

where $\mathbf{w}^1, \mathbf{w}^2, \ldots, \mathbf{w}^k$ is some permutation of $\mathbf{p}, \mathbf{e}^{h_1}, \ldots, \mathbf{e}^{h_{k-1}}$.

There are two cases. First, suppose H is *not* contained in I. In this case, there is some index h that is not in I, and then $dx_{i_1}(\mathbf{w}_{tp}^1) dx_{i_2}(\mathbf{w}_{tp}^2) \cdots dx_{i_k}(\mathbf{w}_{tp}^k) = 0$, as some term in this product is $dx_{i_m}(\mathbf{e}_{tp}^h) = 0$. Hence we see that

$$Z^H(p) = 0 \text{ for } H \text{ not contained in } I.$$

Now suppose $H = I_m$ for some m. Then there can be only one possible nonzero term. For example,

if $m = 1$, it is $+ dx_{i_1}(\mathbf{p}_{tp}) dx_{i_2}(\mathbf{e}_{tp}^{h_1}) \cdots dx_{i_k}(\mathbf{e}_{tp}^{h_{k-1}}) = x_{i_1}$

if $m = 2$, it is $- dx_{i_1}(\mathbf{e}_{tp}^{h_1}) dx_{i_2}(\mathbf{p}_{tp}) \cdots dx_{i_k}(\mathbf{e}_{tp}^{h_{k-1}}) = -x_{i_2}$

$$\vdots$$

if $m = k$, it is $(-1)^{k-1} dx_{i_1}(\mathbf{e}_{tp}^{h_1}) \cdots dx_{i_{k-1}}(\mathbf{e}_{tp}^{h_{k-1}}) dx_{i_k}(\mathbf{p}_{tp})$
$$= (-1)^{k-1} x_{i_k}.$$

Hence, putting these formulas together, we see

$$(*4) \quad \xi(\varphi) = \sum_{m=1}^{k} (-1)^{m-1} \left[\int_0^1 t^{k-1} x_{i_m} A(tp) dt \right] dx_{I_m}.$$

Now let us find $d(\xi(\varphi))$, the first term on the right-hand side of
$(*_1)$. We know that each term $Z^H(p)dx_H$ contributes

$$\left[\sum_{j=1}^{n} \frac{\partial}{\partial x_j} Z^H(p)dx_j \right] dx_H$$

to $d(\xi(\varphi))$, by the definition (and linearity) of d.

We calculate this, using the fact that we can differentiate under the
integral sign. Recall that tp is the point $(tx_1, \ldots, tx_n)$. Then

$$\frac{\partial}{\partial x_j} \left(\int_0^1 t^{k-1} x_{i_m} A(tx_1, \ldots, tx_n)dt \right)$$

$$= \int_0^1 \frac{\partial}{\partial x_j} \left(t^{k-1} x_{i_m} A(tx_1, \ldots, tx_n) \right) dt$$

and this integral is equal to

$$\int_0^1 t^k x_{i_m} A_{x_j}(tx_1, \ldots, tx_n)dt \qquad \text{if } j \neq i_m$$

and is equal to

$$\int_0^1 t^k x_{i_m} A_{x_j}(tx_1, \ldots, tx_n)dt$$

$$+ \int_0^1 t^{k-1} A(tx_1, \ldots, tx_n)dt \qquad \text{if } j = i_m.$$

(Note that we use the chain rule here:

$$\frac{\partial}{\partial x_j}(A(tx_1, \ldots, tx_n)) = A_{x_j}(tx_1, \ldots, tx_n)\frac{\partial}{\partial x_j}(tx_j)$$

$$= t A_{x_j}(tx_1, \ldots, tx_n);$$

and the product rule when $j = i_m$.)

Hence, substituting the value of these integrals into $(*_4)$, we obtain

$(*_5)$ $d(\xi(\varphi))$

$$= \sum_{m=1}^{k} \left[\sum_{j=1}^{n} \left((-1)^{m-1} \int_0^1 t^k x_{i_m} A_{x_j}(tp)dt \right) dx_j \right] dx_{I_m}$$

$$+ \sum_{m=1}^{k} (-1)^{m-1} \left[\int_0^1 t^{k-1} A(tp)dt \right] dx_{i_m} dx_{I_m}.$$

Now in the last of these summands, we have

$$dx_{i_1} dx_{I_1} = dx_{i_1}(dx_{i_2} dx_{i_3} \cdots dx_{i_k}) = dx_{i_1} dx_{i_2} \cdots dx_{i_k} = dx_I$$
$$dx_{i_2} dx_{I_2} = dx_{i_2}(dx_{i_1} dx_{i_3} \cdots dx_{i_k}) = -dx_{i_1} dx_{i_2} \cdots dx_{i_k} = -dx_I$$
$$\vdots$$
$$dx_{i_k} dx_{I_k} = dx_{i_k}(dx_{i_1} dx_{i_2} \cdots dx_{i_{k-1}}) = (-1)^{k-1} dx_{i_1} \cdots dx_{i_k}$$
$$= (-1)^{k-1} dx_I.$$

Indeed, the term $dx_{i_m} dx_{I_m} = (-1)^{m-1} dx_I$.

Putting this into the last summation in $(*_5)$, we see all the terms count $+$, and there are k of them, so

$(*_6)$ $d(\xi(\varphi))$

$$= \sum_{m=1}^{k} \left[\sum_{j=1}^{n} \left((-1)^{m-1} \int_0^1 t^k x_{i_m} A_{x_j}(tp)dt \right) dx_j \right] dx_{I_m}$$

$$+ \left[\int_0^1 k t^{k-1} A(tp)dt \right] dx_I$$

Now let us look at the second term on the right-hand side of $(*_1)$. This is $\xi(d\varphi)$. To compute this, we first compute $d\varphi$, which is easy.

Since $\varphi = A(p)dx_I = A(x_1, \ldots, x_n)dx_I$, we have that

$$d\varphi = \left(\sum_{j=1}^{n} A_{x_j}(x_1, \ldots, x_n)dx_j \right) dx_I$$

$$= \sum_{j=1}^{n} A_{x_j}(x_1, \ldots, x_n)dx_j \, dx_I.$$

Now we simply substitute this into $(*_2)$, remembering that $d\varphi$ is a $(k+1)$-form, so $\ell = k + 1$. We thus see that

$$(*_7) \quad \xi(d\varphi)(\mathbf{v}_p^1, \ldots, \mathbf{v}_p^k)$$

$$= \sum_{j=1}^{n} \int_0^1 t^k A_{x_j}(tp)dx_j \, dx_I(\mathbf{p}_{tp}, \mathbf{v}_{tp}^1, \ldots, \mathbf{v}_{tp}^k)dt.$$

As we have observed, this k-form $\xi(d\varphi)$ is determined by its values $\xi(d\varphi)(\mathbf{e}_p^J)$ on all k-tuples $\mathbf{e}^J$. Thus we shall compute these.

Suppose $J = (j_1, \ldots, j_k)$. Let us consider

$$dx_j \, dx_I(\mathbf{p}_{tp}, \mathbf{v}_{tp}^1, \ldots, \mathbf{v}_{tp}^k).$$

We expand the right-hand side using Lemma 2.2.15:

$$dx_j \, dx_I(\mathbf{p}_{tp}, \mathbf{e}_{tp}^J)$$
$$= dx_j(\mathbf{p}_{tp})dx_I(\mathbf{e}_{tp}^J)$$
$$- dx_j(\mathbf{e}_{tp}^{j_1})dx_{i_1}(\mathbf{p}_{tp})dx_{i_2}(\mathbf{e}_{tp}^{j_2}) \cdots dx_{i_k}(\mathbf{e}_{tp}^{j_k})$$
$$+ \cdots + (-1)^k dx_j(\mathbf{e}_{tp}^{j_k})dx_{i_1}(\mathbf{e}_{tp}^{j_1})dx_{i_2}(\mathbf{e}_{tp}^{j_1}) \cdots dx_{i_k}(\mathbf{p}_{tp}).$$

Note that the first term is

$$\begin{cases} x_j \text{ if } J = I \\ 0 \text{ if } J \neq I \end{cases}$$

and the remaining terms in the summation

$$(-1)^m dx_j(\mathbf{e}_{tp}^{j_m})dx_{i_1}(\mathbf{e}_{tp}^{j_1}) \cdots dx_{i_m}(\mathbf{p}_{tp}) \cdots dx_{i_k}(\mathbf{e}_{tp}^{j_k})$$

have the value

$$\begin{cases} (-1)^m x_{i_m} & \text{if } j_m = j \text{ and } J_m = I_m \\ 0 & \text{otherwise} \end{cases}$$

(as we have a factor of 0 somewhere in the expression, unless $j_m = j$ and $J_m = I_m$). Note that in either case this is

$$(-1)^m x_{i_m} dx_j \, dx_{I_m}.$$

Thus we see from $(*_7)$ that

$$(*_8) \quad \xi(d\varphi)$$

$$= \left[\sum_{j=1}^{n} \int_0^1 t^k x_j A_{x_j}(tp) dt \right] dx_I$$

$$+ \sum_{m=1}^{k} \left[\sum_{j=1}^{n} \left((-1)^m \int_0^1 t^k x_{i_m} A_{x_j}(tp) dt \right) dx_j \right] dx_{I_m}.$$

Now we compute $d(\xi(\varphi)) + \xi(d\varphi)$ by adding the expressions $(*_6)$ and $(*_8)$. Observe that the first summand in $(*_6)$ and the second summand in $(*_8)$ cancel each other out, as every term therein occurs with the opposite sign in each. We thus find

$$d(\xi(\varphi)) + \xi(d\varphi)$$

$$= \left[\int_0^1 \left(kt^{k-1} A(tp) + \sum_{j=1}^{n} t^k x_j A_{x_j}(tp) \right) dt \right] dx_I.$$

We claim that the integrand is $\frac{d}{dt}(t^k A(tp))$. To verify this we simply compute, using first the product rule and then the chain rule:

$$\frac{d}{dt}(t^k A(tp)) = \frac{d}{dt}(t^k A(tx_1, tx_2, \dots, tx_n))$$

$$= \left(\frac{d}{dt}(t^k) \right) A(tx_1, \dots, tx_n)$$

$$+ (t^k) \frac{d}{dt}(A(tx_1, \dots, tx_n))$$

$$
\begin{aligned}
&= kt^{k-1}A(tx_1,\ldots,tx_n) \\
&\quad + t^k\left(A_{x_1}(tx_1,\ldots,tx_n)\cdot x_1\right. \\
&\qquad \left.+\cdots+A_{x_n}(tx_1,\ldots,tx_n)\cdot x_n\right) \\
&= kt^{k-1}A(tx_1,\ldots,tx_n)+\sum_{j=1}^{n}t^k x_j A_{x_j}(tx_1,\ldots,tx_n) \\
&= kt^{k-1}A(tp)+\sum_{j=1}^{n}t^k x_j A_{x_j}(tp).
\end{aligned}
$$

Thus, by the fundamental theorem of calculus,

$$
d(\xi(\varphi))+\xi(d\varphi)=\left[t^k A(tp)\Big|_0^1\right]dx_I = A(p)dx_I = \varphi
$$

verifying $(*_1)$ and completing the proof. $\qquad\square$

In fact this proof gives a formula for ψ with $\varphi = d\psi$.

COROLLARY 2.4.4. *Let $\mathcal{R}$ be a star-shaped region $\mathcal{R}$ in $\mathbb{R}^n$ with center the origin. Let φ be a closed k-form defined on $\mathcal{R}$, $k > 0$. If*

$$
\varphi = \sum_I A^I(p)dx_I
$$

then $\varphi = d\psi$ with

$$
\psi = \sum_I \sum_{m=1}^{k}(-1)^{m-1}\left[\int_0^1 t^{k-1}x_{i_m}A(tp)dt\right]dx_{I_m},
$$

where I_m denotes I with its mth entry deleted.

Proof. For φ closed, $(*_1)$ reduces to $\varphi = d\psi$ where $\psi = \xi(\varphi)$. But then from $(*_3)$ we see that ψ is as claimed. $\qquad\square$

EXAMPLE 2.4.5. We shall work out the formula in Corollary 2.4.4 explicitly for forms in a region $\mathcal{R}$ in $\mathbb{R}^3$ with center the origin.

(a) Let φ be a closed 1-form on $\mathcal{R}$,

$$\varphi = A(x, y, z)dx + B(x, y, z)dy + C(x, y, z)dz.$$

Then

$$\psi = \left(\int_0^1 A(tx, ty, tz)dt \right)(x) + \left(\int_0^1 B(tx, ty, tz)dt \right)(y)$$
$$+ \left(\int_0^1 C(tx, ty, tz)dt \right)(z).$$

(b) Let φ be a closed 2-form on $\mathcal{R}$,

$$\varphi = A(x, y, z)dy\,dz + B(x, y, z)dz\,dx + C(x, y, z)dx\,dy.$$

Then

$$\psi = \left(\int_0^1 t A(tx, ty, tz)dt \right)(y\,dz - z\,dy)$$
$$+ \left(\int_0^1 t B(tx, ty, tz)dt \right)(z\,dx - x\,dz)$$
$$+ \left(\int_0^1 t C(tx, ty, tz)dt \right)(x\,dy - y\,dx).$$

(c) Let φ be a (necessarily) closed 3-form on $\mathcal{R}$,

$$\varphi = A(x, y, z)dx\,dy\,dz.$$

Then

$$\psi = \left(\int_0^1 t^2 A(tx, ty, tz)dt \right)(x\,dy\,dz + y\,dz\,dx + z\,dx\,dy). \quad \diamond$$

REMARK 2.4.6. Since Corollary 2.4.4 gives a formula for ψ, we could have proved Theorem 2.4.3 by simply starting with that formula and verifying that $d\psi = \varphi$. That would have been somewhat simpler than our proof (though it would still have involved quite a bit of computation), but starting with that formula would have been like pulling a rabbit out of a hat. Instead, we decided to show the reader where this particular "rabbit" came from. Moreover (though we do not need it here), the more general construction of $\xi(\varphi)$ for any form φ, not just a closed form, is an important one, and so we decided to present this general construction. ◇

Theorem 2.4.3 is valid for $k > 0$. It is illuminating to see what happens for $k = 0$. Since a 0-form is a smooth function, we use the more common letter f to denote it, rather than φ. We recall that f is closed if and only if it is constant (Corollary 1.2.13). We also observe that on the right-hand side of $(*_1)$, the first term must be 0, as it would have to be a (-1)-form, and there aren't any. Thus it remains only to evaluate the second term $\xi(df)$, which is also a 0-form, i.e., a function.

COROLLARY 2.4.7. *Let $\mathcal{R}$ be a star-shaped region $\mathcal{R}$ in $\mathbb{R}^n$ with center the origin. Let f be a smooth function on $\mathcal{R}$.*

(a) If f is constant, then $\xi(df)(p) = 0$ for every point p of $\mathcal{R}$.

(b) In general, if p is any point of $\mathcal{R}$,

$$\xi(df)(p) = f(p) - f(0).$$

Proof.

(a) If f is constant, then $df = 0$, so $\xi(df) = 0$ as we observed near the beginning of the proof of Theorem 2.4.3.

(b) Fix a point $p = (x_1, \ldots, x_n)$. Then $(*_2)$ gives

$$\xi(df)(p) = \int_0^1 df(\mathbf{p}_{tp})dt.$$

Consider the function $g(t) = f(tp) = f(tx_1, \ldots, tx_n)$. Then, applying the chain rule and recalling the meaning of $dx_1, \ldots, dx_n$ and df, we have

$$g'(t) = \frac{d}{dt}f(tp) = \frac{d}{dt}f(tx_1, \ldots, tx_n)$$

$$= x_1 \frac{\partial}{\partial x_1}f(tx_1, \ldots, tx_n) + \cdots + x_n \frac{\partial}{\partial x_n}f(tx_1, \ldots, tx_n)$$

$$= f_{x_1}(tp)x_1 + \cdots + f_{x_n}(tp)x_n$$

$$= f_{x_1}(tp)dx_1(\mathbf{p}_{tp}) + \cdots + f_{x_n}(tp)dx_n(\mathbf{p}_{tp})$$
$$= df(\mathbf{p}_{tp}).$$

Thus we see that

$$\xi(df)(p) = \int_0^1 g'(t)dt = \left[g(t)\big|_0^1\right]$$
$$= g(1) - g(0) = f(p) - f(0). \qquad \square$$

REMARK 2.4.8. Since $\mathbb{R}^n$ is star-shaped, we see that Theorem 2.4.3 is indeed a generalization of Theorem 1.4.1. ◇

2.5 Exercises

Let $f_1, \ldots, f_4, \varphi_1, \ldots, \varphi_4, \psi_1, \ldots, \psi_4$ be as in the first few exercises of Chapter 1. Let $p = (-1, 1, 2)$ and let $\mathbf{v}_p = \begin{bmatrix} 1 \\ 3 \\ 5 \end{bmatrix}_p$ and $\mathbf{w}_p = \begin{bmatrix} 2 \\ -1 \\ 0 \end{bmatrix}_p$.

1. Evaluate: (a) $\varphi_1(\mathbf{v}_p)$ (b) $\varphi_1(\mathbf{w}_p)$ (c) $\varphi_2(\mathbf{v}_p)$ (d) $\varphi_2(\mathbf{w}_p)$
   (e) $\varphi_3(\mathbf{v}_p)$ (f) $\varphi_3(\mathbf{w}_p)$ (g) $\varphi_4(\mathbf{v}_p)$ (h) $\varphi_4(\mathbf{w}_p)$

2. Evaluate: (a) $(df_1)(\mathbf{v}_p)$ (b) $(df_1)(\mathbf{w}_p)$ (c) $(df_2)(\mathbf{v}_p)$
   (d) $(df_2)(\mathbf{w}_p)$ (e) $(df_3)(\mathbf{v}_p)$ (f) $(df_3)(\mathbf{w}_p)$ (g) $(df_4)(\mathbf{v}_p)$
   (h) $(df_4)(\mathbf{w}_p)$, and verify that Corollary 2.1.11 is true in these cases.

3. Evaluate: (a) $\psi_1(\mathbf{v}_p, \mathbf{w}_p)$ (b) $\psi_2(\mathbf{v}_p, \mathbf{w}_p)$ (c) $\psi_3(\mathbf{v}_p, \mathbf{w}_p)$
   (d) $\psi_4(\mathbf{v}_p, \mathbf{w}_p)$
   Let $\rho_1 = (x^2 + yz)dx\,dy\,dz$ and $\rho_2 = (x + y^2 + z^3)dx\,dy\,dz$. Let
   $p = (3, 2, 4)$ and let $\mathbf{u}_p = \begin{bmatrix} 1 \\ 4 \\ 2 \end{bmatrix}_p$, $\mathbf{v}_p = \begin{bmatrix} 2 \\ 8 \\ 9 \end{bmatrix}_p$, and $\mathbf{w}_p = \begin{bmatrix} 1 \\ 3 \\ -2 \end{bmatrix}_p$.

4. Evaluate: (a) $\rho_1(\mathbf{u}_p, \mathbf{v}_p, \mathbf{w}_p)$ (b) $\rho_2(\mathbf{u}_p, \mathbf{v}_p, \mathbf{w}_p)$

5. Evaluate: (a) $d\psi_1(\mathbf{u}_p, \mathbf{v}_p, \mathbf{w}_p)$ (b) $d\psi_2(\mathbf{u}_p, \mathbf{v}_p, \mathbf{w}_p)$
   (c) $d\psi_3(\mathbf{u}_p, \mathbf{v}_p, \mathbf{w}_p)$ (d) $d\psi_4(\mathbf{u}_p, \mathbf{v}_p, \mathbf{w}_p)$

6. Let $S = \{v_p^1, \ldots, v_p^k\}$ be a k-tuple of vectors in $T_p\mathbb{R}^n$.
   (a) If S is linearly dependent, show that $\varphi(v_p^1, \ldots, v_p^k) = 0$ for any k-form φ defined on a neighborhood of p.
   (b) If S is linearly independent, show that $\varphi(v_p^1, \ldots, v_p^k) \neq 0$ for some k-form φ defined on a neighborhood of p.

7. Use the construction in Section 2.4 to find primitives for the differential forms φ in Exercises 10-12 of Chapter 1.

3 Push-forwards and Pull-backs in $\mathbb{R}^n$

In this chapter we develop the notions of push-forwards and pull-backs. As we will see, the end result is:

SLOGAN 3.0.1. *Points and tangent vectors push forward; functions and differential forms pull back.*

In proper mathematical language, this is

SLOGAN 3.0.2. *Points and tangent vectors are covariant; functions and differential forms are contravariant.*

The reciprocal nature of this behavior is a special case of the general phenomenon:

SLOGAN 3.0.3. *Whatever the variance of objects is, functions on those objects have the opposite variance.*

We now embark on explaining and justifying these slogans.

We shall assume throughout this chapter that we are in the following situation.

SITUATION 3.0.4. $\mathcal{R}$ is a region in $\mathbb{R}^m$, $\mathcal{S}$ is a region in $\mathbb{R}^n$, and $k : \mathcal{R} \longrightarrow \mathcal{S}$ is a smooth function. ◇

We will coordinatize $\mathbb{R}^m$ by $(u_1, \ldots, u_m)$ and $\mathbb{R}^n$ by $(x_1, \ldots, x_n)$. In the case where $m = 3$ we will coordinatize $\mathbb{R}^m$ by (u, v, w) and in the case where $n = 3$ we will coordinatize $\mathbb{R}^n$ by (x, y, z), and similarly for $m = 1, 2$ and $n = 1, 2$.

3.1 Tangent vectors

As a first step, we must carefully examine the notion of a tangent vector. Let us recall that a tangent vector is simply the velocity vector of a curve, and that velocity is the derivative of position.

Differential Forms, Second Edition. http://dx.doi.org/10.1016/B978-0-12-394403-0.00003-7

We coordinatize $\mathbb{R}^1$ by (t) (which we think of as time). By analogy with $\mathcal{R} \subseteq \mathbb{R}^m$, we let $\mathcal{Q}$ be an open interval in $\mathbb{R}^1$ containing the point (t_0).

DEFINITION 3.1.1. Let $r : \mathcal{Q} \longrightarrow \mathcal{R}$ be a smooth function, with $r(t_0) = p$. Thus $r(\mathcal{Q})$ is a smooth curve in $\mathcal{R}$. Then the *tangent vector* to this curve at p is the vector

$$\mathbf{r}'(t_0) \in T_p\mathbb{R}^m$$

defined as follows: if $(u_1, \ldots, u_m) = r(t) = (r_1(t), \ldots, r_m(t))$, then

$$\mathbf{r}'(t_0) = \begin{bmatrix} r_1'(t_0) \\ \vdots \\ r_m'(t_0) \end{bmatrix}_p \in T_p\mathbb{R}^m.$$

Furthermore, a *tangent vector* to $\mathbb{R}^m$ at p is the tangent vector to some curve at p.   ◇

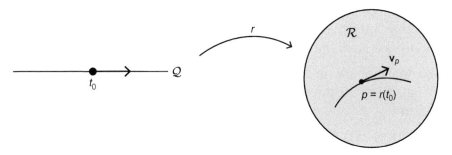

EXAMPLE 3.1.2. Let $r : \mathbb{R}^1 \longrightarrow \mathbb{R}^3$ by $(u, v, w) = r(t) = (t, t^2, t^3)$. Then

$$\mathbf{r}'(t) = \begin{bmatrix} 1 \\ 2t \\ 3t^2 \end{bmatrix},$$

giving, for example, the tangent vectors

$$\mathbf{r}'(-1) = \begin{bmatrix} 1 \\ -2 \\ 3 \end{bmatrix}_{(-1,1,-1)}, \quad \mathbf{r}'(0) = \begin{bmatrix} 1 \\ 0 \\ 0 \end{bmatrix}_{(0,0,0)},$$

$$\mathbf{r}'(1) = \begin{bmatrix} 1 \\ 2 \\ 3 \end{bmatrix}_{(1,1,1)} , \quad \mathbf{r}'(2) = \begin{bmatrix} 2 \\ 4 \\ 12 \end{bmatrix}_{(2,4,8)} . \qquad \diamond$$

REMARK 3.1.3. We should note that we are not only considering the geometric object $r(\mathcal{Q})$ as a smooth curve in $\mathcal{R}$, but that we have chosen a particular parameterization of this curve, namely that given by the function r.

For example, consider the curve C_1 in $\mathbb{R}^2$ given by $(u, v) = r(t) = (\sqrt{1 - t^2}, t)$ for $-1 < t < 1$ and the curve C_2 given by $(u, v) = s(t) = (\cos(t), \sin(t))$ for $-\pi/2 < t < \pi/2$. These both are the same geometric object: the right half of the unit circle in the (u, v)-plane. But C_1 and C_2 are different as parameterized curves. As it happens, $(1, 0) = r(1) = s(0)$ and these two curves have the same tangent vector $\mathbf{r}'(1) = \mathbf{s}'(0) = \begin{bmatrix} 0 \\ 1 \end{bmatrix}_{(1,0)}$ at this point, but this is the only point where they have the same tangent vector. For example, $\left(\frac{1}{\sqrt{2}}, \frac{1}{\sqrt{2}}\right) = r\left(\frac{1}{\sqrt{2}}\right) = s(\pi/4)$ and these two curves have different tangent vectors at this point, $\mathbf{r}'\left(\frac{1}{\sqrt{2}}\right) = \begin{bmatrix} -1 \\ 1 \end{bmatrix}_{\left(\frac{1}{\sqrt{2}}, \frac{1}{\sqrt{2}}\right)}$ and $\mathbf{s}'(\pi/4) = \begin{bmatrix} -\frac{1}{\sqrt{2}} \\ \frac{1}{\sqrt{2}} \end{bmatrix}_{\left(\frac{1}{\sqrt{2}}, \frac{1}{\sqrt{2}}\right)} . \qquad \diamond$

REMARK 3.1.4. It is very useful to adopt the following viewpoint: We consider a curve C_1 parameterized by r_1 passing through the point p when $t = t_1$ and a curve C_2 parametrized by r_2 passing through the point p when $t = t_2$. We say that these two curves are equivalent if $\mathbf{r}_1'(t_1) = \mathbf{r}_2'(t_2)$. Then a tangent vector to $\mathbb{R}^n$ at p is an equivalence class of curves.

Furthermore, in this situation the tangent vector represented by this equivalence class is the vector $\mathbf{r}_1'(t_1) = \mathbf{r}_2'(t_2) \in T_p\mathbb{R}^n$. $\qquad \diamond$

EXAMPLE 3.1.5. For $p = (1, 0)$ the two curves C_1 and C_2 of Remark 3.1.3 are equivalent, and these curves both represent the tangent vector $\begin{bmatrix} 0 \\ 1 \end{bmatrix}_{(1,0)}$.   ◇

REMARK 3.1.6. We will be careful throughout to maintain the notational distinction between the derivative $r'(t)$, a linear transformation, and the vector $\mathbf{r}'(t)$, a tangent vector.   ◇

3.2 Points, tangent vectors, and push-forwards

Let $\mathcal{R}$ be a region in $\mathbb{R}^m$, S be a region in $\mathbb{R}^n$, and $\mathcal{T}$ be a region in $\mathbb{R}^p$. Let $k : \mathcal{R} \longrightarrow S$ and $\ell : S \longrightarrow \mathcal{T}$ be smooth maps.

We begin with a lemma that looks silly, but we state it to make a point (compare Lemma 3.2.5 below).

LEMMA 3.2.1. *Let k_* be the map induced on points by k, given by $k_* = k$, and similarly for ℓ_*. Then:*

(a) If $k = \mathrm{id}$, then $k_* = \mathrm{id}$.

(b) If m is the composition $m = \ell(k)$, then m_* is the composition $m_* = \ell_*(k_*)$.

Proof. Since $k_* = k$ and $\ell_* = \ell$, this is completely trivial.   □

Since $k_* = k$, there is really no necessity to introduce this new notation, and we will not use it in this context further (except in Remark 3.3.2(a) below, where we wish to make a similar point). But we will use the notation k_* in other contexts, where it does denote something new. In fact, we use it in the following context.

DEFINITION 3.2.2. Let $k : \mathcal{R} \longrightarrow S$ be a smooth map. Let p be any point of $\mathcal{R}$ and let $q = k(p)$. Then $k_* : T_p\mathbb{R}^m \longrightarrow T_q\mathbb{R}^n$, the map *induced* on tangent vectors by k, is the map

$$k_*(\mathbf{v}_p) = \big(k'(p)(\mathbf{v}_p)\big)_q$$

where $\mathbf{v}_p$ is any tangent vector to $\mathbb{R}^m$ at p, $\mathbf{v}_p \in T_p\mathbb{R}^m$.

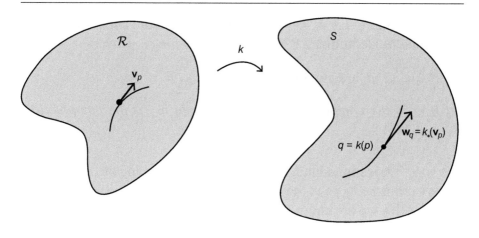

$\diamond$

REMARK 3.2.3. Since the derivative $k'(p)$ is a linear transformation, we immediately see that $k_* : T_p\mathbb{R}^m \longrightarrow T_q\mathbb{R}^n$ is a linear transformation. $\diamond$

REMARK 3.2.4. Let us formulate Definition 3.2.2 in a more concrete way. Let $k : \mathcal{R} \longrightarrow \mathcal{S}$ be given by

$$(x_1, \ldots, x_n) = (k_1(u_1, \ldots, u_m), \ldots, k_n(u_1, \ldots, u_m))$$

and let $q = k(p)$.

Then k' is the derivative matrix

$$k' = \begin{bmatrix} \frac{\partial k_1}{\partial u_1} & \cdots & \frac{\partial k_1}{\partial u_m} \\ \vdots & \vdots & \vdots \\ \frac{\partial k_n}{\partial u_1} & \cdots & \frac{\partial k_n}{\partial u_m} \end{bmatrix}.$$

Thus if

$$\mathbf{v}_p = \begin{bmatrix} a_1 \\ \vdots \\ a_m \end{bmatrix}_p,$$

then

$$k_*(\mathbf{v}_p) = \left(\begin{bmatrix} \frac{\partial k_1}{\partial u_1}(p) & \cdots & \frac{\partial k_1}{\partial u_m}(p) \\ \vdots & \vdots & \vdots \\ \frac{\partial k_n}{\partial u_1}(p) & \cdots & \frac{\partial k_n}{\partial u_m}(p) \end{bmatrix} \begin{bmatrix} a_1 \\ \vdots \\ a_m \end{bmatrix} \right)_q. \qquad \diamond$$

LEMMA 3.2.5. *Let k_* be the map induced on tangent vectors by k, as defined in* Definition 3.2.2, *and similarly for ℓ_*. Then:*

(a) If $k = \mathrm{id}$, then $k_* = \mathrm{id}$.

(b) If m is the composition $m = \ell(k)$, then m_* is the composition $m_* = \ell_*(k_*)$.

Proof. (a) is immediate: If k is the identity mapping on points, then for any point p, $k'(p)$ is the identity linear transformation.

As for (b), we are claiming that

$$m'(p) = \big(\ell(k)\big)'(p) = \ell'(k(p))k'(p).$$

But this is just the chain rule.   □

DEFINITION 3.2.6. Let $k : \mathcal{R} \longrightarrow \mathcal{S}$, let $p \in \mathcal{R}$ be a point, and let $\mathbf{v}_p \in T_p\mathbb{R}^m$ be a tangent vector to $\mathbb{R}^m$ at p.

(a) The point $q = k(p) \in \mathcal{S}$ is the *push-forward* of p via k.

(b) The tangent vector $\mathbf{w}_q = k_*(\mathbf{v}_p) \in T_q\mathbb{R}^n$ is the *push-forward* of $\mathbf{v}_p$ via k.

(c) If $(\mathbf{v}_p^1, \ldots, \mathbf{v}_p^j)$ is a j-tuple of tangent vectors to $\mathbb{R}^m$ at p, its *push-forward* via k is the j-tuple $k_*(\mathbf{v}_p^1, \ldots, \mathbf{v}_p^j) = (k_*(\mathbf{v}_p^1), \ldots, k_*(\mathbf{v}_p^j))$ of tangent vectors to $\mathbb{R}^n$ at q.   ◇

Let us now compute some push-forwards. First we compute the push-forward of some particular vectors. Note we use the usual notation for partial derivatives in the statement of this lemma. Note also we continue to be careful and use the boldface $\mathbf{k}_{u_i}(p)_q$ to denote a tangent vector.

LEMMA 3.2.7. *Let $k : \mathcal{R} \longrightarrow \mathcal{S}$ be a smooth map, given by*

$$(x_1, \ldots, x_n) = (k_1(u_1, \ldots, u_m), \ldots, k_n(u_1, \ldots, u_m)).$$

Let p be any point of $\mathcal{R}$ and let $q = k(p)$. Then

$$k_*(\mathbf{e}_p^i) = \mathbf{k}_{u_i}(p)_q = \begin{bmatrix} \frac{\partial k_1}{\partial u_i}(p) \\ \vdots \\ \frac{\partial k_n}{\partial u_i}(p) \end{bmatrix}_q .$$

Proof. By Definition 3.2.2, $k_*\left(\mathbf{e}_p^i\right) = \left(k'(p)(\mathbf{e}_p^i)\right)_q$. Now $\mathbf{e}_p^i$ is a vector with a 1 in position i and 0 in every other position, so the result of multiplying a matrix by this vector is the ith column of that matrix. The matrix of $k'(p)$ is given explicitly in Remark 3.2.4, and so the result follows immediately. $\qquad\qquad\qquad\square$

Now we compute the push-forward of tangent vectors in general.

LEMMA 3.2.8. *Let $k : \mathcal{R} \longrightarrow \mathcal{S}$ be a smooth map, given by*

$$(x_1, \ldots, x_n) = (k_1(u_1, \ldots, u_m), \ldots, k_n(u_1, \ldots, u_m)).$$

Let p be any point of $\mathcal{R}$ and let $q = k(p)$. Let $\mathcal{Q}$ be an open interval in $\mathbb{R}^1$ containing the point p. Let $r : \mathcal{Q} \longrightarrow \mathcal{R}$ be a parameterized curve with $r(t_0) = p$, and consider its tangent vector $\mathbf{r}'(t_0) \in T_p\mathbb{R}^m$. Write $(u_1, \ldots, u_m) = (r_1(t), \ldots, r_m(t))$. Then

$$\begin{aligned} k_*(\mathbf{r}'(t_0)) &= k'(p)\left(\mathbf{r}'(t_0)\right)_q \\ &= \begin{bmatrix} \sum_{i=1}^m \frac{\partial k_1}{\partial u_i}(p)r_i'(t_0) \\ \vdots \\ \sum_{i=1}^m \frac{\partial k_n}{\partial u_i}(p)r_i'(t_0) \end{bmatrix}_q . \end{aligned}$$

Proof. This is just Definition 3.2.2, Definition 3.1.1, and Remark 3.2.4. $\qquad\qquad\qquad\square$

REMARK 3.2.9. We say that points and tangent vectors "push forward" because they go the "same way" that the function k does. The function k goes from $\mathcal{R}$ to $\mathcal{S}$, and under the induced maps, points and tangent vectors similarly go from $\mathcal{R}$ to $\mathcal{S}$, i.e., they go in the "same direction." $\qquad\qquad\qquad\diamond$

REMARK 3.2.10. Let us reconsider Lemma 3.2.8 from the viewpoint of Remark 3.1.4. The curve C in $\mathcal{R}$ parameterized by r represents the tangent vector $\mathbf{v}_p = \mathbf{r}'(t_0) \in T_p\mathbb{R}^m$, where $p = r(t_0)$. By exactly the same logic, the curve D in $\mathcal{S}$ parameterized by $k \circ r$ represents the tangent vector $\mathbf{w}_q = (kr)'(t_0) \in T_q\mathbb{R}^n$, where $q = kr(t_0) = k(p)$. Then if we let $D = k_*(C)$, we have that if C represents the tangent vector $\mathbf{v}_p$, then $k_*(C)$ represents the tangent vector $k_*(\mathbf{v}_p)$, and furthermore that this tangent vector is given by $k_*(\mathbf{v}_p) = (k'(p)(\mathbf{v}_p))_q$.   ◇

REMARK 3.2.11. A tangent vector to a curve C is itself a push-forward of a tangent vector to $\mathbb{R}^1$, as we now see: Coordinatize $\mathbb{R}^1$ by t, i.e., let $\mathbb{R}^1 = \{(t)\}$, and consider the constant vector field $\mathbf{i} = \mathbf{e}_1 = [1]_t$ at every point $t \in \mathbb{R}^1$. Let $\mathcal{Q}$ be an open interval in $\mathbb{R}^1$ containing the point t_0. Let $r : \mathcal{Q} \longrightarrow \mathcal{R}$ be a smooth map and let $p = r(t_0)$. Thus $r(\mathcal{Q})$ is a smooth curve in $\mathcal{R}$. The tangent vector to this curve at p is the push-forward $r_*([1]_{t_0}) \in T_p\mathbb{R}^m$.

Let us see this explicitly. Let $r(t) = (r_1(t), \ldots, r_m(t))$. Then $r'(t_0)$ is the linear transformation from $T_{t_0}\mathbb{R}^1$ to $T_p\mathbb{R}^m$ with matrix

$$r'(t_0) = \begin{bmatrix} r_1'(t_0) \\ \vdots \\ r_m'(t_0) \end{bmatrix}.$$

Thus

$$r_*(\mathbf{i}_{t_0}) = \left(r'(t_0)\right)(\mathbf{i}_{t_0}) = \left(\begin{bmatrix} r_1'(t_0) \\ \vdots \\ r_m'(t_0) \end{bmatrix}\right)([1]_{t_0})$$

$$= \begin{bmatrix} r_1'(t_0) \\ \vdots \\ r_m'(t_0) \end{bmatrix}_p \in T_p\mathbb{R}^m$$

as in Definition 3.1.1.   ◇

MNEMONIC 3.2.12. In computing the rectangular matrix k' as in Remark 3.2.4, we need to remember that in this matrix

Columns ↔ Independent variables,

Rows ↔ Dependent variables

and not the other way around. In order to easily remember this, we find the mnemonic

Chocolate Is Really Delicious

to be particularly useful. ◇

3.3 Differential forms and pull-backs

Now we come to "pull-backs." These go in the "wrong direction." As this may be a notion that the reader is unfamiliar with, we shall take special care in its explication.

We suppose that we are in the same situation as in the preceding section. That is, we let $\mathcal{R}$ be a region in $\mathbb{R}^m$, $\mathcal{S}$ be a region in $\mathbb{R}^n$, and $\mathcal{T}$ be a region in $\mathbb{R}^p$, and we let $k : \mathcal{R} \longrightarrow \mathcal{S}$ and $\ell : \mathcal{S} \longrightarrow \mathcal{T}$ be smooth maps.

We begin by seeing how to pull back (smooth) functions.

DEFINITION 3.3.1. Let $\mathcal{F}(\mathcal{S})$ be the set of smooth functions $\mathcal{F}(\mathcal{S}) = \{f : \mathcal{S} \longrightarrow \mathbb{R}^1\}$ and similarly for $\mathcal{F}(\mathcal{R})$. Let $k : \mathcal{R} \longrightarrow \mathcal{S}$ be a smooth map. $k^* : \mathcal{F}(\mathcal{S}) \longrightarrow \mathcal{F}(\mathcal{R})$, the map *induced* on functions by k, is the map

$$k^*(f) = f(k)$$

where the right-hand side denotes composition. Thus $k^*(f) \in \mathcal{F}(\mathcal{R})$, the *pull-back* of f via k, is the function defined by

$$\big(k^*(f)\big)(p) = \big(f(k)\big)(p) = f(k(p))$$

for any point $p \in \mathcal{R}$. ◇

REMARK 3.3.2.

(a) Recall we have the map k_* induced by k on points, and this is just $k_* = k$ (compare Lemma 3.2.1). Then in this notation Definition 3.3.1 reads

$$k^*(f) = f(k_*).$$

(b) Note that Definition 3.3.1 really makes sense. We are trying to define a function $k^*(f)$ on $\mathcal{R}$, so we must specify its value at any point p of $\mathcal{R}$, and this definition does so.

(c) To emphasize what's going on here, let's note that we have the following situation to contend with. We start with a point p in $\mathcal{R}$ and a function S defined on N. Then it doesn't make sense to evaluate f at p, as f isn't defined there. What does make sense is to try one of the following strategies: Either push forward p to a point $q = k(p)$ on S, where f is defined, and evaluate f there, or pull back f to a new function $k^*(f)$ on $\mathcal{R}$, where p lives, and evaluate this function there. The first strategy gives us the value $f(q) = f(k(p))$, and the very definition of $k^*(f)$ is that the second strategy should give us the *same* answer, which is exactly what Definition 3.3.1 says.

(d) In case you're thinking that this is all too complicated, and that we should simply push functions forward, just as we push points forward, let me note that this *won't work*. Suppose we start with a function f defined on $\mathcal{R}$, and we want to get a new function $k_*(f)$ on S. How should that be defined at a point q of S? If $q = f(p)$, then you might want to define $k_*(f)(q)$ to be $f(p)$. But there are two problems with this definition: First, the point q might *not* be in the image of k, i.e., there might not be any such point p and we could not make this definition. Second, q might be the image of *more than one* which would also be a difficulty. For, if $q = k(p_1) = k(p_2)$, what should $k_*(f)(q)$ be defined as, $f(p_1)$ or $f(p_2)$?

(e) We hope this argument has convinced you that pull-backs of functions, although surprising at first glance, are actually very natural (while any attempt to push functions forward is not only unnatural, but also undoable).

(f) We are going to use the same strategy to pull back differential forms. Everything we have said in this remark will apply to that situation, so you should pause here to make sure you really understand it.   ◇

REMARK 3.3.3. We denote push-forwards by k_* (lower-star) and pull-backs by k^* (upper-star). This is standard mathematical notation. ◇

We have just seen how to pull back functions, i.e., 0-forms. Recall our notation from Chapter 1: $\Omega^j(S) = \{k\text{-forms on } S\}$ and similarly for $\Omega^j(\mathcal{R})$. (Thus, in Definition 3.3.1, $\mathcal{F}(S) = \Omega^0(S)$, and similarly for $\mathcal{F}(R)$.)

Now we see how to pull back 1-forms.

DEFINITION 3.3.4. Let $k : \mathcal{R} \longrightarrow S$ be a smooth map. $k^* : \Omega^1(S) \longrightarrow \Omega^1(\mathcal{R})$, the map *induced* on 1-forms by k, is the map

$$k^*(\varphi) = \varphi(k_*)$$

where the right-hand side denotes composition. Thus $k^*(\varphi) \in \Omega^1(\mathcal{R})$, the *pull-back* of φ via k, is the 1-form defined by

$$\left(k^*(\varphi)\right)(\mathbf{v}_p) = \left(\varphi(k_*)\right)(\mathbf{v}_p) = \varphi(k_*(\mathbf{v}_p))$$

for any tangent vector $\mathbf{v}_p$ at any point $p \in \mathcal{R}$. ◇

REMARK 3.3.5.

(a) Note how analogous Definition 3.3.4 is to Definition 3.3.1. In particular, recall that 1-forms are functions on tangent vectors, so again we are pulling back a function by pushing forward the place where it is evaluated.

(b) In particular, note that Definition 3.3.4 is formally *identical* to the reformulation of Definition 3.3.1 given in Remark 3.3.2(a).

(c) In fact, the analog of all of Remark 3.3.2 applies to this situation as well. Note that $k^*(\varphi)$, as defined in Definition 3.3.4, really is a function on tangent vectors to points in $\mathcal{R}$; that is, it is a 1-form on $\mathcal{R}$ whose value $\left(k^*(\varphi)\right)(\mathbf{v}_p)$ on a tangent vector $\mathbf{v}_p$ at a point $p \in \mathcal{R}$ is given by $\varphi(k_*(\mathbf{v}_p))$. Note that this expression really does make sense, as the push-forward $k_*(\mathbf{v}_p)$, which was defined in Definition 3.3.4, is a tangent vector at the point $k(p) \in S$. Thus it makes sense to evaluate the 1-form $\varphi \in \Omega^1(S)$ on it. We encourage the reader to reread Remark 3.3.2 with 1-forms in mind. ◇

Having pulled back 1-forms, we now pull back j-forms. Here there is absolutely nothing new going on. If we simply remember that a j-form is a function on j-tuples of tangent vectors (all tangent at the same point), the generalization of Definition 3.3.4 becomes clear.

DEFINITION 3.3.6. Let $k : \mathcal{R} \longrightarrow \mathcal{S}$ be a smooth map. $k^* : \Omega^j(\mathcal{S}) \longrightarrow \Omega^j(\mathcal{R})$, the map *induced* on j-forms by k, is the map

$$k^*(\varphi) = \varphi(k_*)$$

where the right-hand side denotes composition. Thus $k^*(\varphi) \in \Omega^j(\mathcal{R})$, the *pull-back* of φ via k, is the j-form defined by

$$\left(k^*(\varphi) \right) \left((\mathbf{v}_p^1, \ldots, \mathbf{v}_p^j) \right) = \left(\varphi(k_*) \right) \left((\mathbf{v}_p^1, \ldots, \mathbf{v}_p^j) \right)$$
$$= \varphi \left(k_*(\mathbf{v}_p^1), \ldots, k_*(\mathbf{v}_p^j) \right)$$

for any j-tuple of tangent vectors $(\mathbf{v}_p^1, \ldots, \mathbf{v}_p^j)$ at any point $p \in \mathcal{R}$. ◇

We have developed Lemma 3.2.5 for push-forwards. Let us first see that we have a complete analog of Lemma 3.2.5 for pull-backs. Note, however, that in part (b) the order of composition is *reversed*.

LEMMA 3.3.7. *Let $\mathcal{R}$ be a region in $\mathbb{R}^m$, $\mathcal{S}$ be a region in $\mathbb{R}^n$, and $\mathcal{T}$ be a region in $\mathbb{R}^p$. Let $k : \mathcal{R} \longrightarrow \mathcal{S}$ and $\ell : \mathcal{S} \longrightarrow \mathcal{T}$ be smooth maps. Then:*

(a) If $k = \mathrm{id}$, then $k^* = \mathrm{id}$.

(b) If m is the composition $m = \ell(k)$, then m^* is the composition $m^* = k^*(\ell^*)$.

Proof. Part (a) follows immediately from Lemma 3.2.5(a):

$$\left(\mathrm{id}^*(\varphi) \right) \left((\mathbf{v}_p^1, \ldots, \mathbf{v}_p^j) \right) = \left(\varphi(\mathrm{id}_*) \right) \left((\mathbf{v}_p^1, \ldots, \mathbf{v}_p^j) \right)$$
$$= \varphi \left(\mathrm{id}_*(\mathbf{v}_p^1), \ldots, \mathrm{id}_*(\mathbf{v}_p^j) \right)$$
$$= \varphi \left(\mathbf{v}_p^1, \ldots, \mathbf{v}_p^j \right)$$

so $\mathrm{id}^* = \mathrm{id}$.

Part (b) follows immediately from Lemma 3.2.5(b):

$$
\begin{aligned}
\left(m^*(\varphi)\right)\left(\mathbf{v}_p^1, \ldots, \mathbf{v}_p^j\right) &= \left(\varphi(m_*)\right)\left((\mathbf{v}_p^1, \ldots, \mathbf{v}_p^j)\right) \\
&= \varphi\left(m_*(\mathbf{v}_p^1), \ldots, m_*(\mathbf{v}_p^j)\right) \\
&= \varphi\left(\ell_*(k_*(\mathbf{v}_p^1)), \ldots, \ell_*(k_*(\mathbf{v}_p^j))\right) \\
&= \left(\ell^*(\varphi)\right)\left(k_*(\mathbf{v}_p^1), \ldots, k_*(\mathbf{v}_p^j)\right) \\
&= \left(k^*(\ell^*(\varphi))\right)\left(\mathbf{v}_p^1, \ldots, \mathbf{v}_p^j\right)
\end{aligned}
$$

so $m^* = k^*(\ell^*)$ $\qquad\square$

We now embark on a theoretical discussion of properties of the pull-back. But one point of this discussion is to enable us to easily derive concrete formulas for pull-backs, which we will then proceed to do.

LEMMA 3.3.8. *In the situation of* Definition 3.3.6, *let ψ and ρ be j-forms on S and let a and b be real numbers. Then*

$$
k^*(f\psi + g\rho) = ak^*(\psi) + bk^*(\rho).
$$

Proof. We simply compute

$$
\begin{aligned}
\left(k^*(a\psi + b\rho)\right)\left((\mathbf{v}_p^1, \ldots, \mathbf{v}_p^j)\right) &= \left((a\psi + b\rho)(k_*)\right)\left((\mathbf{v}_p^1, \ldots, \mathbf{v}_p^j)\right) \\
&= \left(a\psi + b\rho\right)\left(k_*(\mathbf{v}_p^1), \ldots, k_*(\mathbf{v}_p^j)\right) \\
&= a\psi\left(k_*(\mathbf{v}_p^1), \ldots, k_*(\mathbf{v}_p^j)\right) \\
&\quad + b\rho\left(k_*(\mathbf{v}_p^1), \ldots, k_*(\mathbf{v}_p^j)\right) \\
&= a\left(k^*(\psi)\right)\left((\mathbf{v}_p^1, \ldots, \mathbf{v}_p^j)\right) \\
&\quad + b\left(k^*(\rho)\right)\left((\mathbf{v}_p^1, \ldots, \mathbf{v}_p^j)\right). \quad\square
\end{aligned}
$$

From our definition of pull-backs, we see that if φ is a differential form on S, then $k^*(\varphi)$ is a function on tangent vectors at points of M. We need somewhat more for $k^*(\varphi)$ to be a differential form—it must be multilinear, alternating, and smooth. We prove these properties now.

LEMMA 3.3.9. *In the situation of* Definition 3.3.6, $\psi = k^*(\varphi)$ *is multilinear, alternating, and smooth, and hence a j-form on $\mathcal{R}$.*

Proof. Multilinearity: This is a direct consequence of the facts that φ is linear (by the definition of a j-form) and that k_* is linear (Remark 3.2.3). Suppose that for some i, $\mathbf{v}_p^i = a\mathbf{u}_p + b\mathbf{w}_p$ for real numbers a and b. Then

$$\psi\left((\mathbf{v}_p^1, \ldots, a\mathbf{u}_p + b\mathbf{w}_p, \ldots, \mathbf{v}_p^j)\right)$$

$$= k^*(\varphi)\left((\mathbf{v}_p^1, \ldots, a\mathbf{u}_p + b\mathbf{w}_p, \ldots, \mathbf{v}_p^j)\right)$$

$$= \varphi\left(k_*(\mathbf{v}_p^1), \ldots, k_*(a\mathbf{u}_p + b\mathbf{w}_p), \ldots, k_*(\mathbf{v}_p^j)\right)$$

$$= \varphi\left(k_*(\mathbf{v}_p^1), \ldots, ak_*(\mathbf{u}_p) + bk_*(\mathbf{w}_p), \ldots, k_*(\mathbf{v}_p^j)\right)$$

$$= a\varphi\left(k_*(\mathbf{v}_p^1), \ldots, k_*(\mathbf{u}_p), \ldots, k_*(\mathbf{v}_p^j)\right)$$

$$+ b\varphi\left(k_*(\mathbf{v}_p^1), \ldots, k_*(\mathbf{w}_p), \ldots, k_*(\mathbf{v}_p^j)\right)$$

$$= ak^*(\varphi)\left((\mathbf{v}_p^1, \ldots, \mathbf{u}_p, \ldots, \mathbf{v}_p^j)\right)$$

$$+ bk^*(\varphi)\left((\mathbf{v}_p^1, \ldots, \mathbf{w}_p, \ldots, \mathbf{v}_p^j)\right)$$

$$= a\psi\left((\mathbf{v}_p^1, \ldots, \mathbf{u}_p, \ldots, \mathbf{v}_p^j)\right) + b\psi\left((\mathbf{v}_p^1, \ldots, \mathbf{w}_p, \ldots, \mathbf{v}_p^j)\right)$$

Alternation: This is a direct consequence of the fact that φ is alternating. Suppose that for some $i_1 \neq i_2$, $\mathbf{v}_p^{i_1} = \mathbf{v}_p^{i_2}$, in which case $k_*(\mathbf{v}_p^{i_1}) = k_*(\mathbf{v}_p^{i_2})$. Then

$$\psi\left((\mathbf{v}_p^1, \ldots, \mathbf{v}_p^{i_1}, \ldots, \mathbf{v}_p^{i_2}, \ldots, \mathbf{v}_p^j)\right)$$

$$= k^*(\varphi)\left((\mathbf{v}_p^1, \ldots, \mathbf{v}_p^{i_1}, \ldots, \mathbf{v}_p^{i_2}, \ldots, \mathbf{v}_p^j)\right)$$

$$= \varphi\left((\mathbf{v}_p^1, \ldots, k_*(\mathbf{v}_p^{i_1}), \ldots, k_*(\mathbf{v}_p^{i_2}), \ldots, \mathbf{v}_p^j)\right)$$

$$= 0.$$

Smoothness: We know that φ is a sum of terms of the form $A^I dx_I$, with each A^I a smooth function, and, by the linearity of k^* and the fact that a sum of smooth functions is smooth, it suffices to show that $\psi_I = k^*(A^I dx_I)$ is smooth for each I.

Fix a particular value I_0 of I.

Referring to Definition 2.2.1(3), let $\mathbf{F}^1, \ldots, \mathbf{F}^j$ be smooth vector fields on $\mathcal{R}$, and consider the function $e(p) = \psi_{I_0}(\mathbf{F}_p^1, \ldots, \mathbf{F}_p^j)$. Then

$$
\begin{aligned}
e(p) &= \psi_{I_0}(\mathbf{F}_p^1, \ldots, \mathbf{F}_p^j) \\
&= \left(k^*(A^{I_0} dx_{I_0}) \right)(\mathbf{F}_p^1, \ldots, \mathbf{F}_p^j) \\
&= (A^{I_0} dx_{I_0}) \left(k_*(\mathbf{F}_p^1), \ldots, k_*(\mathbf{F}_p^j) \right) \\
&= A^{I_0}(k(p)) dx_{I_0} \left(k_*(\mathbf{F}_p^1), \ldots, k_*(\mathbf{F}_p^j) \right).
\end{aligned}
$$

Now $A^{I_0}(k(p))$ is a smooth function of p, and, referring to Definition 2.2.11, we see that $dx_{I_0}(k_*(\mathbf{F}_p^1), \ldots, k_*(\mathbf{F}_p^j))$ is a sum of terms each of which is $\pm$ a product of j factors, each of those factors being the i_1th entry in the vector $k_*(\mathbf{F}_p^{i_2})$ for some i_1 and i_2. But each of those entries is a smooth function of p as well, and hence $e(p)$ is a smooth function of p, as required. $\qquad \square$

COROLLARY 3.3.10. *In the situation of Definition 3.3.6,*

$$
k^* : \Omega^j(\mathcal{S}) \longrightarrow \Omega^j(\mathcal{R})
$$

is a linear transformation.

Proof. Once we know that the pull-back of a j-form on $\mathcal{S}$ is a j-form on $\mathcal{R}$, which is what Lemma 3.3.9 tells us, this is just a restatement of Lemma 3.3.8. $\qquad \square$

LEMMA 3.3.11. *Let ψ be a j-form on $\mathcal{S}$ and let f be a smooth function on $\mathcal{S}$. Then*

$$
k^*(f\psi) = k^*(f) k^*(\psi).
$$

Proof. We want to show that $k^*(f\psi) = k^*(f)k^*(\psi)$. By definition, this means that for any point p of $\mathcal{R}$, and any j-tuple of vectors
$(\mathbf{v}_p^1, \ldots, \mathbf{v}_p^j)$ all tangent to $\mathbb{R}^m$ at p,

$$k^*(f\psi)\left((\mathbf{v}_p^1, \ldots, \mathbf{v}_p^j)\right) = (k^*(f))(p)k^*(\psi)\left((\mathbf{v}_p^1, \ldots, \mathbf{v}_p^j)\right).$$

Recall that, by definition, $(k^*(f))(p) = k(f(p))$. Then we have

$$\begin{aligned}
k^*(f\psi)\left((\mathbf{v}_p^1, \ldots, \mathbf{v}_p^j)\right) &= (f\psi)\left((k_*(\mathbf{v}_p^1), \ldots, k_*(\mathbf{v}_p^j))\right) \\
&= f(k(p))\psi\left((k_*(\mathbf{v}_p^1), \ldots, k_*(\mathbf{v}_p^j))\right) \\
&= f(k(p))k^*(\psi)\left((\mathbf{v}_p^1, \ldots, \mathbf{v}_p^j)\right) \\
&= (k^*(f))(p)k^*(\psi)\left((\mathbf{v}_p^1, \ldots, \mathbf{v}_p^j)\right)
\end{aligned}$$

as claimed. $\square$

COROLLARY 3.3.12. *Let ψ and ρ be j-forms on S and let f and
g be smooth functions on S. Then*

$$k^*(f\psi + g\rho) = k^*(f)k^*(\psi) + k^*(g)k^*(\rho).$$

Proof. This is immediate from Lemma 3.3.11 and Lemma 3.3.8.
                                                                                                       $\square$

REMARK 3.3.13. We have stated Corollary 3.3.12 explicitly in
order to observe that it is a strengthening of Lemma 3.3.8, as we
may regard Lemma 3.3.8 as the special case of Corollary 3.3.12
where f and g are the constant functions $f(q) = a$ and $g(q) = b$
respectively.   ◇

Let us now develop a concrete formula for the pull-back of a 1-form.
First we deal with a special case.

LEMMA 3.3.14. *Let $k : \mathcal{R} \longrightarrow S$ be given by*

$$\begin{aligned}
(x_1, \ldots, x_n) = k(u_1, \ldots, u_m) &= (k_1(u_1, \ldots, u_m), \ldots, \\
&\quad k_n(u_1, \ldots, u_m)).
\end{aligned}$$

Let φ be the 1-form on S given by

$$\varphi = dx_j$$

Then $k^*(\varphi)$ is the 1-form on $\mathcal{R}$ given by

$$k^*(\varphi) = \sum_{i=1}^{m} \frac{\partial k_j}{\partial u_i}(u_1, \ldots, u_m) du_i.$$

Proof. Let $p = (u_1, \ldots, u_m)$ and $q = k(p) = (k_1(p), \ldots, k_n(p)) = (x_1, \ldots, x_n)$.

We know that $k^*(\varphi)$ must be of the form

$$k^*(\varphi) = \sum_{i=1}^{m} B^i(u_1, \ldots, u_m) du_i$$

(as any 1-form on $\mathcal{R}$ can be written in that way) and we need to determine the functions $B^1, \ldots, B^m$.

We also know that, if $k^*(\varphi)$ is written in this way,

$$B^i(p) = k^*(\varphi)(e^i_p).$$

But by the definition of the pull-back,

$$k^*(dx_j)(e^i_p) = dx_j(k_*(e^i_p)).$$

We computed $k_*(e^i_p)$ in Lemma 3.3.8. Substituting that answer here, we find

$$k^*(dx_j)(e^i_p) = dx_j(k_*(e^i_p)) = dx_j\left(\begin{bmatrix} \frac{\partial k_1}{\partial u_i}(p) \\ \vdots \\ \frac{\partial k_n}{\partial u_i}(p) \end{bmatrix}_q\right)$$

Recalling that dx_j "picks out" the jth entry of a tangent vector, we obtained the claimed formula. $\square$

Now we find a formula for the pull-back of a 1-form in general.

COROLLARY 3.3.15. *Let* $k : \mathcal{R} \longrightarrow \mathcal{S}$ *be given by*

$$(x_1, \ldots, x_n) = k(u_1, \ldots, u_m) = (k_1(u_1, \ldots, u_m), \ldots,$$
$$k_n(u_1, \ldots, u_m)).$$

Let φ *be the* 1*-form on* $\mathcal{S}$ *given by*

$$\varphi = \sum_{j=1}^{n} A^j(x_1, \ldots, x_n) dx_j.$$

Then $k^*(\varphi)$ *is the* 1*-form on* $\mathcal{R}$ *given by*

$$k^*(\varphi) = \sum_{i=1}^{m} \left[\sum_{j=1}^{n} A^j(k(u_1, \ldots, u_n)) \frac{\partial k_j}{\partial u_i}(u_1, \ldots, u_m) \right] du_i.$$

Proof.   We have

$$k^*(\varphi) = k^* \left(\sum_{j=1}^{n} A^j(x_1, \ldots, x_n) dx_j \right)$$

$$= \sum_{j=1}^{n} k^* \left(A^j(x_1, \ldots, x_n) dx_j \right)$$

$$= \sum_{j=1}^{n} k^*(A^j(x_1, \ldots, x_n)) k^*(dx_j)$$

$$= \sum_{i=1}^{m} \left[\sum_{j=1}^{n} A^j(k(u_1, \ldots, u_n)) \frac{\partial k_j}{\partial u_i}(u_1, \ldots, u_m) \right] du_i$$

by Lemma 3.3.8, Lemma 3.3.11, and Lemma 3.3.14.   □

EXAMPLE 3.3.16. Let $k : \mathbb{R}^2 \longrightarrow \mathbb{R}^3$ be given by

$$k(u, v) = (u^2 + v, 2u - v, u + v^3) = (x, y, z)$$

and let φ be the 1-form on $\mathbb{R}^3$ given by

$$\varphi = (x + y + 1)dx + (2z - y)dy + xdz.$$

Then

$$
\begin{aligned}
k^*(\varphi) &= ((u+1)^2(2u) + (v^3 + 2v)(2) + (u^2 + v)(1))du \\
&\quad +((u+1)^2(1) + (v^3 + 2v)(-1) + (u^2 + v)(3v^2))dv \\
&= (2u^3 + 2v^3 + 5u^2 + 5v + 2)du + (3u^2v^2 + 2v^3 + u^2 \\
&\quad +2u - 2v + 1)dv.
\end{aligned}
$$
$\diamond$

METHOD 3.3.17. Corollary 3.3.15 looks complicated, but its result is actually simple to state:

For a 1-form φ, we obtain $k^*(\varphi)$ by "substitution."

Let us explain this. We use the notation of Corollary 3.3.15 (in an abbreviated form, i.e., suppressing the arguments).

We begin with

$$\varphi = A^1 dx_1 + \cdots + A^n dx_n.$$

Now $x_i = k_i$, so by the definition of the exterior derivative,

$$dx_i = \frac{\partial k_i}{\partial u_1} du_1 + \cdots + \frac{\partial k_i}{\partial u_m} du_m$$

for each i.

Then, "substituting" this into φ, we obtain

$$
\begin{aligned}
&A^1 \left(\frac{\partial k_1}{\partial u_1} du_1 + \cdots + \frac{\partial k_1}{\partial u_m} du_m \right) \\
&+ \cdots \\
&+ A^n \left(\frac{\partial k_n}{\partial u_1} du_1 + \cdots + \frac{\partial k_n}{\partial u_m} du_m \right).
\end{aligned}
$$

Gathering terms in each du_i together, we see that this sum is

$$
\begin{aligned}
&\left(A^1 \frac{\partial k_1}{\partial u_1} + \cdots + A^n \frac{\partial k_n}{\partial u_1} \right) du_1 \\
&+ \cdots \\
&+ \left(A^1 \frac{\partial k_1}{\partial u_m} + \cdots + A^n \frac{\partial k_1}{\partial u_m} \right) du_m,
\end{aligned}
$$

which is exactly the expression for $k^*(\varphi)$ given by Corollary 3.1.15. $\diamond$

EXAMPLE 3.3.18. Let us redo Example 3.3.16 using this method. Let $k(u, v) = (u^2 + v, 2u - v, u + v^3) = (x, y, z)$ and let $\varphi = (x + y + 1)dx + (2z - y)dy + xdz$. Then

$$x = u^2 + v, \quad y = 2u - v, \quad z = u + v^3$$

and so

$$dx = 2u\,du + dv, \quad dy = 2du - dv, \quad dz = du + 3v^2dv.$$

Substituting, we obtain

$$
\begin{aligned}
k^*(\varphi) &= (u + 1)^2(2u\,du + dv) + (v^3 + 2v)(2du - dv) \\
&\quad + (u^2 + v)(du + 3v^2dv) \\
&= (2u^3 + 2v^3 + 5u^2 + 5v + 2)du \\
&\quad + (3u^2v^2 + 2v^3 + u^2 + 2u - v + 1)dv
\end{aligned}
$$

as we obtained before.   ◇

We now obtain a formula for the pull-back of a j-form in general.

LEMMA 3.3.19. *For any n-by-j matrix M and any j-tuple $I = (i_1, \ldots, i_j)$ of integers between 1 and n, let $P_I(M)$ be the j-by-j matrix whose rows are the rows $i_1, \ldots, i_j$ (in that order) of M.*

For any j-tuple $H = (h_1, \ldots, h_j)$ of integers between 1 and m, let $M_H(p)$ be the n-by-j matrix

$$M_H(p) = \left[k_{h_1}(p) \,|\ldots|\, k_{h_j}(p) \right]$$

where the subscripts denote partial derivatives.
Then

$$k^*(dx_I) = \sum_I \det(P_I(M_H))du_H.$$

Proof. We know that

$$k^*(dx_I) = \sum_H B^H du_H$$

for some set of functions $\{B^H\}$, and furthermore that, for each H,

$$B^H(p) = \left(k^*(dx_I)\right)(e_p^H) = \left(k^*(dx_I)\right)(e_p^{h_1}, \ldots, e_p^{h_j})$$
$$= dx_I(k_*(e_p^{h_1}), \ldots, k_*(e_p^{h_j}))$$
$$= dx_I(M_H(p)) = \det(P_I(M_H(p)))$$

by Theorem 2.2.19.   □

COROLLARY 3.3.20. *Let φ be the j-form on S given by*

$$\varphi = \sum_I A^I dx_I.$$

Then

$$k^*(\varphi) = \sum_H \left[\sum_I A^I(k)\det(P_I(M_H)) \right] du_H.$$

Proof. Using Corollary 3.3.12 and Lemma 3.3.19, and reversing the order of summation, this follows immediately.   □

There is one special case of Corollary 3.3.20 that is so important we point it out separately.

COROLLARY 3.3.21. *Let $\mathcal{R}$ be a region in $\mathbb{R}^n$, let S be a region in $\mathbb{R}^n$, and let $k : \mathcal{R} \longrightarrow S$ be a smooth map, $(x_1, \ldots, x_n) = k(u_1, \ldots, u_n)$. Let $\varphi = A dx_1 \cdots dx_n$ be an n-form on S. Then*

$$k^*(\varphi) = A(k)\det(k')du_1 \cdots du_n.$$

Proof. In this case, the matrix $P_I(M_H)$ is just k', the derivative (or Jacobian matrix) of k.   □

We will see how to evaluate the pull-back of a j-form for $j > 1$ by "substitution" in the next section.

REMARK 3.3.22. Let us notice we have a certain sort of dual behavior here. On the one hand, we can push forward a single tangent vector, but not a smooth vector field. On the other hand, we can pull back a 1-form, i.e., a smooth cotangent vector field, but not a single cotangent vector.   ◇

REMARK 3.3.23. Suppose we are in the (very) important but (very) special situation where the smooth map k has a smooth inverse k^{-1}. Then we can also push forward a smooth vector field and we can also pull back a single cotangent vector.

Given a tangent vector $\mathbf{v}_p$ we can of course always push it forward via k to obtain a tangent vector $\mathbf{w}_q = k_*(\mathbf{v}_p)$, where $q = k(p)$, (so that $p = k^{-1}(q)$). But in this situation, given a tangent vector $\mathbf{w}_q$ we can also push it forward via k^{-1} to obtain a tangent vector $\mathbf{u}_p = (k^{-1})_*(\mathbf{w}_q)$. Note that these two push-forwards are inverses of each other, i.e.,

$$(k^{-1})_* = (k_*)^{-1}$$

as

$$(k^{-1})_*(k_*(\mathbf{v}_p)) = (k^{-1}k)_*(\mathbf{v}_p) = (\mathrm{id})_*(\mathbf{v}_p) = \mathrm{id}(\mathbf{v}_p) = \mathbf{v}_p$$

and similarly

$$k_*(k_*^{-1}(\mathbf{w}_q)) = (kk^{-1})_*(\mathbf{w}_q) = (\mathrm{id})_*(\mathbf{w}_p) = \mathrm{id}(\mathbf{w}_q) = \mathbf{w}_q$$

both by Lemma 3.2.5.

Similarly, given a differential form φ on $\mathcal{S}$ we can of course always pull it back via k to obtain a differential form $k^*(\varphi)$ on $\mathcal{R}$. But in this situation, given a differential form ψ on $\mathcal{R}$ we can also pull it back via k^{-1} to obtain a differential form $\rho = (k^{-1})^*(\psi)$ on $\mathcal{S}$. By exactly the same computation, this time using Lemma 3.3.7, we once again obtain that these two pull-backs are inverses of each other,

$$(k^{-1})^* = (k^*)^{-1}.$$ ◇

REMARK 3.3.24. We have adopted the language of modern algebraic and differential topology here. But we should remark that this is opposite to the language of classical tensor analysis. As we have stated, differential forms are contravariant as they pull back, rather than push forward. But in the language of tensor analysis, differential forms are called alternating covariant tensors. Thus the reader should beware!   ◇

3.4 Pull-backs, products, and exterior derivatives

Our objective in this section is two-fold: first, to show that pull-backs commute with exterior products (Theorem 3.4.1), and second, to show that pull-backs commute with exterior derivatives (Theorem 3.4.8). We will also use Theorem 3.4.1 to compute pullbacks of j-forms (Method 3.4.4).

Again we suppose that we have a smooth map $k : \mathcal{R} \longrightarrow \mathcal{S}$ from a region $\mathcal{R}$ in $\mathbb{R}^m$ to a region $\mathcal{S}$ in $\mathbb{R}^n$.

THEOREM 3.4.1. *Let φ_1 be an i-form and let φ_2 be a j-form on $\mathcal{S}$. Then*

$$k^*(\varphi_1\varphi_2) = k^*(\varphi_1)k^*(\varphi_2).$$

Proof. Recall Definition 2.2.13, applied to our situation: $\varphi_1\varphi_2$ is the $m = i + j$ form given as follows: Let C be a set of left coset representatives of $S_i \times S_j$ in S_m. Then

$$(\varphi_1\varphi_2)(\mathbf{w}_q^1, \ldots, \mathbf{w}_q^m) = \sum_{\sigma \in C} \text{sign}(\sigma)\varphi_1(\mathbf{w}_q^{\sigma(1)}, \ldots, \mathbf{w}_q^{\sigma(i)})$$
$$\times \varphi_2(\mathbf{w}_q^{\sigma(i+1)}, \ldots, \mathbf{w}_q^{\sigma(m)})$$

where q is any point of $\mathcal{S}$ and $(\mathbf{w}_q^1, \ldots, \mathbf{w}_q^m)$ is any m-tuple of tangent vectors to $\mathbb{R}^n$ at q.

Let p be any point of $\mathcal{R}$ and let $(\mathbf{v}_p^1, \ldots, \mathbf{v}_p^m)$ be any m-tuple of tangent vectors to $\mathbb{R}^m$ at p.

We compute:

$$k^*(\varphi_1\varphi_2)(\mathbf{v}_p^1, \ldots, \mathbf{v}_p^m)$$
$$= (\varphi_1\varphi_2)(k_*(\mathbf{v}_p^1), \ldots, k_*(\mathbf{v}_p^m))$$
$$= \sum_{\sigma \in C} \text{sign}(\sigma)\varphi_1(k_*(\mathbf{v}^{\sigma(1)}), \ldots, k_*(\mathbf{v}^{\sigma(i)}))$$
$$\times \varphi_2(k_*(\mathbf{v}^{\sigma(i+1)}), \ldots, k_*(\mathbf{v}^{\sigma(m)}))$$
$$= \sum_{\sigma \in C} \text{sign}(\sigma)k^*(\varphi_1)(\mathbf{v}^{\sigma(1)}, \ldots, \mathbf{v}^{\sigma(i)})$$
$$\times k^*(\varphi_2)(\mathbf{v}^{\sigma(i+1)}, \ldots, \mathbf{v}^{\sigma(m)})$$
$$= (k^*(\varphi_1)k^*(\varphi_2))(\mathbf{v}_p^1, \ldots, \mathbf{v}_p^m)$$

and so

$$k^*(\varphi_1\varphi_2) = k^*(\varphi_1)k^*(\varphi_2)$$

as claimed.   □

REMARK 3.4.2. Note that Theorem 3.4.1 is a generalization of Lemma 3.3.11, which handled the case $i = 0$ (since a 0-form is just a smooth function).   ◇

REMARK 3.4.3.

(a) We phrase Theorem 3.4.1 as stating

Pull-back commutes with exterior product.

(b) Equivalently, Theorem 3.4.1 states that the following diagram commutes:

$$\Omega^i(\mathcal{R}) \times \Omega^j(\mathcal{R}) \xleftarrow{k^* \times k^*} \Omega^i(\mathcal{S}) \times \Omega^j(\mathcal{S})$$
$$\downarrow \qquad\qquad\qquad\qquad \downarrow$$
$$\Omega^{i+j}(\mathcal{R}) \xleftarrow{\quad k^* \quad} \Omega^{i+j}(\mathcal{S})$$

where the vertical maps are products.   ◇

METHOD 3.4.4. Theorem 3.4.1 gives us a method of pulling back j-forms for any j. Recall from Theorem 2.2.10 that any j-form φ can be written as

$$\varphi = \sum_I A^I dx_I$$

where I ranges over ordered j-tuples of elements of $\{1, \ldots, n\}$. By linearity,

$$k^*(\varphi) = \sum_I k^*(A^I dx_I).$$

Thus it suffices to be able to pull back a single term.

Consider a term $A^{I_0} dx_{I_0}$. Let $I_0 = \{i_1, \ldots, i_j\}$. Then, by Theorem 3.4.1,

$$k^*(A^{I_0} dx_{I_0}) = k^*(A^{I_0} dx_{i_1} \cdots dx_{i_j}) = k^*(A^{I_0})k^*(dx_{i_1}) \cdots k^*(dx_{i_j}).$$

We know how to pull back each of the factors in this product, as each factor is either the pull-back of a function, given by Definition 3.3.1, or the pull-back of a 1-form dx_j, given by Lemma 3.3.14. So we simply pull them all back and then multiply them.

In applying this method, we may use Lemma 3.3.14 directly or we may use "substitution" as in Method 3.3.17. ◇

EXAMPLE 3.4.5. Let $k(u, v) = (u^2 + v, 2u - v, u + v^3) = (x, y, z)$ as in Example 3.3.16, and let $\varphi = z\, dy\, dz + (y + 1)dz\, dx - dx\, dy$. Then again

$$x = u^2 + v, \quad y = 2u - v, \quad z = u + v^3$$

and so again

$$dx = 2u\, du + dv, \quad dy = 2du - dv, \quad dz = du + 3v^2 dv.$$

Substituting,

$$
\begin{aligned}
k^*(\varphi) &= (u + v^3)(2du - dv)(du + 3v^2 dv) \\
&\quad + (2u - v + 1)(du + 3v^2 dv)(2u\, du + dv) \\
&\quad - (2u\, du + dv)(2du - dv) \\
&= (u + v^3)\left\{(6v^2 + 1)du\, dv\right\} \\
&\quad + (2u - v + 1)\left\{(1 - 6uv^2)du\, dv\right\} \\
&\quad - \left\{(-2u - 2)du\, dv\right\} \\
&= \left\{(u + v^3)(6v^2 + 1) + (2uv + 1)(1 - 6uv^2)\right. \\
&\quad \left. + (2u + 2)\right\} du\, dv.
\end{aligned}
$$

◇

Now we turn our attention to exterior derivatives. We build up to our main result with two preliminary results.

LEMMA 3.4.6. *Let φ be a 0-form, i.e., a smooth function, on S. Then*

$$dk^*(\varphi) = k^*(d\varphi).$$

Proof. For clarity, we set $A = \varphi$.

We want to show that the two 1-forms $d(k^*(A))$ and $k^*(dA)$ are equal. Now 1-forms are functions on tangent vectors, so to show two

1-forms are equal is to show they have the same value on each tangent vector. Thus we make the following claim:

$$(*) \qquad d(k^*(A))(\mathbf{v}_p) = (k^*(dA))(\mathbf{v}_p)$$

for any tangent vector $\mathbf{v}_p$ to $\mathbb{R}^m$ at any point p of $\mathcal{R}$.

Let us begin by recalling the meaning of $dg(\mathbf{u}_z)$, where g is a smooth function and $\mathbf{u}_z$ is a tangent vector at the point z. (We are choosing unbiased letters as we will be applying this in two different situations.) Let $r(t)$ be any parameterized smooth curve with $r(0) = z$ and $r'(0) = \mathbf{u}_z$. Then, by Theorem 2.1.8, $dg(\mathbf{u}_z) = e'(0)$ where $e(t)$ is the composition $e(t) = g(r(t))$.

With this in mind let us consider the left-hand side of $(*)$. Let $r(t)$ be any parameterized curve with $r(0) = p$ and $r'(0) = \mathbf{v}_p$. Then

$$(*_1) \qquad d(k^*(A))(\mathbf{v}_p) = \big((k^*(A))(r)\big)'(0) = \big((A(k))(r)\big)'(0).$$

Let $s(t)$ be the composition $s(t) = k(r(t))$. Then, by the chain rule and the definition of k_* (Definition 3.2.2), $s'(0) = k'(r(0))r'(0) = k'(p)(\mathbf{v}_p) = k_*(\mathbf{v}_p)$. With this in mind let us consider the right-hand side of $(*)$. Then

$$(*_2) \;\; (k^*(dA))(\mathbf{v}_p) = dA(k_*(\mathbf{v}_p)) = \big(A(s)\big)'(0) = \big(A(k(r))\big)'(0).$$

But $\big((A(k))(r)\big) = \big(A(k(r))\big)$ (composition of functions is associative), so the left-hand and right-hand sides of $(*)$ are equal, as claimed. $\qquad \square$

LEMMA 3.4.7. *Let $j \geq 0$ and let $I = (i_1, \ldots, i_j)$ be any ordered j-tuple of integers between 1 and n. Then $k^*(dx_I)$ is a closed j-form on $\mathcal{R}$.*

Proof. For completeness we have let $j \geq 0$.

First consider the case $j = 0$. Then $I = \{\}$ and $dx_I = 1$, the constant function $f = 1$ on $\mathcal{S}$. But then $k^*(dx_I)$ is the constant function $k^*(f) = 1$ on $\mathcal{R}$, so $d(k^*(f)) = 0$.

The interesting cases are the cases with $j > 0$. We prove them by induction on j.

To begin the induction we consider the case $j = 1$, so $dx_I = dx_i$ for some integer i. For clarity, we let $A(x_1, \ldots, x_n) = x_i$. Then, as we observed in Remark 1.2.2, $dx_i = dA$. Now

$$k^*(dA) = d(k^*(A))$$

by Lemma 3.4.6 (as A is a 0-form) and then

$$d(k^*(dA)) = d(d(k^*(A)) = 0$$

by Poincaré's Lemma (Theorem 1.2.7), so $dA = dx_i$ is closed.

Now suppose the Lemma is true for any $(j-1)$-tuple and consider a j-tuple $I = (i_1, \ldots, i_j)$. Let $J = (i_2, \ldots, i_j)$. Then J is a $(j-1)$-tuple and $dx_I = dx_{i_1} dx_J = dA dx_J$ where $A(x_1, \ldots, x_n) = x_i$ as above. Then

$$k^*(dx_I) = k^*(dA dx_J) = k^*(dA)k^*(dx_J) \qquad \text{by Theorem 3.4.1}$$
$$= d(k^*(A))k^*(dx_J) \qquad \text{by Lemma 3.4.6.}$$

Then, applying the Leibniz rule (Theorem 1.2.5),

$$d(k^*(dx_I)) = d(d(k^*(A))k^*(dx_J) - d(k^*(A))d(k^*(dx_J))$$
$$= 0 - 0 = 0$$

as the first term on the right-hand side is 0 by Poincaré's Lemma, and the second term on the right-hand side is 0 by the inductive hypothesis. □

THEOREM 3.4.8. *Let φ be a j-form on S. Then*

$$dk^*(\varphi) = k^*(d\varphi).$$

Proof. By Theorem 2.2.10, φ can be written as

$$\varphi = \sum_I A^I dx_I$$

where I ranges over ordered j-tuples of elements of $\{1, \ldots, n\}$.

By the linearity of both k^* and d, it suffices to prove the theorem when φ consists of a single term. Thus we let $\varphi = A^{I_0} dx_{I_0}$. On the one hand, using the Leibniz rule (Theorem 1.2.5), Poincaré's Lemma, and Theorem 3.4.1 we have

$$
\begin{aligned}
k^*(d\varphi) &= k^*(dA^{I_0} dx_{I_0} - A^{I_0} d(dx_{I_0})) \\
&= k^*(dA^{I_0} dx_{I_0}) \\
&= k^*(dA^{I_0}) k^*(dx_{I_0})
\end{aligned}
$$

and on the other hand, using Theorem 3.4.1, the Leibniz Rule, and Lemma 3.4.7 we have

$$
\begin{aligned}
d(k^*(\varphi)) &= d(k^*(A^{I_0} dx_{I_0})) \\
&= d(k^*(A^{I_0}) k^*(dx_{I_0})) \\
&= d(k^*(A^{I_0})) k^*(dx_{I_0}) - k^*(A^{I_0}) d(k^*(dx_{I_0})) \\
&= d(k^*(A^{I_0})) k^*(dx_{I_0}).
\end{aligned}
$$

But by Lemma 3.4.6, these are equal.   □

REMARK 3.4.9.

(a) We phrase Theorem 3.4.8 as stating

Pull-back commutes with exterior differentiation.

(b) Equivalently, Theorem 3.4.8 states that the following diagram commutes:

$$
\begin{array}{ccc}
\Omega^j(\mathcal{R}) & \overset{k^*}{\longleftarrow} & \Omega^j(\mathcal{S}) \\
d\downarrow & & \downarrow d \\
\Omega^{j+1}(\mathcal{R}) & \overset{k^*}{\longleftarrow} & \Omega^{j+1}(\mathcal{S})
\end{array}
$$

◇

Theorem 3.4.8 has the following immediate but important consequence.

COROLLARY 3.4.10.

(a) Let φ be a closed j-form on $\mathcal{S}$. Then $k^*(\varphi)$ is a closed j-form on $\mathcal{R}$.

(b) Let φ be an exact j-form on $\mathcal{S}$. Then $k^*(\varphi)$ is an exact j-form on $\mathcal{R}$.

Proof.

(a) If $d(\varphi) = 0$ then $d(k^*(\varphi)) = k^*(d(\varphi)) = k^*(0) = 0$.

(b) If $\varphi = d\psi$ then $k^*(\varphi) = k^*(d\psi) = d(k^*(\psi))$. $\qquad\qquad$ □

3.5 Smooth homotopies and the Converse of Poincaré's Lemma, III

Recall Poincaré's Lemma "$d^2 = 0$" (Theorem 1.2.7), which had as an immediate consequence that every exact differential form is closed (Corollary 1.2.14).

Recall also that we proved partial converses to Poincaré's Lemma as Theorem 1.4.1 and Theorem 2.4.3. In this section we consider a much more general situation, and apply it to obtain a more general partial converse to Poincaré's Lemma.

Throughout this section we let I denote the unit interval $I = [0, 1]$ (except that it will also denote a multi-index; since these two contexts are very different, there should not be much danger of confusion). We parameterize I by t.

DEFINITION 3.5.1. Let N be a region in $\mathbb{R}^n = \{(x_1, \ldots, x_n)\}$. Then $N \times I \subset \mathbb{R}^{n+1} = \{(x_1, \ldots, x_n, t)\}$ is defined by

$$N \times I = \{(x_1, \ldots, x_n, t) | (x_1, \ldots, x_n) \in N, \ t \in I\}.$$

For $t \in I$ we let $i_t : N \longrightarrow N \times I$ be the inclusion

$$i_t(x_1, \ldots, x_n) = (x_1, \ldots, x_n, t).$$

Also, we let $\pi : N \times I \longrightarrow N$ be the projection

$$\pi(x_1, \ldots, x_n, t) = (x_1, \ldots, x_n). \qquad\qquad \diamond$$

DEFINITION 3.5.2. Let N be a region in $\mathbb{R}^n$. A set of ℓ-forms $\{\theta_t\}_{t \in I}$ on N is a *smooth family of ℓ-forms on N*, if, writing

$\theta_t = \sum_I A_I(x_1, \ldots, x_n, t)dx_I$, each function $A_I(x_1, \ldots, x_n, t)$ is a smooth function on $N \times I$.

In this situation we define the *integral of the family* $\{\theta_t\}_{t \in I}$ to be the ℓ-form $\Theta = \text{Int}\left(\{\theta_t\}\right)$ on N given by

$$\text{Int}\left(\sum_I A_I(x_1, \ldots, x_n, t)dx_I\right)$$
$$= \sum_I \left[\int_0^1 A_I(x_1, \ldots, x_n, t)dt\right] dx_I. \qquad \diamond$$

The following construction is crucial. We present it first, and then show its application.

LEMMA 3.5.3. *Let N be a region in $\mathbb{R}^n$. Then there is a linear map, for every $\ell \geq 0$,*

$$\tilde{\xi} : \Omega^\ell(N \times I) \longrightarrow \Omega^{\ell-1}(N)$$

with the property that, for any ℓ-form φ on $N \times I$,

$$i_1^*(\varphi) - i_0^*(\varphi) = \tilde{\xi}(d\varphi) + d(\tilde{\xi}(\varphi)).$$

Proof. First we need to establish some notation. We denote points in N by p, and points in $N \times I$ by (p, t). We let $i_t : N \longrightarrow N \times I$ be the map given by $i_t(p) = (p, t)$; we think of i_t as including N in $N \times I$ at "level t."

We let $\tilde{\mathbf{f}}$ be the constant vector field on $N \times I$ whose value at the point (p, t) is

$$\tilde{\mathbf{f}}_{(p,t)} = \begin{bmatrix} 0 \\ \vdots \\ 0 \\ 1 \end{bmatrix}_{(p,t)}.$$

Conceptually, $\tilde{\xi}$ is easy to define.

The case $\ell = 0$ is trivial: If φ is an ℓ-form on $N \times I$ for $\ell = 0$, then $\Omega^{\ell-1}(N) = \{0\}$, so we must have $\tilde{\xi}(\varphi) = 0$.

Now for the interesting case: Let φ be an ℓ-form on $N \times I$ for some $\ell > 0$. Then

$$\tilde{\xi}(\varphi) = \text{Int}\left(\{i_t^*(\varphi_{\tilde{\mathbf{f}}})\}\right).$$

Let us examine this definition. First, φ is an ℓ-form on $N \times I$. Then its contraction $\varphi_{\tilde{\mathbf{f}}}$ (as defined in Definition 2.2.23) is an $(\ell-1)$-form on $N \times I$, and then, for each $t \in I$, the pullback $i_t^*(\varphi_{\tilde{\mathbf{f}}})$ is an $(\ell-1)$-form on N, so the integral of this family, as defined in Definition 3.5.2, is indeed an $(\ell-1)$-form on N.

We need a bit more notation.

If $\mathbf{v}$ is the constant vector field on N whose value at the point p is

$$\mathbf{v}_p = \begin{bmatrix} a_1 \\ \vdots \\ a_n \end{bmatrix}_p,$$

we let $\tilde{\mathbf{v}}$ be the constant vector field on $N \times I$ whose value at the point (t, p) is

$$\tilde{\mathbf{v}}_{(p,t)} = \begin{bmatrix} a_1 \\ \vdots \\ a_n \\ 0 \end{bmatrix}_{(p,t)},$$

and we note that

$$\tilde{\mathbf{v}}_{(p,t)} = (i_t)_*(\mathbf{v}_p),$$

the push-forward of the tangent vector $\mathbf{v}_p$ by the map i_t.

Let us reexamine the above definition of $\tilde{\xi}$. In the case $\ell = 0$, there is nothing more to say. Suppose $\ell > 0$. Since $\tilde{\xi}(\theta)$ is an $(\ell-1)$-form, we can evaluate it on any $(\ell-1)$-tuple of tangent vectors. Conversely, specifying its value on any arbitrary $(\ell-1)$-tuple of tangent vectors defines it.

Then the above definition shows that $\tilde{\xi}(\theta)$ satisfies, and in fact is defined by, the formula:

$$\tilde{\xi}(\varphi)(\mathbf{v}_p^1, \ldots, \mathbf{v}_p^{\ell-1}) = \int_0^1 \varphi(\tilde{\mathbf{f}}_{(p,t)}, \tilde{\mathbf{v}}_{(p,t)}^1, \ldots, \tilde{\mathbf{v}}_{(p,t)}^{\ell-1}) \, dt.$$

Actually, we will find it convenient in the course of the proof to have $\tilde{\mathbf{f}}_{(p,t)}$ as the last argument, rather than the first argument, so we will instead use the formula:

$$\tilde{\xi}(\varphi)(\mathbf{v}_p^1, \ldots, \mathbf{v}_p^{\ell-1}) = (-1)^{\ell-1} \int_0^1 \varphi(\tilde{\mathbf{v}}_{(p,t)}^1, \ldots, \tilde{\mathbf{v}}_{(p,t)}^{\ell-1}, \tilde{\mathbf{f}}_{(p,t)}) dt.$$

Since $\tilde{\xi}$ is clearly linear, it suffices to verify the claim of the lemma when φ consists of a single term. There are two cases:

(a) $$\varphi = A(p, t) dx_I,$$

(b) $$\varphi = A(p, t) dx_I dt.$$

Here $A(p, t)$ is a smooth function on $N \times I$. Also, I denotes a multi-index, and we assume the indices in I are in increasing order.

Also by linearity, it suffices to show that the left-hand and right-hand sides take the same value on any ℓ-tuple of constant vector fields of the form $\mathbf{e}^H$ (whose value at the point p is $\mathbf{e}_p^H$). We assume the multi-indices in H are in increasing order as well.

Let us first compute the left-hand side.

In case (a), $i_1^*(A(p, t) dx_I) = A(p, 1) dx_I$ and $i_0^*(A(p, t) dx_I) = A(p, 0) dx_I$, so in this case

$$i_1^*(A(p, t) dx_I) - i_0^*(A(p, t) dx_I) = \big(A(p, 1) - A(p, 0)\big) dx_I$$

and hence

$$\big(i_1^*(A(p, t) dx_I) - i_0^*(A(p, t) dx_I)\big) (\mathbf{e}_p^H)$$
$$= \big(A(p, 1) - A(p, 0)\big) dx_I(\mathbf{e}_p^H).$$

In case (b), $i_1^*(A(p, t) dx_I dt) = 0$ and $i_0^*(A(p, t) dx_I dt) = 0$, so in this case

$$i_1^*(A(p, t) dx_I dt) - i_0^*(A(p, t) dx_I dt) = 0$$

and hence

$$\big(i_1^*(A(p, t) dx_I) - i_0^*(A(p, t) dx_I)\big) (\mathbf{e}_p^H) = 0.$$

Now for the right-hand side. We have two terms on the right-hand side, which, for convenience, we will label T_1 and T_2.

First we consider case (a).

For T_1:

$$d\varphi = d(A(p, t)dx_I)$$

$$= \left(\sum_{i=1}^{n} A_{x_i}(p, t)dx_i dx_I \right) + A_t(p, t)dt\, dx_I$$

$$= \left(\sum_{i=1}^{n} A_{x_i}(p, t)dx_i dx_I \right) + (-1)^\ell A_t(p, t)dx_I dt$$

and then

$$\left(\tilde{\xi}(d\varphi) \right)(\mathbf{e}_p^H) = \left(\sum_{i=1}^{n} (-1)^{((\ell+1)-1)} \right.$$

$$\int_0^1 A_{x_i}(p, t)dx_i dx_I(\tilde{\mathbf{e}}_{(p,t)}^H, \tilde{\mathbf{f}}_{(p,t)})dt \Bigg)$$

$$+(-1)^\ell (-1)^{((\ell+1)-1)}$$

$$\times \int_0^1 \left[A_t(p, t)dx_I dt(\tilde{\mathbf{e}}_{(p,t)}^H, \tilde{\mathbf{f}}_{(p,t)}) \right] dt.$$

Now in every term of the first line in this expression, the integrand is 0, as $dx_i dx_I$ evaluates to 0 on any $(\ell+1)$-tuple $(\tilde{\mathbf{e}}_{(p,t)}^H, \tilde{\mathbf{f}}_{(p,t)})$ containing the constant vector field $\mathbf{e}_{(p,t)}^{n+1}$. As for the term in the second line, $dx_I dt(\tilde{\mathbf{e}}_{(p,t)}^H, \tilde{\mathbf{f}}_{(p,t)}) = dx_I(\mathbf{e}_p^H)$. Thus for T_1 we obtain:

$$\left(\tilde{\xi}(d\varphi) \right)(\mathbf{e}_p^H) = \int_0^1 \left[A_t(p, t)dx_I dt(\tilde{\mathbf{e}}_{(p,t)}^H) \right] dt$$

$$= \left(\int_0^1 A_t(p, t)dt \right) dx_I(\mathbf{e}_p^H)$$

$$= (A(p, t)|_0^1)(\mathbf{e}_p^H)$$

$$= \left(A(p, 1) - A(p, 0) \right) dx_I(\mathbf{e}_p^H).$$

For T_2: $\tilde{\xi}(A(p, t)dx_I) = 0$, as the integrand is always 0 (again, dx_I evaluates to 0 on any ℓ-tuple containing $\tilde{\mathbf{f}}_{(p,t)}$), so $d(\tilde{\xi}(A(p, t)dx_I) = d(0) = 0$. Thus for T_2 we obtain:

$$(d(\tilde{\xi}(\varphi))(\mathbf{e}_p^H) = 0.$$

Adding these two expressions we find that

$$\left((\tilde{\xi}d + d\tilde{\xi})(\varphi)\right)(\mathbf{e}_p^H) = \left((i_1^* - i_0^*)(\varphi)\right)(\mathbf{e}_p^H),$$

as claimed, in this case.

Next we consider case (b). Again we label the two terms on the right-hand side T_1 and T_2.

For T_1:

$$d\varphi = d(A(p,t)dx_I dt)$$

$$= \left(\sum_{i=1}^{n} A_{x_i}(p,t)dx_i dx_I dt\right) + A_t(p,t)dtdx_I dt$$

$$= \left(\sum_{i=1}^{n} A_{x_i}(p,t)dx_i dx_I dt\right)$$

and then

$$\left(\tilde{\xi}(d\varphi)\right)(\mathbf{e}_p^H) = (-1)^{((\ell+1)-1)}$$

$$\times \sum_{i=1}^{n}\int_0^1 \left[A_{x_i}(p,t)dx_i dx_I dt(\tilde{\mathbf{e}}_{(p,t)}^H, \tilde{\mathbf{f}}_{(p,t)})\right] dt$$

$$= (-1)^{\ell}\sum_{i=1}^{n}\int_0^1 \left[A_{x_i}(p,t)dx_i dx_I(\tilde{\mathbf{e}}_{(p,t)}^H)\right] dt$$

$$= (-1)^{\ell}\sum_{i=1}^{n}\left[\int_0^1 A_{x_i}(p,t)dt\right] dx_i dx_I(\tilde{\mathbf{e}}_{(p,t)}^H).$$

For T_2:

$$(d(\tilde{\xi}(\varphi))(\mathbf{e}_p^H) = d\left((-1)^{\ell-1}\int_0^1 A(p,t)dx_I dt(\tilde{\mathbf{e}}_{(p,t)}^H, \tilde{\mathbf{f}}_{(p,t)})dt\right)$$

$$= (-1)^{\ell-1}\sum_{i=1}^{n}\frac{\partial}{\partial x_i}\left[\int_0^1 A(p,t)dt\right] dx_i dx_I(\tilde{\mathbf{e}}_{(p,t)}^H)$$

$$= (-1)^{\ell-1}\sum_{i=1}^{n}\left[\int_0^1 A_{x_i}(p,t)dt\right] dx_i dx_I(\tilde{\mathbf{e}}_{(p,t)}^H).$$

We note that the signs in T_1 and T_2 are opposite. Thus, adding these two expressions we find that

$$\left((\tilde{\xi}d + d\tilde{\xi})(\varphi)\right)(\mathbf{e}_p^H) = 0 = \left((i_1^* - i_0^*)(\varphi)\right)(\mathbf{e}_p^H),$$

as claimed, in this case as well, completing the proof. $\qquad\square$

COROLLARY 3.5.4. *In the situation of Lemma* 3.5.3, $\mathrm{supp}(\tilde{\xi}(\varphi)) \subseteq \pi(\mathrm{supp}(\varphi))$. *In particular, if φ has compact support, then $\tilde{\xi}(\varphi)$ has compact support.*

Proof. Referring to the definition of $\tilde{\xi}(\varphi)$, we simply observe that, if $S = \mathrm{supp}(\tilde{\xi}(\varphi))$, then $p \notin \pi(S)$ implies that $i_t(p) \notin S$ for every $t \in I$. $\qquad\square$

REMARK 3.5.5. Since $\tilde{\xi}$ is determined by its effect on forms consisting of a single term, an examination of its definition in the proof of Lemma 3.5.3 shows that we can define $\tilde{\xi}$ on ℓ-forms by the formulas

$$\tilde{\xi}(A(p,t)dx_I) = 0,$$

$$\tilde{\xi}(A(p,t)dx_I dt) = (-1)^{\ell-1}\left(\int_0^1 A(p,t)dt\right)dx_I$$

and extend to $\tilde{\xi} : \Omega^\ell(N \times I) \longrightarrow \Omega^{\ell-1}(N)$ by linearity. (Note that in the first line in this definition, I is a multi-index of ℓ individual indices, while in the second line it is a multi-index of $\ell - 1$ individual indices.) $\qquad\diamond$

DEFINITION 3.5.6. Let M be a region in $\mathbb{R}^m$ and let N be a region in $\mathbb{R}^n$. Let $f_0 : M \longrightarrow N$ and $f_1 : M \longrightarrow N$ be smooth maps. A *smooth homotopy* between f_0 and f_1 is a smooth map $F : M \times I \longrightarrow N$ with

$$F(x_1, \ldots, x_m, 0) = f_0(x_1, \ldots, x_m) \text{ and } F(x_1, \ldots, x_m, 1)$$
$$= f_1(x_1, \ldots, x_m)$$

for every $(x_1, \ldots, x_m) \in M$.

If there exists a smooth homotopy between f_0 and f_1, then f_0 and f_1 are said to be *smoothly homotopic*. $\qquad\diamond$

REMARK 3.5.7. In the situation of Definition 3.5.6, define $f_t(x_1, \ldots, x_m) = F(x_1, \ldots, x_m, t) = F(i_t(x_1, \ldots, x_m))$ for $t \in I$. We usually think of the smooth homotopy F as providing a smooth deformation of f_0 to f_1 through the maps f_t. We think of t as "time" and so we start with f_0 at time 0, pass through f_t at time t, and end with f_1 at time 1.   ◇

THEOREM 3.5.8. *Let M be a region in $\mathbb{R}^m$ and let N be a region in $\mathbb{R}^n$. Let $f_0 : M \longrightarrow N$ and $f_1 : M \longrightarrow N$ be smoothly homotopic smooth maps. Then there is a linear map, defined for every $j \geq 0$,*

$$\Xi : \Omega^j(N) \longrightarrow \Omega^{j-1}(M)$$

such that, for any k-form φ on N,

$$f_1^*(\varphi) - f_0^*(\varphi) = d(\Xi(\varphi)) + \Xi(d\varphi).$$

Proof. We define this map as follows: Let $F : M \times I \longrightarrow N$ be a homotopy. Then

$$\Xi(\varphi) = \tilde{\xi}(F^*(\varphi))$$

where $\tilde{\xi}$ is the map in Lemma 3.5.3.   □

COROLLARY 3.5.9. *Let M be a region in $\mathbb{R}^m$ and let N be a region in $\mathbb{R}^n$. Let $f_0 : M \longrightarrow N$ and $f_1 : M \longrightarrow N$ be smoothly homotopic smooth maps. Let φ be a closed k-form on N. Then $f_1^*(\varphi) - f_0^*(\varphi)$ is exact.*

Proof. We need to show that $f_1^*(\varphi) - f_0^*(\varphi) = d\psi$ for some $(k-1)$-form ψ. We simply let $\psi = \Xi(\varphi)$ as in the conclusion of Theorem 3.5.8 (noting that $\Xi(d\varphi) = 0$ as $d\varphi = 0$, since φ is closed, by hypothesis).   □

(We have chosen the notation $\tilde{\xi}$ and Ξ in this section because we are generalizing Theorem 2.4.3, and we used the notation ξ in the proof of that theorem.)

DEFINITION 3.5.10. Let M be a region in $\mathbb{R}^m$. Then M is *smoothly contractible* if the identity map from M to itself is smoothly homotopic to the constant map taking M to some (and hence any) point of M.   ◇

Now we have our next generalization of the converse of Poincaré's Lemma.

COROLLARY 3.5.11. *(Converse of Poincaré's Lemma) Let M be a smoothly contractible region in $\mathbb{R}^m$. Let φ be a closed k-form defined on M, $k > 0$. Then φ is exact.*

Proof. Let $f_1 : M \longrightarrow M$ be the identity map and $f_0 : M \longrightarrow M$ be the map taking M to a point. Then $f_1^*(\varphi) = \varphi$ for any φ. Also, f_0 is a constant map, so for $k > 0$, $f_0^*(\varphi) = 0$ for any φ. Thus, by Corollary 3.5.9,

$$f_1^*(\varphi) = f_0^*(\varphi) + d(\xi(\varphi)),$$
$$\varphi = 0 + d(\xi(\varphi)),$$
$$\varphi = d(\xi(\varphi))$$

is exact. $\qquad\qquad\qquad\qquad\qquad\qquad\qquad\qquad\qquad\qquad\qquad$ □

REMARK 3.5.12. We may ask what happens for $k = 0$. Note that a smoothly contractible region M is certainly connected.

Corollary 3.5.9 still shows that, for any closed 0-form φ, $\varphi - f_1^*(\varphi)$ is exact. Since the only exact 0-form is 0, this states that $\varphi - f_1^*(\varphi) = 0$. But every closed 0-form on M is constant, so if φ is closed, $f_1^*(\varphi) = \varphi$. Thus in this case we simply get the equality $0 = 0$, i.e., Corollary 3.5.9 simply gives no information in case $k = 0$. $\qquad\qquad$ ◇

REMARK 3.5.13. If M is any star-shaped region in $\mathbb{R}^m$, then M is smoothly contractible. Let p_0 be any fixed point in M. Then the map $F : M \times I \longrightarrow M$ given by

$$F(p, t) = p_0 + t(p - p_0)$$

is a homotopy from the identity map on M to the constant map taking M to the point p_0. Thus we see that Corollary 3.5.11 is a generalization of Theorem 2.4.3. $\qquad\qquad$ ◇

EXAMPLE 3.5.14. For any nonnegative real number $a < 1$ and any positive real number $A > 1$, or $A = \infty$, let M be the region in $\mathbb{R}^2$ given, in polar coordinates, by

$$M = \{re^{i\theta} \mid a < r < A, \ -\pi < \theta < \pi\}.$$

M is the region consisting of all points in the plane lying strictly between the circles of radius a and A except for points lying on the negative x-axis. M is smoothly contractible, as we see from the map

$$F : M \times I \longrightarrow M \text{ given by } F(re^{i\theta}, t) = r^t e^{it\theta},$$

but M is not star-shaped except when $a = 0$.   ◇

(Psychologically speaking, in the examples in this section it is easiest to think of starting with the identity map f_1 and going "backwards in time" from $t = 1$ to $t = 0$ to deform it to a constant map f_0.)

EXAMPLE 3.5.15. Usually it is easier to "see" that a region is smoothly contractible than to give an explicit smooth homotopy that accomplishes the contraction. For example, we have the following contractible "amoeba," where we have indicated with dotted lines the stages of the contraction.

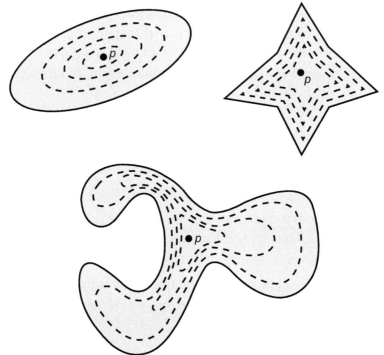

◇

3.6 Exercises

Let

$$f_1 = x^2 - y^2,$$
$$f_2 = 2xy,$$
$$\varphi_1 = xy\,dx + x\,dy,$$
$$\varphi_2 = 2y\,dx + x\,dy,$$
$$\varphi_3 = (x^2 + y^2)\,dy,$$
$$\psi_1 = (xy - y)\,dx\,dy,$$
$$\psi_2 = (x + y^2)\,dx\,dy.$$

Let $k : \mathbb{R} \longrightarrow \mathbb{R}^2$ be given by $(x, y) = k(t) = (t, t^2)$.

1. Find: (a) $k^*(f_1)$ (b) $k^*(f_2)$ (c) $k^*(\varphi_1)$ (d) $k^*(\varphi_2)$ (e) $k^*(\varphi_3)$.

2. Show that: (a) $k^*(df_1) = d(k^*(f_1))$ (b) $k^*(df_2) = d(k^*(f_2))$, thereby verifying Theorem 3.4.8 in these cases.

Let $k : \mathbb{R}^2 \longrightarrow \mathbb{R}^2$ be given by $(x, y) = k(u, v) = (u^2 + 1, uv)$.

3. Find: (a) $k^*(f_1)$ (b) $k^*(f_2)$ (c) $k^*(\varphi_1)$ (d) $k^*(\varphi_2)$ (e) $k^*(\varphi_3)$ (f) $k^*(\psi_1)$ (g) $k^*(\psi_2)$.

4. Show that: (a) $k^*(df_1) = d(k^*(f_1))$ (b) $k^*(df_2) = d(k^*(f_2))$ (c) $k^*(d\varphi_1) = d(k^*(\varphi_1))$ (d) $k^*(d\varphi_2) = d(k^*(\varphi_2))$ (e) $k^*(d\varphi_3) = d(k^*(\varphi_3))$, thereby verifying Theorem 3.4.8 in these cases.

5. Show that: (a) $k^*(\varphi_1\varphi_2) = k^*(\varphi_1)k^*(\varphi_2)$ (b) $k^*(\varphi_1\varphi_3) = k^*(\varphi_1)k^*(\varphi_3)$ (c) $k^*(\varphi_2\varphi_3) = k^*(\varphi_2)k^*(\varphi_3)$, thereby verifying Theorem 3.4.1 in these cases.

6–8. Do Exercises 3–5 with $k : \mathbb{R}^n \longrightarrow \mathbb{R}^2$ given by $(x, y) = k(u, v) = (u^2v, u - v)$.

9. Let M be the region in $\mathbb{R}^3 = \{(u, v, w)\}$ that is the complement of the w-axis $\{(0, 0, w)\}$ and the circle in the (u, v)-plane $\{(u, v, 0) \mid u^2 + v^2 = 4\}$. Let α be the 1-form

$\alpha = \frac{-v}{u^2+v^2}du + \frac{u}{u^2+v^2}dv$ and let β be the 1-form of Chapter 1, Exercise 15. Note that α and β are both defined on M. Let $a : M \longrightarrow \mathbb{R}^2 - \{(0, 0)\}$ by $(x, y) = a(u, v, w) = (u, v)$ and let $b : M \longrightarrow \mathbb{R}^2 - \{(0, 0)\}$ by $(x, y) = b(u, v, w) = (u^2 + v^2 - 4, w^2)$. Show that $\alpha = a^*(\varphi^1)$ and $\beta = b^*(\varphi^1)$. Note that, since φ^1 is closed, this shows that α and β are closed, by Corollary 3.4.10(a).

4 Smooth Manifolds

In this chapter we develop the notion of a manifold. More precisely, we define and investigate smooth manifolds with and without boundary. We also develop the notion of an orientation of a manifold. While a manifold is an abstract notion, we will be devoting much of our attention to the more concrete notion of a submanifold of Euclidean space.

Before we begin our rigorous development, we give an intuitive notion of what a manifold is.

Intuitively, the notion of a manifold is simple. An n-dimensional manifold M is a space that locally (i.e., near any of its points) "looks like" a piece of $\mathbb{R}^n$. (Here "piece" means open set.) Let us illustrate this by some examples, which will make this idea clear.

$n = 0$: The space $\mathbb{R}^0$ is a single point. Thus a 0-manifold is one in which every point has a neighborhood that looks like a single point. In other words, a 0-manifold is just a collection of isolated points.

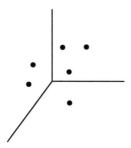

$n = 1$: The space $\mathbb{R}^1$ is the real line. Thus a 1-manifold is one in which every point has a neighborhood that looks like an open line segment. Thus, a 1-manifold is just a curve, or a union of curves. For example, we have the trefoil knot:

Differential Forms, Second Edition. http://dx.doi.org/10.1016/B978-0-12-394403-0.00004-9

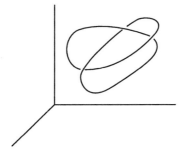

We need to be a bit clearer about what we mean by "looks like," and we will be below. For now let us observe that we can obtain a neighborhood of any point on the curve by "bending" a line segment:

$n = 2$: The space $\mathbb{R}^2$ is the plane. Thus a 2-manifold is one in which every point has a neighborhood that looks like an open bit of the plane. This is just a surface (or a union of surfaces). For example, we have the 2-sphere or the 2-torus:

Again, we can obtain a neighborhood of any point on either of these by "bending" a disk in the plane.

We not only need the notion of a manifold, but also the more general notion of a manifold with boundary. An n-dimensional manifold with boundary M is a space that locally (i.e., near any of its points) "looks like" a piece of a closed half-space in $\mathbb{R}^n$. (Again "piece" means open set.) The points p that have a neighborhood "looking like" a neighborhood of a point in the interior of a half-space constitute the interior of M, and the interior of M is an n-manifold. The points p that have a neighborhood "looking like" a neighborhood of a point on the boundary of a half-space constitute the boundary ∂M of M, and ∂M is an $(n-1)$-manifold. We allow ∂M to be empty, and we regard the empty set $\{\}$ as a manifold of any dimension. Again we illustrate with some examples.

$n = 0$: The boundary of a 0-manifold with boundary is always empty, so a 0-manifold with boundary is just the same as a 0-manifold.

$n = 1$: A typical 1-manifold with boundary is a curve (including its endpoints) and its boundary consists of the endpoints.

$n = 2$: Here is a typical 2-manifold with boundary, the boundary being the curve that bounds it.

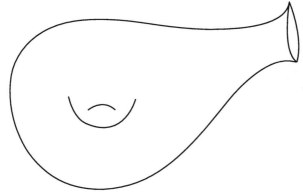

We should emphasize that a manifold (with boundary) is an abstract object. The above examples are pictures of manifolds (possibly with boundary) that are subsets—indeed, submanifolds (possibly with boundary) of Euclidean space. The theory we develop will work in general, but we will usually be considering submanifolds of Euclidean spaces, as we want to be able to do concrete computations, and so we need concrete representations of our manifolds.

Actually, we need more structure than just that of a manifold. We want to do calculus on manifolds, i.e., we want to be able to define derivatives, and hence obtain tangent vectors and differential forms. We also—and this is our principal objective here—want to be able to integrate differential forms. In order to do this we need a notion of smoothness, so we will be developing the notion of a smooth manifold.

Once we have the notion of a smooth manifold, we will proceed to develop the notions of tangent vectors and differential forms on a smooth manifold. This will lead us naturally to the notion of a smooth vector bundle.

Also, once we have the notion of a smooth manifold, we will be able to state and prove the converse of Poincaré's Lemma in its most general form.

This will all suffice for us to be able to do differential calculus on smooth manifolds. But in the next chapter we will also want to be able to do integral calculus on smooth manifolds, i.e., to be able to integrate differential forms, and for that we will need the additional notion of an orientation for a smooth manifold, which we develop as well. We also include a section that develops a very useful technical tool, partitions of unity.

4.1   The notion of a smooth manifold

We wish to consider not only the notion of a manifold, but also the more general notion of a manifold with boundary. Thus there are two ways we could proceed. The most logically economical way would be to define the more general notion of a manifold with boundary at the start, and note how it specializes to the notion of a manifold. However, we feel that that approach would be rather confusing. Thus instead we will adopt the approach of first defining the notion of a manifold, and then defining the notion of a manifold with boundary.

In order to properly define a manifold we need some notions from point-set topology. We define these notions and state some of the basic results, but it would take us too far afield to prove them, so we will forbear from doing so.

DEFINITION 4.1.1. A topological space X is *Hausdorff* if any two distinct points of X have distinct open neighborhoods. In other words, given any two distinct points p and q of X, there are open neighborhoods $\mathcal{U}$ and $\mathcal{V}$ of p and q respectively with $\mathcal{U} \cap \mathcal{V} = \{\}$.   ◇

DEFINITION 4.1.2. A topological space X is *compact* if any open cover of X has a finite subcover. In other words, given any collection $\{\mathcal{U}_i \mid i \in I\}$ of open sets $\mathcal{U}_i$ (where I is any indexing set) with $X = \cup_{i \in I} \mathcal{U}_i$, there is a finite subset $I_0 \subseteq I$ with $X = \cup_{i \in I_0} \mathcal{U}_i$.   ◇

It is easy to recognize compact subsets of Euclidean space.

THEOREM 4.1.3. *Let $X \subseteq \mathbb{R}^n$ for some n. Then X is compact if and only if it is closed and bounded.*

EXAMPLE 4.1.4. Every closed ball in $\mathbb{R}^n$ is compact. That is, let $p_0 \in \mathbb{R}^n$ be arbitrary and let R be any real number. Then $\{p \in \mathbb{R}^n \mid \|p - p_0\| \leq R\}$ is compact.   ◇

For most of our work we will be considering compact manifolds, but we do need a more general notion for some of our work.

DEFINITION 4.1.5. A topological space X is *paracompact* if any open cover of X has a locally finite open refinement. In other words, given any collection $\{\mathcal{U}_i \mid i \in I\}$ of open sets $\mathcal{U}_i$ (where I is any indexing set) with $X = \cup_{i \in I} \mathcal{U}_i$, there is a subset $\mathcal{U}_i' \subseteq \mathcal{U}_i$ for each i, with $\mathcal{U}_i'$ an open subset of X, such that $X = \cup_{i \in I} \mathcal{U}_i'$ and such that for any point $p \in X$, $\{i \in I \mid p \in \mathcal{U}_i'\}$ is finite.   ◇

We say that a set is "countable" if it is finite or countably infinite.

THEOREM 4.1.6. *Let X be a Hausdorff space in which every point has a neighborhood homeomorphic to an open subset of $\mathbb{R}^n$ for some fixed n. The following are equivalent:*

(a) X is separable, i.e., X has a countable basis of open sets.

(b) X has countably many components and each component of X is paracompact.

(c) There is a countable collection $\{\mathcal{C}_1, \mathcal{C}_2, \ldots\}$ of compact subsets of X with $X = \cup_i \mathcal{C}_i$.

(d) There is a countable collection $\{\mathcal{D}_1, \mathcal{D}_2, \ldots\}$ of compact subsets of X with $\mathcal{D}_i \subseteq \text{int}(\mathcal{D}_{i+1})$ for each i with $X = \cup_i \mathcal{D}_i$.

EXAMPLE 4.1.7. For any n, $\mathbb{R}^n$ is paracompact, as $\mathbb{R}^n$ satisfies condition (d) of Theorem 4.1.6, with $\mathcal{D}_i = \{p \in \mathbb{R}^n \mid \|p\| \leq i\}$. ◇

EXAMPLE 4.1.8. Let X be a topological space consisting of a discrete set of points. X is always paracompact. If X has countably many points, X is separable. If X has uncountably many points, X is not separable.   ◇

DEFINITION 4.1.9. Fix $n \geq 0$. A topological space M is an *n*-dimensional *manifold* (or *n-manifold*) if it is a separable Hausdorff topological space such that every point p of M has an open neighborhood homeomorphic to an open subset of $\mathbb{R}^n$.

A *curve* is a 1-manifold and a *surface* is a 2-manifold.   ◇

REMARK 4.1.10. We require that M be Hausdorff in order to eliminate the following pathological sort of example: Begin with $\mathbb{R}^1$, which
is certainly a 1-manifold, and split 0 into two points 0^+ and 0^-, such
that the deleted neighborhoods of each are the same as the deleted
neighborhoods of 0 in $\mathbb{R}^1$.

Then each point of this space has a neighborhood homeomorphic to
an open subset of $\mathbb{R}^1$ (indeed, to $\mathbb{R}^1$ itself), so would be a 1-manifold,
except that we require our manifolds to be Hausdorff, which this space
is not, as 0^+ and 0^- do not have disjoint neighborhoods.   ◇

REMARK 4.1.11. Most of our discussion will concern compact
manifolds. But we require that M be separable in order that it not
be "too big" (cf. Example 4.1.8).   ◇

Expanding on Definition 4.1.9, we see that it is equivalent to the
following:

DEFINITION 4.1.12. Fix $n \geq 0$. An n-dimensional *manifold M*
is a separable Hausdorff topological space with a collection $\mathcal{A} =$
$\{k_i : \mathcal{O}_i \longrightarrow \mathcal{U}_i\}_{i \in I}$, where k_i is a homeomorphism from an open
set $\mathcal{O}_i$ of $\mathbb{R}^n$ to an open set $\mathcal{U}_i$ of M, and such that every point p
of M is in at least one $\mathcal{U}_i$ (i.e., $\{\mathcal{U}_i\}$ is an open cover of M). The
open sets $\mathcal{U}_i$ are called *coordinate charts*. (To be precise, the data of
coordinate charts also include the homeomorphisms k_i and the sets
$\mathcal{O}_i$, but we simply abbreviate all these data by $\mathcal{U}_i$.) The collection $\mathcal{A}$
is called an *atlas*.   ◇

Our next order of business is to define smooth manifolds.

A smooth manifold is one on which we can do calculus, i.e., where
we have a notion of what it means for a function to be differentiable.
Let us see what this involves.

We are trying to define the notion of a smooth function on M.
We only know what differentiability means on $\mathbb{R}^n$, so we need to
relate this new notion to that old one. Let us try to do so as follows:
Consider a continuous function $f : M \longrightarrow \mathbb{R}$. Let p be a point of M.
Pick a coordinate chart $\mathcal{U}_1$ containing p, so we have $k_1 : \mathcal{O}_1 \longrightarrow \mathcal{U}_1$
with $\mathcal{O}_1$ an open subset of $\mathbb{R}^n$. Then we can consider the composition

$$f \circ k_1 : \mathcal{O}_1 \longrightarrow \mathbb{R}$$

$$\mathcal{O}_1 \xrightarrow{k_1} M \xrightarrow{f} \mathbb{R}.$$

Let x be the point of $\mathcal{O}_1$ with $k_1(x) = p$. We would like to say that f is smooth at p if $f \circ k_1$ is smooth at x. The latter makes sense, as x is a point of $\mathbb{R}^n$.

Unfortunately, there is a problem with this: The point p might also be in a second coordinate chart $\mathcal{U}_2$, with $k_2 : \mathcal{O}_2 \longrightarrow \mathcal{U}_2$. Then we could also consider $f \circ k_2 : \mathcal{O}_2 \longrightarrow \mathbb{R}$

$$\mathcal{O}_2 \xrightarrow{k_2} M \xrightarrow{f} \mathbb{R}.$$

If y is the point of $\mathbb{R}^n$ with $k_2(y) = p$, we could equally well say that f is smooth at p if $f \circ k_2$ is smooth at y. But how do we know that this definition (using $\mathcal{U}_2$) is the same as the first definition (using $\mathcal{U}_1$)? In fact, *we don't*.

Thus in order to make sense of the notion of differentiability, we'll have to introduce some more structure on M—the structure of a smooth manifold.

DEFINITION 4.1.13. A *diffeomorphism* $\ell : \mathcal{O}_1 \longrightarrow \mathcal{O}_2$ between two open subsets of $\mathbb{R}^n$ is a smooth homeomorphism ℓ whose inverse $\ell^{-1} : \mathcal{O}_2 \longrightarrow \mathcal{O}_1$ is also smooth. ◇

DEFINITION 4.1.14. Let $k_1 : \mathcal{O}_1 \longrightarrow \mathcal{U}_1$ and $k_2 : \mathcal{O}_2 \longrightarrow \mathcal{U}_2$ be two coordinate charts on M. They are said to be *compatible* if

$$k_2^{-1} \circ k_1 : k_1^{-1}(\mathcal{U}_1 \cap \mathcal{U}_2) \longrightarrow k_2^{-1}(\mathcal{U}_1 \cap \mathcal{U}_2)$$

is a diffeomorphism. ◇

REMARK 4.1.15. Note that $k_1^{-1}(\mathcal{U}_1 \cap \mathcal{U}_2) \subseteq \mathcal{O}_1$ and $k_2^{-1}(\mathcal{U}_1 \cap \mathcal{U}_2) \subseteq \mathcal{O}_2$ are both subsets of $\mathbb{R}^n$, so this makes sense. Also, $k_2^{-1} \circ k_1$ is certainly a homeomorphism, so the point of this definition is that both $k_2^{-1} \circ k_1$ and $k_1 \circ k_2^{-1}$ are smooth. (Of course, this is only a condition if $\mathcal{U}_1 \cap \mathcal{U}_2 \neq \emptyset$; if $\mathcal{U}_1 \cap \mathcal{U}_2 = \emptyset$, k_1 and k_2 are automatically compatible.) ◇

DEFINITION 4.1.16. A *smooth atlas* $\mathcal{A} = \{k_i : \mathcal{O}_i \longrightarrow \mathcal{U}_i\}_{i \in I}$ on M is an atlas $\mathcal{A}$ as in Definition 4.1.12, where furthermore any two coordinate charts $\mathcal{U}_{i_1}$ and $\mathcal{U}_{i_2}$ in $\mathcal{A}$ are compatible.

Two smooth atlases $\mathcal{A}^1$ and $\mathcal{A}^2$ on M are said to be *equivalent* if $\mathcal{U}_i^1$ and $\mathcal{U}_j^2$ are compatible for every pair of coordinate charts $\mathcal{U}_i^1 \in \mathcal{A}^1$ and $\mathcal{U}_j^2 \in \mathcal{A}^2$. ◇

DEFINITION 4.1.17. A *smooth structure* on an n-dimensional manifold M is an equivalence class of smooth atlases on M.

A manifold M together with a smooth structure on it is called a *smooth manifold*. ◇

We will often abbreviate smooth n-dimensional manifold to smooth n-manifold.

We should remark that a smooth manifold is often defined slightly differently. Here is the alternative definition:

DEFINITION 4.1.18. A *smooth manifold M* is a manifold together with a maximal smooth atlas. ◇

REMARK 4.1.19. It is easy to see that these notions are the same: Note that if $\mathcal{A}^1$ and $\mathcal{A}^2$ are any two equivalent smooth atlases, their union $\mathcal{A}^1 \cup \mathcal{A}^2$ is also a smooth atlas. From this we see that any equivalence class of smooth atlases determines a maximal atlas, and vice versa. We prefer our definition because an individual smooth atlas often comprises only a finite number of coordinate charts (and in case M is compact, we *always* can get an atlas consisting of only finitely many coordinate charts), whereas a maximal atlas is inherently something infinite. ◇

LEMMA 4.1.20. *Let M be a (smooth) n-manifold. Then any open subset N of M is a (smooth) n-manifold.*

Proof. Let $\mathcal{A} = \{k_i : \mathcal{O}_i \longrightarrow \mathcal{U}_i\}_{i \in I}$ be a (smooth) atlas on M. Set $\mathcal{U}_i' = (\mathcal{U}_i \cap N)$, $\mathcal{O}_i' = k_i^{-1}(\mathcal{U}')$, and $k_i' = k_i|\mathcal{O}_i'$ (the restriction of k_i to $\mathcal{O}_i'$). Then $\mathcal{A}' = \{k_i' : \mathcal{O}_i' \longrightarrow \mathcal{U}_i'\}_{i \in I}$ is a (smooth) atlas on N. □

EXAMPLE 4.1.21.

(a) $\mathbb{R}^n$ is a smooth n-manifold. It has a smooth atlas consisting of a single chart id : $\mathbb{R}^n \longrightarrow \mathbb{R}^n$.

(b) Any region $\mathcal{R} \subseteq \mathbb{R}^n$ is a smooth n-manifold. This follows immediately from the first part of this example and Lemma 4.1.20.

(c) $\mathbf{R}^n$ is a smooth n-manifold. It has a smooth atlas consisting of the single coordinate chart $k : \mathbb{R}^n \longrightarrow \mathbf{R}^n$ defined by

$$k(x_1, \ldots, x_n) = x_1 e^1 + \cdots + x_n e^n = \begin{bmatrix} x_1 \\ \vdots \\ x_n \end{bmatrix},$$

where $\mathcal{E} = \{e^1, \ldots, e^n\}$ is the standard basis of $\mathbf{R}^n$.

More generally, any n-dimensional vector space V is a smooth n-manifold. Let $\mathcal{B} = \{b^1, \ldots, b^n\}$ be any basis of V. Then V has the smooth atlas consisting of the single coordinate chart $k : \mathbb{R}^n \longrightarrow \mathbf{R}^n$ defined by

$$k(x_1, \ldots, x_n) = x_1 b^1 + \cdots + x_n b^n.$$

(The map k depends on the choice of basis, but the smooth structure does not.)

(d) $\text{Mat}_{n \times n}(\mathbb{R}) = \{n\text{-by-}n \text{ matrices with entries in } \mathbb{R}\}$ is a smooth n^2-manifold with a single coordinate chart $k : \mathbb{R}^{n^2} \longrightarrow \text{Mat}_{n \times n}(\mathbb{R})$ defined by $k(x_1, \ldots, x_{n^2}) = $ the matrix with $x_1, \ldots, x_n$ as its first row, $x_{n+1}, \ldots, x_{2n}$ as its second row, $\ldots, x_{(n-1)n+1}, \ldots, x_{n^2}$ as its last row. Let $GL_n(\mathbb{R}) = \{\text{nonsingular } n\text{-by-}n \text{ matrices}\} \subset \text{Mat}_{n \times n}(\mathbb{R})$. Then $GL_n(\mathbb{R})$ is an open subset of $\text{Mat}_{n \times n}(\mathbb{R})$ as $GL_n(\mathbb{R})$ is the complement of the closed subset $(\det)^{-1}(0)$ (where det is the determinant function). Thus $GL_n(\mathbb{R})$ is also smooth n^2-manifold.

Similarly, for any n-dimensional vector space V, $\text{End}(V) = \{\text{linear transformations} : V \longrightarrow V\}$ is a smooth n^2-manifold, and $\text{Aut}(V) = \{\text{invertible linear transformations} : V \longrightarrow V\}$ is also a smooth n^2-manifold.

(e) Let $T\mathbb{R}^n$ be the union of the tangent spaces $T_p\mathbb{R}^n$ for all points $p \in \mathbb{R}^n$. Then we can give $T\mathbb{R}^n$ a topology and indeed a smooth structure making it a smooth $2n$-manifold by giving it a single coordinate chart $k : \mathbb{R}^{2n} \longrightarrow T\mathbb{R}^n$ with

$$k(x_1, \ldots, x_{2n}) = \begin{bmatrix} x_{n+1} \\ \vdots \\ x_{2n} \end{bmatrix}_{(x_1, \ldots, x_n)}.$$

Similarly, if $\mathcal{R} \subseteq \mathbb{R}^n$ is any open set, $T\mathcal{R}$, the union of the tangent spaces $T_p\mathbb{R}^n$ for all points $p \in \mathcal{R}$, is an open subset of $T\mathbb{R}^n$ and hence is also a smooth $2n$-manifold.   ◇

We thus see from Example 4.1.21 parts (a) and (b) that in our previous discussion of tangent vectors, forms, etc., we have been dealing with a very special kind of smooth manifold. It is our objective to generalize this discussion to arbitrary smooth n-manifolds.

But before we do so, let us look at some other examples.

EXAMPLE 4.1.22. Let $\mathcal{R}$ be a region in $\mathbb{R}^n$ and let $f : \mathcal{R} \longrightarrow \mathbb{R}^k$ be an arbitrary function. Then the graph G of f, $G(f) = \{(p, q) \in \mathbb{R}^n \times \mathbb{R}^k \mid q = f(p)\}$ is a smooth n-manifold with atlas consisting of a single chart $k : \mathcal{R} \longrightarrow G(f)$ defined by $k(p) = (p, f(p))$.   ◇

EXAMPLE 4.1.23.

(a) The unit circle $S^1 = \{(x, y) \mid x^2 + y^2 = 1\}$ in $\mathbb{R}^2$ is a smooth manifold. Here is an atlas:

$$\mathcal{A} = \{k_1 : \mathcal{O}_1 \longrightarrow \mathcal{U}_1, k_2 : \mathcal{O}_2 \longrightarrow \mathcal{U}_2\},$$

where

$$\mathcal{O}_1 = \{t \mid 0 < t < 2\pi\}, \quad \mathcal{U}_1 = S^1 - \{(0, -1)\},$$
$$k_1 : \mathcal{O}_1 \longrightarrow S^1 \text{ by } k_1(t) = (\cos(t + 3\pi/2), \sin(t + 3\pi/2));$$
$$\mathcal{O}_2 = \{t \mid 0 < t < 2\pi\}, \quad \mathcal{U}_2 = S^1 - \{(0, 1)\},$$
$$k_2 : \mathcal{O}_1 \longrightarrow S^1 \text{ by } k_2(t) = (\cos(t + \pi/2), \sin(t + \pi/2)).$$

Then

$$\mathcal{O} = k_1^{-1}(\mathcal{U}_1 \cap \mathcal{U}_2) = k_2^{-1}(\mathcal{U}_1 \cap \mathcal{U}_2)$$
$$= \{t \mid 0 < t < \pi\} \cup \{t \mid \pi < t < 2\pi\},$$
$$\ell = k_2^{-1} \circ k_1 : \mathcal{O} \longrightarrow \mathcal{O} \text{ by } \begin{cases} \ell(t) = t + \pi \text{ if } 0 < t < \pi, \\ \ell(t) = t - \pi \text{ if } \pi < t < 2\pi. \end{cases}$$

(b) The unit sphere $S^2 = \{(x, y, z) \mid x^2 + y^2 + z^2 = 1\}$ in $\mathbb{R}^3$ is a smooth manifold. Here is an atlas, where we use polar coordinates (r, θ) for points in $\mathbb{R}^2$:

$$\mathcal{A} = \{k_1 : \mathcal{O}_1 \longrightarrow \mathcal{U}_1, k_2 : \mathcal{O}_2 \longrightarrow \mathcal{U}_2\},$$

where

$$\begin{aligned}
\mathcal{O}_1 &= \{(r, \theta) \mid r < \pi\}, \quad \mathcal{U}_1 = S^2 - \{(0, 0, -1)\}, \\
&k_1 : \mathcal{O}_1 \longrightarrow S^1 \text{ by} \\
&k_1(r, \theta) = (\sin(r)\cos(\theta), \sin(r)\sin(\theta), \cos(r)); \\
\mathcal{O}_2 &= \{(r, \theta) \mid r < \pi\}, \quad \mathcal{U}_2 = S^1 - \{(0, 0, 1)\}, \\
&k_2 : \mathcal{O}_1 \longrightarrow S^1 \text{ by} \\
&k_2(r, \theta) = (\sin(r)\cos(\theta), -\sin(r)\sin(\theta), -\cos(r)).
\end{aligned}$$

Then

$$\mathcal{O} = k_1^{-1}(\mathcal{U}_1 \cap \mathcal{U}_2) = k_2^{-1}(\mathcal{U}_1 \cap \mathcal{U}_2) = \{(r, \theta) \mid 0 < r < \pi\},$$
$$\ell = k_2^{-1} \circ k_1 : \mathcal{O} \longrightarrow \mathcal{O} \text{ by } \ell(r, \theta) = (\pi - r, -\theta). \quad \diamond$$

Example 4.1.23 was a lot of work in some simple cases. We shall now quote, without proof, a theorem that allows us to easily obtain examples of smooth manifolds.

DEFINITION 4.1.24. Let $F : \mathbb{R}^N \longrightarrow \mathbb{R}^k$ be a smooth function. A point $p \in \mathbb{R}^N$ is a *regular point* of F if the derivative matrix $F'(p)$ of F at p has rank k.

A point $q \in \mathbb{R}^k$ is a *regular value* of F if every point $p \in \mathbb{R}^N$ with $F(p) = q$ is a regular value of F. $\quad \diamond$

THEOREM 4.1.25 (*Implicit function theorem*). *Let M be a subset of $\mathbb{R}^N$. Then M is a smooth submanifold of $\mathbb{R}^N$ of dimension $n = N - k$ if and only if for every point p of M there exist an open neighborhood $\mathcal{N} \subseteq \mathbb{R}^N$ of p in $\mathbb{R}^N$ and a smooth function $F : \mathcal{N} \longrightarrow \mathbb{R}^k$ such that $q = F(p)$ is a regular value of F and $M \cap \mathcal{N} = F^{-1}(q)$.*

EXAMPLE 4.1.26.

(a) Let $f : \mathbb{R}^n \longrightarrow \mathbb{R}^k$. Coordinatize $\mathbb{R}^n$ by $(x_1, \ldots, x_n)$ and $\mathbb{R}^k$ by $(y_1, \ldots, y_k)$ and let $f = (f_1, \ldots, f_k)$, i.e., $y_i = f_i(x_1, \ldots, x_n)$ for

$i = 1, \ldots, k$. Let $N = n + k$ and coordinatize $\mathbb{R}^N$ by $(z_1, \ldots, z_N)$. Let $F : \mathbb{R}^N \longrightarrow \mathbb{R}^k$ by

$$F(z_1, \ldots, z_N) = (z_{n+1} - f_1(z_1, \ldots, z_n), \ldots, z_{n+k} - f_k(z_1, \ldots, z_n)).$$

Then the derivative matrix $F'(p)$ is the k-by-N block matrix

$$F'(p) = \left[-f'(p) \,|\, I \right],$$

where I is the k-by-k identity matrix. Thus $F'(p)$ always has rank k, so every point q is a regular value of F. In particular, $q = (0, \ldots, 0)$ is a regular value of F. But $F^{-1}(0, \ldots, 0) = G(f)$, the graph of f (cf Example 4.1.22). Thus $G(f)$ is a smooth submanifold of $\mathbb{R}^N$.

(b) Let $F : \mathbb{R}^{n+1} \longrightarrow \mathbb{R}^1$ by

$$F(x_1, \ldots, x_{n+1}) = x_1^2 + \cdots + x_{n+1}^2.$$

Then $S^n = F^{-1}(1)$ is the unit sphere in $\mathbb{R}^{n+1}$.

$$F'(x_1, \ldots, x_{n+1}) = [2x_1 \ldots 2x_{n+1}]$$

and this matrix has rank 1 except at the point $(0, \ldots, 0)$. But this point is not in $F^{-1}(1)$, so 1 is a regular value of F. Thus S^n is a smooth submanifold of $\mathbb{R}^{n+1}$. Note S^n is compact.

Note also that $F^{-1}(0)$ is the single point $(0, \ldots, 0)$, which, for $n > 0$, is not an n-manifold.

(c) Let $F : \mathbb{R}^3 \longrightarrow \mathbb{R}^1$ by

$$F(x, y, z) = x^2 + y^2 - z^2.$$

An analysis very similar to that in part (b) shows that $F^{-1}(r)$ is a smooth surface for $r \neq 0$. For $r > 0$ it is a hyperboloid of 1 sheet and for $r < 0$ it is a hyperboloid of 2 sheets. Note that neither of these is compact.

For $r = 0$ it is a cone, which is not a smooth surface as the point $(0, 0, 0)$ does not have a good neighborhood. But $F^{-1}(0) - (0, 0, 0)$ is a smooth surface.

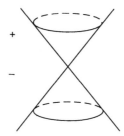

(d) The equation $y^2 = x^3 + x$ defines a smooth curve in $\mathbb{R}^2$. It is $f^{-1}(0)$ where $f(x, y) = x^3 + x - y^2$. Observe that $f'(x, y) = [3x^2 + 1 \quad -2y]$ has rank 1 everywhere. This curve is pictured in the figure below. Note it is not compact.

The equation $y^2 = x^3 - x$ also defines a smooth curve in $\mathbb{R}^2$. It is $f^{-1}(0)$ where $f(x, y) = x^3 - x - y^2$. Observe that $f'(x, y) = [3x^2 - 1 \quad -2y]$ has rank 1 except at the points $(\pm \sqrt{1/3}, 0)$ and $f(\pm \sqrt{1/3}, 0) \neq 0$. This curve is pictured in the figure below. Note that it has two components, one of which is compact and the other of which is not.

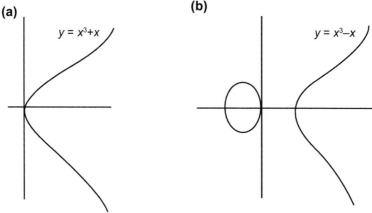

The equation $y^2 = x^3$ does not define a smooth curve in $\mathbb{R}^2$. It is $f^{-1}(0)$ where $f(x, y) = x^3 - y^2$. Observe that $f'(x, y) = [3x^2 \quad -2y]$ has rank 0 at the point $(0, 0)$ and $f(0, 0) = 0$. This curve is not smooth at $(0, 0)$–it has a "cusp" at $(0, 0)$. However, $f^{-1}(0) - (0, 0)$ is a smooth curve.

(e) Let $F : \mathbb{R}^3 \longrightarrow \mathbb{R}^2$ by

$$F(x, y, z) = (x^2 + y^2 + z^2, z - x\sqrt{3}).$$

Then $(4, 4\sqrt{3})$ is a regular value of F, so $F^{-1}(4, 4\sqrt{3})$ is a smooth curve in $\mathbb{R}^3$. It is easy to see that it is a circle.

Now let $F : \mathbb{R}^3 \longrightarrow \mathbb{R}^3$ by

$$F(x, y, z) = (x^2 + y^2 + z^2, z - x\sqrt{3}, y - 2x).$$

Then $(4, 4\sqrt{3}, 8)$ is a regular value of F, so $F^{-1}(4, 4\sqrt{3}, 8)$ is a smooth 0-manifold in $\mathbb{R}^3$. It is easy to see that it is the union of two points, $F^{-1}(4, 4\sqrt{3}, 8) = \{(-4, 0, 0), (-3, 2, \sqrt{3})\}$.

(f) Let us identify $\mathbb{R}^2$ with the complex plane $\mathbb{C}$. Let $F : \mathbb{C}^2 \longrightarrow \mathbb{C} \times \mathbb{R}$ be defined by

$$F(z_1, z_2) = (z_1^2 - z_2^3, \|z_1\|^2 + \|z_2\|^2).$$

Then $(0, 1)$ is a regular point and hence $F^{-1}(0, 1)$ is a smooth curve. Note that this curve is contained in the unit sphere S^3 in $\mathbb{C}^2$.   ◇

REMARK 4.1.27. Roughly speaking, Theorem 4.1.25 says that a "level curve" of a "nice" function from $\mathbb{R}^2$ to $\mathbb{R}^1$ is indeed a smooth curve, a "level surface" of a "nice" function from $\mathbb{R}^3$ to $\mathbb{R}^1$ is indeed a smooth surface, and that the intersection of the "level surfaces" of a "nice" pair of functions from $\mathbb{R}^3$ to $\mathbb{R}^1$ is a smooth curve, etc.   ◇

REMARK 4.1.28. Note that M as given by the implicit function theorem, Theorem 4.1.25, is always locally closed in $\mathbb{R}^N$, i.e., in the notation of that theorem, for every point p of M, $M \cap \mathcal{N}$ is a closed subset of $\mathcal{N}$. Consequently, if the function F is defined on all of $\mathbb{R}^N$ (as in almost all of the examples above), M is closed in $\mathbb{R}^N$. In this case, M is compact if and only if it is bounded.   ◇

REMARK 4.1.29. In a way, Example 4.1.26 was misleadingly simple. As Theorem 4.1.25 states, M is a smooth submanifold of $\mathbb{R}^n$ if for every point p of M there is a "nice" function F defined on a neighborhood of p. In this example there was a single such F defined on all of M. But in general this is not always the case.   ◇

In our considerations of manifolds above, we went from "global" to "local." By that we mean that we started with a smooth manifold, a global object, and went to coordinate patches, a collection of local objects. But we should observe that we can go in the other direction, starting with a suitable collection of coordinate patches and obtaining a manifold from them. This idea will be important below, when a priori we will only have a collection of local objects and we will obtain a global object from them.

First we make the global to local direction more precise.

COROLLARY 4.1.30. *Let M be a smooth manifold and let $\mathcal{A} = \{k_i : \mathcal{O}_i \longrightarrow \mathcal{U}_i\}$ be a smooth atlas on M. For any ordered pair (i_1, i_2) let $\mathcal{O}_{i_1 i_2} = k_{i_1}^{-1}(\mathcal{U}_{i_1} \cap \mathcal{U}_{i_2}) \subseteq \mathcal{O}_{i_1}$. Let $\ell_{i_2 i_1} = k_{i_2}^{-1} \circ k_{i_1} : \mathcal{O}_{i_1 i_2} \longrightarrow \mathcal{O}_{i_2 i_1}$. Then M is homeomorphic to the space obtained by identifying all points $p_1 \in \mathcal{O}_{i_1}$ and $p_2 \in \mathcal{O}_{i_2}$ such that $p_2 = \ell_{i_2 i_1}(p_1)$. Furthermore, each $\ell_{i_2 i_1}$ is a diffeomorphism from an open set in $\mathbb{R}^n$ to an open set in $\mathbb{R}^n$, and $\{\ell_{i_2 i_1}\}$ is compatible in the sense that $\ell_{i_3 i_1} = \ell_{i_3 i_2} \circ \ell_{i_2 i_1}$ whenever the composition is defined, for every triple (i_1, i_2, i_3).*

Proof. The given identification is nothing other than the observation that if $p \in \mathcal{U}_1 \cap \mathcal{U}_2$ with $p = k_1(q_1)$ and $p = k_2(q_2)$ then $q_2 = k_{i_2}^{-1} \circ k_{i_1}(q_1) = \ell_{i_2 i_1}(p_1)$.

Also, $\ell_{i_3 i_1} = k_{i_3}^{-1} \circ k_{i_1} = k_{i_3}^{-1} \circ k_{i_2} \circ k_{i_2}^{-1} \circ k_{i_1} = \ell_{i_3 i_2} \circ \ell_{i_2 i_1}$. □

Now we give the local to global direction.

THEOREM 4.1.31. *Let $\{\mathcal{O}_i\}$ be a collection of open sets in $\mathbb{R}^n$. Suppose that for each ordered pair (i_1, i_2) there is an open subset (which may be empty) $\mathcal{O}_{i_1 i_2}$ of $\mathcal{O}_{i_1}$ and a diffeomorphism $\ell_{i_2 i_1} : \mathcal{O}_{i_1 i_2} \longrightarrow \mathcal{O}_{i_2 i_1}$, with the collection of maps $\{\ell_{i_2 i_1}\}$ satisfying the compatibility relations $\ell_{i_3 i_1} = \ell_{i_3 i_2} \circ \ell_{i_2 i_1}$ whenever the composition is defined, for every triple (i_1, i_2, i_3).*

Let M be the space obtained by identifying points in $\{\mathcal{O}_i\}$ where $q_1 \in \mathcal{O}_1$ is identified with $q_2 \in \mathcal{O}_2$ if $q_2 = \ell_{i_2 i_1}(q_1)$. Suppose that M is a separable Hausdorff space. Then M is an n-manifold. Further, if $\mathcal{U}_i$ is the image of $\mathcal{O}_i$ under this identification, and $k_i : \mathcal{O}_i \longrightarrow \mathcal{U}_i$ is the identification map, then $\mathcal{A} = \{k_i : \mathcal{O}_i \longrightarrow \mathcal{U}_i\}$ is a smooth atlas on M, giving M a smooth structure. Also, $\ell_{i_2 i_1} = k_{i_2}^{-1} \circ k_{i_1}$ for every (i_1, i_2).

Proof. The compatibility conditions give that the above identification is an equivalence relation, and then it is immediate from the definition that every point in M has a neighborhood homeomorphic to an open subset of $\mathbb{R}^n$, so, given that M is separable and Hausdorff, M is an n-manifold. Then it is immediate that $\mathcal{A}$ is a smooth atlas on M. $\square$

As we stated above, the motivation for our definition of a smooth manifold was to be able to define a smooth function. We do that now.

DEFINITION 4.1.32. Let M be a smooth manifold. A function $f : M \longrightarrow \mathbb{R}$ is *smooth* if for every coordinate chart $k : \mathcal{O} \longrightarrow \mathcal{U}$ the composition $f \circ k : \mathcal{O} \longrightarrow \mathbb{R}$ is smooth. $\diamond$

More generally we can define a smooth map between two smooth manifolds.

DEFINITION 4.1.33. Let M_1 and M_2 be smooth manifolds. A function $f : M_1 \longrightarrow M_2$ is *smooth* if it has the following property: Let p_1 be any point of M_1 and let $p_2 = f(p_1)$. Then M_1 has a coordinate chart $k_1 : \mathcal{O}_1 \longrightarrow \mathcal{U}_1$ and M_2 has a coordinate chart $k_2 : \mathcal{O}_2 \longrightarrow \mathcal{U}_2$ with $p_1 \in \mathcal{U}_1, p_2 \in \mathcal{U}_2, f(\mathcal{U}_1) \subseteq \mathcal{U}_2$, and with the composition $k_2^{-1} \circ f \circ k_1 : \mathcal{O}_1 \longrightarrow \mathcal{O}_2$ a smooth function. $\diamond$

(Note that in Definition 4.1.12 and the sequel, $\{k_i : \mathcal{O}_i \longrightarrow \mathcal{U}_i\}$ were all charts on the same manifold, whereas here we have charts on two different manifolds. We have decided to use the same notation in these two slightly different contexts, as the alternative would be to introduce extra new notation which would just be more cumbersome and confusing. We thus call the reader's attention to the fact that we have these two different uses, and that these two uses will recur below.)

REMARK 4.1.34. The definition of smoothness is independent of the choice of coordinate charts in each of these definitions. $\diamond$

REMARK 4.1.35. The case of 0-manifolds is a simple one, but it requires us to reconsider our previous definitions. As we have seen, a connected 0-manifold is a point. The condition of separability implies that a manifold can have at most countably many components, so a general 0-manifold M is a finite or countably infinite set of discrete points. (M is compact when this set is finite.) The notion of an atlas for a 0-manifold is a trivial one, so a smooth 0-manifold M is just a 0-manifold M. Also, there is no condition on a map between points to be smooth, so a smooth map $k : M \longrightarrow N$ between 0-manifolds M and N is any function from M to N. Then k is a diffeomorphism if and only if it is a homeomorphism, which is the case if and only if it is an isomorphism of sets. (In particular, if $M = \{p\}$ and $N = \{q\}$ each consists of a single point, the unique function $k : M \longrightarrow N$, i.e., the function given by $k(p) = q$, is a diffeomorphism.)   ◇

Now let us consider the more general notion of a manifold with boundary.

To begin with we introduce some (nonstandard) notation.

NOTATION 4.1.36.

(a) $\mathbb{R}^0 = \mathbb{R}^0_- = \mathbb{R}^0_{--} = \{0\}$, a single point.

(b) For $n > 0$, $\mathbb{R}^n = \{(x_1, \ldots, x_n)\}$, $\mathbb{R}^n_- = \{(x_1, \ldots, x_n) \mid x_1 \leq 0\}$, and $\mathbb{R}^n_{--} = \{(x_1, \ldots, x_n) \mid x_1 < 0\}$. For $n = 1$, we also let $\mathbb{R}^1_+ = \{x_1 \mid x_1 \geq 0\}$ and $\mathbb{R}^1_{++} = \{x_1 \mid x_1 > 0\}$.

(c) We identify $\mathbb{R}^{n-1}$ with the subset $\{(0, x_2, \ldots, x_n)\}$ of $\mathbb{R}^n$. (Note this is not the "usual" $\mathbb{R}^{n-1}$.) Thus, $\mathbb{R}^{n-1}$ is included in $\mathbb{R}^n_-$, and indeed is the complement of $\mathbb{R}^{n-1}_{--}$ in $\mathbb{R}^{n-1}$.

(d) We set $\mathbb{R}^{-1} = \{\}$, the empty set.   ◇

Of course, $\mathbb{R}^n_{--}$ is homeomorphic to $\mathbb{R}^n$, but it is convenient to introduce it. Also, we will see that it is convenient to have $\mathbb{R}^1_{++}$ available.

We now have the following definition, generalizing Definition 4.1.9.

DEFINITION 4.1.37. Fix $n \geq 0$. A topological space M is an n-dimensional *manifold with boundary* (or *n-manifold with boundary*) if it is a separable Hausdorff topological space such that every point p of M has an open neighborhood homeomorphic to an open subset of $\mathbb{R}^n_-$ (or, for $n = 1$, also of $\mathbb{R}^1_+$).

The points of M that have neighborhoods homeomorphic to neighborhoods of points in $\mathbb{R}^n_{--}$ (or, for $n = 1$, also in $\mathbb{R}^1_{++}$) form the *interior* of M, denoted $Int(M)$, while the points of M that have neighborhoods homeomorphic to neighborhoods in $\mathbb{R}^n_-$ (or, for $n = 1$, also in $\mathbb{R}^n_+$) of points in $\mathbb{R}^{n-1}$ form the *boundary* of M, denoted ∂M. ◇

This is equivalent to the following definition (cf Definition 4.1.12).

DEFINITION 4.1.38. Fix $n \geq 0$. An n-dimensional *manifold with boundary* M is a Hausdorff topological space with a collection $\mathcal{A} = \{k_i : \mathcal{O}_i \longrightarrow \mathcal{U}_i\}_{i \in I}$, where k_i is a homeomorphism from an open set $\mathcal{O}_i$ of $\mathbb{R}^n_-$ (or, for $n = 1$, of $\mathbb{R}^1_+$) to an open set $\mathcal{U}_i$ of M, and such that every point p of M is in at least one $\mathcal{U}_i$ (i.e., $\{\mathcal{U}_i\}$ is an open cover of M). The open sets $\mathcal{U}_i$ are called *coordinate charts*. (To be precise, the data of coordinate charts also include the homeomorphisms k_i and the sets $\mathcal{O}_i$, but we simply abbreviate all these data by $\mathcal{U}_i$.) The collection $\mathcal{A}$ is called an *atlas*.

The points q of M with $q = k_i(p)$ for $p \in \mathcal{O}_i \cap \mathbb{R}^n_{--}$ (or, for $n = 1, \mathbb{R}^1_{++}$) form the *interior* of M, denoted $Int(M)$, while the points q of M with $q = k_i(p)$ for $p \in \mathcal{O}_i \cap \mathbb{R}^{n-1}$ form the *boundary* of M, denoted ∂M.   ◇

REMARK 4.1.39. By definition, the empty set $\{\}$ is a manifold of any dimension. If M is a manifold with boundary, then M is a manifold if and only if $\partial M = \{\}$.   ◇

The following is common terminology. Note this use of the word "closed" is unconnected with the topological meaning of the word "closed."

DEFINITION 4.1.40. A *closed* manifold M is a compact manifold (without boundary).   ◇

LEMMA 4.1.41. *If M is an n-manifold with boundary, its boundary ∂M is an $(n-1)$-manifold.*

Proof. If M has atlas $\mathcal{A} = \{k_i : \mathcal{O}_i \longrightarrow \mathcal{U}_i\}_{i \in I}$, let $\mathcal{O}'_i = \mathcal{O}_i \cap \mathbb{R}^{n-1}$, let k'_i be the restriction of k_i to $\mathcal{O}'_i$, and let $\mathcal{U}'_i = k'_i(\mathcal{O}'_i) \subset \mathcal{U}_i$. Then $\mathcal{A}' = \{k'_i : \mathcal{O}'_i \longrightarrow \mathcal{U}'_i\}_{i \in I}$ is an atlas for ∂M.   □

Items 4.1.13 through 4.1.20 generalize virtually without change, so we merely refer the reader to them. We will simply state the following (cf Definition 4.1.17).

DEFINITION 4.1.42. A *smooth structure* on an n-dimensional manifold M with boundary is an equivalence class of smooth atlases on M.

A manifold M with boundary together with a smooth structure on it is called a *smooth manifold with boundary*.   ◇

We have stated this explicitly in order to make the following observation.

LEMMA 4.1.43. *Let M be an n-manifold with boundary with smooth structure given by the atlas $\mathcal{A}$. Let $\mathcal{A}'$ be the atlas on ∂M constructed in the proof of Lemma 4.1.41. Then $\mathcal{A}'$ is a smooth structure on ∂M.*

Proof. If $k_j^{-1} \circ k_i$ is smooth for every $i, j \in I$, then $k_j'^{-1} \circ k_i'$ is smooth for every $i, j \in I$.   □

DEFINITION 4.1.44. The smooth structure on ∂M given by Lemma 4.1.43 is the smooth structure on ∂M *induced* by the smooth structure on M.   ◇

Here are a few examples of manifolds with boundary.

EXAMPLE 4.1.45.

(a) Any closed interval $I = [a, b] \subset \mathbb{R}$ is a smooth 1-manifold with boundary $\partial I = \{a, b\}$.

(b) Any closed ball in $\mathbb{R}^n$, i.e., any set $\{p \in \mathbb{R}^n \mid \|p - p_0\| \leq R\}$ for some fixed point $p_0 \in \mathbb{R}^n$ and any real number $R > 0$, is a smooth n-manifold with boundary $\{p \in \mathbb{R}^n \mid \|p - p_0\| = R\}$.

Similarly any closed annulus $\{p \in \mathbb{R}^n \mid r < \|p - p_0\| \leq R\}$, where $R > r > 0$, is a smooth n-manifold with boundary $\{p \in \mathbb{R}^n \mid \|p - p_0\| = r\} \cup \{p \in \mathbb{R}^n \mid \|p - p_0\| = R\}$.

(c) The closed "northern hemisphere" of the unit sphere in $\mathbb{R}^{n+1}$, given by $M = \{(x_1, \ldots, x_n, x_{n+1}) \mid x_1^2 + \cdots + x_{n+1}^2 = 1, x_{n+1} \geq 0\}$, is a smooth n-manifold with $\partial M = \{(x_1, \ldots, x_n, 0) \mid x_1^2 + \cdots + x_n^2 = 1\}$.

(d) The subset $\{(x, y, z) \mid z = x^2 + y^2, z \leq 1\}$ of $\mathbb{R}^3$ (part of a paraboloid) is a smooth surface with boundary the circle $\{(x, y, 1) \mid x^2 + y^2 = 1\}$.   ◇

Definitions 4.1.32 and 4.1.33 generalize to manifolds with boundary without change. However, there is a subtlety in these extensions that is hidden in the language. Consider Definition 4.1.32. It requires the map $g = f \circ k : \mathcal{O} \longrightarrow \mathbb{R}$ to be smooth. In the case of a manifold with boundary, the subset $\mathcal{O} \subseteq \mathbb{R}^n$ is open as a subset of a half-space in $\mathbb{R}^n$, but may or may not be open as a subset of $\mathbb{R}^n$ itself. In case it is not, we need to recall what it means for $g : \mathcal{O} \longrightarrow \mathbb{R}$ to be smooth: There is an open subset $\tilde{\mathcal{O}} \subseteq \mathbb{R}^n$ with $\mathcal{O} \subseteq \tilde{\mathcal{O}}$ and an extension $\tilde{g} : \tilde{\mathcal{O}} \longrightarrow \mathbb{R}$ of $g : \mathcal{O} \longrightarrow \mathbb{R}$ with $\tilde{g}$ smooth. A similar consideration applies to Definition 4.1.33.

Actually, we will often be looking at a slightly more general situation, that of a "piecewise smooth" manifold with boundary). A piecewise smooth manifold is exactly what its name implies–it is a manifold with boundary that is a union of pieces, each of which is a smooth manifold with boundary. Like (almost) every other author, we will not go through the whole theory of these carefully–it is routine and quite tedious, and the result is simply everything generalizes in a quite straightforward way. But we will introduce a particular example, one that will be extremely useful to us.

DEFINITION 4.1.46. A *brick* B in $\mathbb{R}^n$ is a subset of $\mathbb{R}^n$ of the form

$$B = \{(x_1, \ldots, x_n) \mid a_1 \leq x_1 \leq b_1, \ldots, a_n \leq x_n \leq b_n\}$$

for fixed real numbers $a_1 < b_1, \ldots, a_n < b_n$.   ◇

We observe that ∂B is piecewise smooth, being a union of smooth "faces." Then B itself is not a smooth manifold with boundary, but rather a smooth manifold "with corners," in an obvious sense.

4.2 Tangent vectors and differential forms

In this section we develop the notions of tangent vectors and differential forms on an arbitrary smooth manifold. We will see that this is a straightforward generalization of our earlier work.

We begin with tangent vectors. As before, a tangent vector is the tangent to a smooth curve. Let us make this precise. Recalling Remark 3.1.4, we make the following definition.

DEFINITION 4.2.1. Let M be a smooth manifold and let p be a point of M. Pick a coordinate chart $\mathcal{U}_1$ containing p, so we have $k_1 : \mathcal{O}_1 \longrightarrow \mathcal{U}_1$ with $\mathcal{O}_1$ an open subset of $\mathbb{R}^n$. Consider a curve C_1 parameterized by r_1 passing through the point p when $t = t_1$ and a curve C_2 parametrized by r_2 passing through the point p when $t = t_2$. We say that these two curves are equivalent if $(k_1^{-1} r_1)'(t_1) = (k_1^{-1} r_2)'(t_2)$. Then a *tangent vector* to M at p is an equivalence class of curves.

The *tangent space* $T_p M$ of M at p is the union of tangent vectors to M at p. $\diamond$

REMARK 4.2.2. The first thing we have to see is that this definition makes sense. A priori it depends on the choice of coordinate chart. So suppose we have a second coordinate chart $\mathcal{U}_2$ containing p, so we have $k_2 : \mathcal{O}_2 \longrightarrow \mathcal{U}_2$ with $\mathcal{O}_2$ an open subset of $\mathbb{R}^n$. Then, using this second coordinate chart, curves C_1 and C_2 as above are equivalent if $(k_2^{-1} r_1)'(t_1) = (k_2^{-1} r_2)'(t_2)$. Let $q_1 = k_1^{-1}(p)$ and $q_2 = k_2^{-1}(p)$. Let $\ell = k_2^{-1} \circ k_1$. Then

$$(k_2^{-1} r_1)'(t_1) = (\ell k_1^{-1} r_1)'(t_1) = \ell'(k_1^{-1} r_1)'(t_1)$$

and similarly

$$(k_2^{-1} r_2)'(t_2) = (\ell k_1^{-1} r_2)'(t_2) = \ell'(k_1^{-1} r_2)'(t_2).$$

But, since these two charts are compatible, ℓ is a diffeomorphism from an open subset of $\mathbb{R}^n$ to an open subset of $\mathbb{R}^n$, so ℓ' is an isomorphism $\ell' : T_{q_1}\mathbb{R}^n \longrightarrow T_{q_2}\mathbb{R}^n$. Hence

$$(k_1^{-1} r_1)'(t_1) = (k_1^{-1} r_2)'(t_2)$$

if and only if

$$(k_2^{-1} r_1)'(t_1) = (k_2^{-1} r_2)'(t_2)$$

and so this equivalence relation is independent of the choice of coordinate chart. $\diamond$

LEMMA 4.2.3. *Let M be a smooth manifold and let p be a point of M. Pick a coordinate chart $\mathcal{U}$ containing p, so we have $k : \mathcal{O} \longrightarrow \mathcal{U}$ with $\mathcal{O}$ an open subset of $\mathbb{R}^n$. Let $q = k^{-1}(p)$. Then any tangent vector $\mathbf{v}_p$ is given by $\mathbf{v}_p = k_*(\mathbf{w}_q)$ for a unique tangent vector $\mathbf{w}_q$.*

Proof. Let $\mathbf{v}_p$ be the equivalence class of the curve C parameterized by r with $r(t_0) = p$. Let D be the curve in $\mathcal{O} \subseteq \mathbb{R}^n$ parameterized by $s = k^{-1} \circ r$, with $s(t_0) = q$, and let $\mathbf{w}_q$ be the equivalence class of the curve D. Then $r = k \circ s$, so by the definition of the push-forward, $\mathbf{v}_p = k_*(\mathbf{w}_q)$.

Also, suppose $\mathbf{u}_q$ is any tangent vector with $k_*(\mathbf{u}_q) = k_*(\mathbf{w}_q)$. By Definition 4.2.1, this equality just means that $(k^{-1} \circ k)_*(\mathbf{u}_q) = (k^{-1} \circ k)_*(\mathbf{w}_q)$, i.e., that $\mathbf{u}_q = \mathbf{w}_q$. Thus $\mathbf{w}_q$ is unique. $\quad\square$

This lemma enables us to define a vector space structure on T_pM.

DEFINITION 4.2.4. Let M be a smooth manifold and let p be a point of M. Pick a coordinate chart $\mathcal{U}$ containing p, so we have $k : \mathcal{O} \longrightarrow \mathcal{U}$ with $\mathcal{O}$ an open subset of $\mathbb{R}^n$. Let $q = k^{-1}(p)$.

T_pM has vector space structure as follows: Let $\mathbf{v}_p^1$ and $\mathbf{v}_p^2$ be any two elements of T_pM, i.e., any two tangent vectors to M at p, and let c^1 and c^2 be any two scalars. Let $\mathbf{w}_q^1$ and $\mathbf{w}_q^2$ be the unique elements of $T_q\mathbb{R}^n$ with $k_*(\mathbf{w}_q^1) = \mathbf{v}_q^1$ and $k_*(\mathbf{w}_q^2) = \mathbf{v}_q^2$. Then

$$c^1\mathbf{v}^1 + c^2\mathbf{v}^2 = k_*(c^1\mathbf{w}^1 + c^2\mathbf{w}^2). \qquad \diamond$$

COROLLARY 4.2.5. *Let M be a smooth manifold and let p be a point of M. Pick a coordinate chart $\mathcal{U}$ containing p, so we have $k : \mathcal{O} \longrightarrow \mathcal{U}$ with $\mathcal{O}$ an open subset of $\mathbb{R}^n$. Let $q = k^{-1}(p)$. Then $k_* : T_q\mathbb{R}^n \longrightarrow T_pM$ is a vector space isomorphism.*

Proof. By Lemma 4.2.3, k_* is an isomorphism of sets (i.e, a one-to-one and onto map), and in Definition 4.2.4 we have given T_pM a vector space structure so as to make k_* a linear map. $\quad\square$

REMARK 4.2.6. Note that in Lemma 4.2.3 and Definition 4.2.4, and hence in Corollary 4.2.5 as well, we could use any coordinate chart. Once again we have to show our results are independent of the choice of coordinate chart, and the proof of that is the same as the proof in Remark 4.2.2 that the definition of tangent vector in Definition 4.2.1 is independent of the choice of coordinate chart. ◇

Having defined tangent vectors, it is now easy to define smooth vector fields.

DEFINITION 4.2.7. Let $V = \{\mathbf{v}_p | p \in M\}$ be a set of tangent vectors at every point of M. Then V is a *smooth vector field* on M if for every coordinate chart $k : \mathcal{O} \longrightarrow \mathcal{U}$, $W = \{\mathbf{w}_q = k_*^{-1}(\mathbf{v}_p) | q \in \mathcal{O}\}$ is a smooth vector field on $\mathcal{O} \subseteq \mathbb{R}^n$. ◇

Having defined tangent vectors and vector fields on an arbitrary smooth manifold M, we may now define differential forms on M. First we define 0-forms and then we define j-forms for $j > 0$.

DEFINITION 4.2.8. A 0-form on the smooth manifold M is a function f on M with the following property:
Let $k : \mathcal{O} \longrightarrow \mathcal{U}$ be any coordinate chart. Let $f|_\mathcal{U}$ denote the restriction of f to $\mathcal{U}$. Then $g = f \circ k = k^*(f|_\mathcal{U})$ is a smooth function $g : \mathcal{O} \longrightarrow \mathbb{R}$. ◇

LEMMA 4.2.9. *Let M be a smooth manifold. Then a 0-form on M is a smooth function on M.*

Proof. The condition in Definition 4.2.8 is exactly the condition that f be a smooth function on M (cf. Definition 4.1.32). □

DEFINITION 4.2.10. A j-form on the smooth manifold M is a function φ on j-tuples of tangent vectors to M, all based at the same point of M, with the following property:
Let $k : \mathcal{O} \longrightarrow \mathcal{U}$ be any coordinate chart. Let $\varphi|_\mathcal{U}$ denote the restriction of f to j-tuples of tangent vectors all based at some point p of $\mathcal{U}$. Then $\psi = k^*(\varphi|_\mathcal{U})$ is a j-form ψ on $\mathcal{O}$. ◇

REMARK 4.2.11. Let us see what this definition means. Suppose we have a j-tuple of tangent vectors $(\mathbf{v}_p^1, \ldots, \mathbf{v}_p^j)$ to M at p. Now on the one hand $\varphi(\mathbf{v}_p^1, \ldots, \mathbf{v}_p^j)$ makes sense, since φ is a function on these j-tuples. On the other hand, let $q = k^{-1}(p)$. By Lemma 4.2.3,

$\mathbf{v}_p^1 = k_*(\mathbf{w}_q^1), \ldots, \mathbf{v}_p^j = k_*(\mathbf{w}_q^j)$, for unique tangent vectors $\mathbf{w}_q^1, \ldots,$
$\mathbf{w}_q^j$. Then, by the very definition of pull-back,

$$\varphi(\mathbf{v}_p^1, \ldots, \mathbf{v}_p^j) = \varphi(k_*(\mathbf{w}_q^1), \ldots, k_*(\mathbf{w}_q^j))$$
$$= k^*(\varphi)(\mathbf{w}_q^1, \ldots, \mathbf{w}_q^j).$$

Thus we are requiring in this situation that

$$\varphi(\mathbf{v}_p^1, \ldots, \mathbf{v}_p^j) = \psi(\mathbf{w}_q^1, \ldots, \mathbf{w}_q^j)$$

for some j-form ψ on $\mathcal{O}$. This requirement makes sense as we already
know how to define j-forms on regions in $\mathbb{R}^n$. $\diamond$

REMARK 4.2.12. We may similarly check that these definitions are
independent of the choice of coordinate chart, as any two coordinate
charts are compatible. $\diamond$

Now we turn to manifolds with boundary. We must define the notion
of a tangent vector to a point p of a manifold M with boundary. This
only depends on a neighborhood of p, so if p is in the interior of
M, there is no change. But for a point p on the boundary of M, we
must make a slight change. Actually, this is not so much a change as
a reinterpretation, similar to that in our definition of smoothness of a
function f at a boundary point of M.

Let p be a point on the boundary of M. Then Definition 4.2.1 goes
through without change, except that $\mathcal{O}_1$ is an open subset of a half-
space of $\mathbb{R}^n$. Also, the curves C_1 and C_2 passing through p may go
through p or may start or end at p. That is, the curve C_1 may be
given by $r_1 : (t_1 - \delta, t_1 + \delta) \longrightarrow \mathcal{O}_1$, by $r_1 : [t_1, t_1 + \delta) \longrightarrow \mathcal{O}_1$,
or by $r_1 : (t_1 - \delta, t_1] \longrightarrow \mathcal{O}_1$, for some $\delta > 0$, and similarly for the
curve C_2. In these latter two cases, the definition of differentiability
implies that the map $k_1^{-1} r_1$ extends to a map from an open interval in $\mathbb{R}$
containing t_1 to an open subset of $\mathbb{R}^n$ containing $\mathcal{O}_1$, and similarly
for $k_1^{-1} r_2$, and then we again have that these two curves C_1 and C_2
are equivalent if $(k_1^{-1} r_1)'(t_1) = (k_1^{-1} r_2)'(t_2)$. Once again a tangent
vector to M at p is an equivalence class of curves, and the tangent
space $T_p M$ of M at p is the union of tangent vectors to M at p.

DEFINITION 4.2.13. Let M be a smooth manifold with boundary and let p be a point on the boundary of M. Let $k : \mathcal{O} \longrightarrow \mathcal{U}$ be a coordinate patch with $p \in \mathcal{U}$.

Let C be a curve parameterized by $r : (t_0 - \delta, t_0] \longrightarrow \mathcal{O}$, with $r(t_0) = p$, and suppose that r does *not* extend to a smooth map from some open interval containing t_0 to $\mathcal{O}$. Then the tangent vector which is the equivalence class of C is an *outward-pointing* tangent vector to M at p.

Let C be a curve parameterized by $r : [t_0, t_0 + \delta) \longrightarrow \mathcal{O}$, with $r(t_0) = p$, and suppose that r does *not* extend to a smooth map from some open interval containing t_0 to $\mathcal{O}$. Then the tangent vector which is the equivalence class of C is an *inward-pointing* tangent vector to M at p. ◇

REMARK 4.2.14. Note that there are three mutually exclusive possibilities for a tangent vector $\mathbf{v}_p$ to M at a point p of ∂M: Either $\mathbf{v}_p$ points outward, $\mathbf{v}_p$ is tangent to ∂M, or $\mathbf{v}_p$ points inward. ◇

Once we have defined tangent vectors then the rest of our constructions, and in particular the definition of differential forms, go through without change in the case of a manifold with boundary.

REMARK 4.2.15. We need to point out a subtle but important distinction. Suppose that M is a smooth submanifold of N. A k-form τ on M is a function on k-tuples *of tangent vectors to M* at each point of M. On the one hand, τ need not be defined on k-tuples of tangent vectors to N at each point of M, and on the other hand two k-forms ρ_1 and ρ_2 on N that yield the same values on k-tuples of tangent vectors to M define the same k-form on M.

This distinction may be clearer if we give a specific example. Let M be the circle of radius a around the origin in $\mathbb{R}^2$, for any fixed but arbitrary $a > 0$. Let $N = \mathbb{R}^2 - \{(0, 0)\}$. For any point $p = (x, y)$ of M, let $\mathbf{u}_p$ be the unit counterclockwise tangent vector to M at

$$p, \mathbf{u}_p = \begin{bmatrix} \frac{-y}{x^2+y^2} \\ \frac{x}{x^2+y^2} \end{bmatrix}_p .$$ Any tangent vector $\mathbf{v}_p$ to M at p is $\mathbf{v}_p = r\mathbf{u}_p$

for some unique real number r. We may then define a 1-form τ_1 on M by $\tau_1(r\mathbf{u}_p) = r$. Of course, we may also have the identically zero

1-form on M τ_2 defined by $\tau_2(\mathbf{v}_p) = 0$ for every $\mathbf{v}_p$ as above. Note that τ_1 and τ_2 are only defined on tangent vectors to M, not on all tangent vectors to N at points of M.

Now we have previously defined 1-forms φ^1 (in Definition 1.2.17) and ω (in Definition 1.2.21(1)) on N. We observe that $\varphi^1(\mathbf{v}_p) = \tau_1(\mathbf{v}_p)$ for every tangent vector $\mathbf{v}_p$ to M. We have tended to use (and will continue to use) φ^1 for two reasons: First, that we often care about φ^1 as a 1-form on N, and second, that even when we only care about 1-forms on M, φ^1 is easier to write down than τ_1 is. But we emphasize that this latter point is merely a convenience. We also observe that $\omega(\mathbf{v}_p) = 0$ for every tangent vector $\mathbf{v}_p$ to M, so that $\omega(\mathbf{v}_p) = \tau_2(\mathbf{v}_p)$. In fact, we observe that, for any real numbers s_1 and s_2 the forms $\rho_1 = s_1\varphi^1$ and $\rho_2 = s_1\varphi^1 + s_2\omega$ both define the same 1-form, the 1-form $s_1\tau_1$, on M.

Note that what we are doing is restricting the domains of the forms involved. So we may rephrase our last conclusion as follows: Let $i : M \longrightarrow N$ be the inclusion. Then, for any real numbers s_1 and s_2, $i^*(s_1\varphi^1 + s_2\omega) = s_1\tau_1$. $\diamond$

4.3 Further constructions

In this section we first consider exterior differentiation and exterior products of differential forms on arbitrary smooth manifolds with boundary, and we then consider push-forwards of tangent vectors and pull-backs of differential forms under arbitrary smooth maps between smooth manifolds with boundary. Our strategy will be the same as in the last section. We will relate these to the situation of open sets in $\mathbb{R}^n$, where we already know how to do things.

In general, we will see that things work "just the same as before." But we stress to the (impatient) reader that we needed to do things in $\mathbb{R}^n$ first before we could do them in general. Furthermore, while our constructions here are general, and the things we construct are important in themselves, in order to do concrete computations we must work in "local coordinates," i.e., we must transfer the situation to $\mathbb{R}^n$ by using coordinate charts and compute there.

Suppose we have an arbitrary smooth map $F : M_1 \longrightarrow M_2$ between smooth manifolds M_1 and M_2.

We first see how to push tangent vectors forward. This is the key step from which everything else follows.

DEFINITION 4.3.1. Let M_1 and M_2 be smooth manifolds and $F : M_1 \longrightarrow M_2$ be a smooth map. Let $p_1 \in M_1$ and let $p_2 = F(p_1) \in M_2$. Let $\mathbf{v}_1$ be a tangent vector to M_1 at p_1. Then the *push-forward* $\mathbf{v}_2 = F_*(\mathbf{v}_1)$ is the tangent vector to M_2 at p_2 defined as follows:

Let $k_1 : \mathcal{O}_1 \longrightarrow \mathcal{U}_1$ with $p_1 \in \mathcal{U}_1$ and let $k_2 : \mathcal{O}_2 \longrightarrow \mathcal{U}_2$ with $p_2 \in \mathcal{U}_2$. Let $q_1 = k_1^{-1}(p_1)$ and $q_2 = k_2^{-1}(p_2)$. Let $\mathbf{v}_1 = (k_1)_*(\mathbf{w}_1)$. Then $\mathbf{v}_2 = (k_2)_*(\mathbf{w}_2)$ where

$$\mathbf{w}_2 = G_*(\mathbf{w}_1) = G'(q_1)(\mathbf{w}_1) \text{ where } G = k_2^{-1} \circ F \circ k_1,$$

i.e.,

$$F_* = (k_2)_* \circ G_* \circ (k_1)_*^{-1}. \qquad \diamond$$

REMARK 4.3.2. Let us emphasize what we are doing in Definition 4.3.1. We are trying to define F_*. We already know how to define G_*, as G is a map between open subsets of Euclidean space; indeed G_* is just the derivative map G'. We also know how to define $(k_1)_*$ and $(k_2)_*$, as k_1 and k_2 are coordinate maps. Then Definition 4.3.1 states that F_* is defined precisely by requiring that the following diagram commute:

$$
\begin{array}{ccc}
T\mathcal{U}_1 & \xrightarrow{\;F_*\;} & T\mathcal{U}_2 \\[4pt]
{\scriptstyle (k_1)_*}\uparrow & & \uparrow{\scriptstyle (k_2)_*} \\[4pt]
T\mathcal{O}_1 & \xrightarrow{\;G_*\;} & T\mathcal{O}_2
\end{array}
\qquad \diamond
$$

THEOREM 4.3.3.

(1) Let M be a smooth manifold and let $F : M \longrightarrow M$ be the identity map. Then $F_* : TM \longrightarrow TM$ is the identity map.

(2) Let $F_1 : M_1 \longrightarrow M_2$ and $F_2 : M_2 \longrightarrow M_3$ be smooth maps between smooth manifolds. Let $F : M_1 \longrightarrow M_3$ be the composition $F = F_2 \circ F_1$. Then

$$F_* = (F_2)_* \circ (F_1)_* : TM_1 \longrightarrow TM_3.$$

Proof.

(1) is obvious.

For (2), let $p_1 \in M_1$ be an arbitrary point, let $p_2 = F_1(p_1) \in M_2$, and let $p_3 = F_2(p_2) = F(p_1) \in M_3$. Let $k_1 : \mathcal{O}_1 \longrightarrow \mathcal{U}_1 \subseteq M_1$, $k_2 : \mathcal{O}_2 \longrightarrow \mathcal{U}_2 \subseteq M_2$, and $k_3 : \mathcal{O}_3 \longrightarrow \mathcal{U}_3 \subseteq M_3$ be coordinate maps, with $p_1 \in \mathcal{U}_1$, $p_2 \in \mathcal{U}_2$, and $p_3 \in \mathcal{U}_3$. Let $q_1 = k_1^{-1}(p_1) \in \mathcal{O}_1$, $q_2 = k_2^{-1}(p_2) \in \mathcal{O}_2$, and $q_3 = k_3^{-1}(p_3) \in \mathcal{O}_3$.

Finally, let $G_1 : \mathcal{O}_1 \longrightarrow \mathcal{O}_2$ be the composition $G_1 = k_2^{-1} \circ F_1 \circ k_1$, let $G_2 : \mathcal{O}_2 \longrightarrow \mathcal{O}_3$ be the composition $G_2 = k_3^{-1} \circ F_2 \circ k_2$, and let $G : \mathcal{O}_1 \longrightarrow \mathcal{O}_3$ be the composition $G = G_2 \circ G_1 = k_3^{-1} \circ F \circ k_1$. Then by Definition 4.3.1 and Lemma 3.2.5,

$$
\begin{aligned}
F_* &= (k_3)_* \circ G_* \circ (k_1)_*^{-1} \\
&= (k_3)_* \circ (G_2 \circ G_1)_* \circ (k_1)_*^{-1} \\
&= (k_3)_* \circ (G_2)_* \circ (G_1)_* \circ (k_1)_*^{-1} \\
&= (k_3)_* \circ (G_2)_* \circ (k_2)_*^{-1} \circ (k_2)_* \circ (G_1)_* \circ (k_1)_*^{-1} \\
&= (F_2)_* \circ (F_1)_*. \qquad\qquad\qquad\qquad\qquad\qquad\quad \square
\end{aligned}
$$

The proof of Theorem 4.3.3 appears to be just chasing symbols around. But there is one part of the proof that has real content. The heart of this proof is Lemma 3.2.5, which told us that in the situation of maps between open subsets of Euclidean space, $G_* = (G_2 \circ G_1)_* = (G_2)_* \circ (G_1)_*$. (Recall the proof of that was just the chain rule, as $G_* = G'$.) The rest of this proof is just transferring that equality to the more general situation of maps between smooth manifolds.

Now once we have generalized the notion of the induced map on tangent vectors from maps between regions in Euclidean space to maps between smooth manifolds, the remainder of our constructions generalizes in exactly the same way. So we can now essentially repeat some of our earlier definitions and results, with language and proofs that are virtually identical, and so we omit the proofs.

We have the following generalization of Definitions 3.3.1, 3.3.4, and 3.3.6.

DEFINITION 4.3.4. Let $F : M_1 \longrightarrow M_2$ be a smooth map between smooth manifolds.

(1) Let f be a 0-form on M_2, i.e., a smooth function $f : M_2 \longrightarrow \mathbb{R}$. Then the *pull-back* $F^*(f)$ is the 0-form on M_1, i.e., the smooth function, $F^*(f) = f \circ F : M_1 \longrightarrow \mathbb{R}$.

(2) For $j > 0$ let φ be a j-form on M_2. Then the *pull-back* $F^*(\varphi)$ is the j-form on M_1 defined by $(F^*(\varphi))(\mathbf{v}_p^1, \ldots, \mathbf{v}_p^j) = \varphi(F_*(\mathbf{v}_p^1), \ldots, F_*(\mathbf{v}_p^j))$. ◇

THEOREM 4.3.5.

(1) Let M be a smooth manifold and let $F : M \longrightarrow M$ be the identity map. Then for any j, $F^* : \Omega^j(M) \longrightarrow \Omega^j(M)$ is the identity map.

(2) Let $F_1 : M_1 \longrightarrow M_2$ and $F_2 : M_2 \longrightarrow M_3$ be smooth maps between smooth manifolds. Let $F : M_1 \longrightarrow M_3$ be the composition $F = F_2 \circ F_1$. Then for any j,

$$F^* = (F_1)^* \circ (F_2)^* : \Omega^j(M_3) \longrightarrow \Omega^j(M_1).$$

LEMMA 4.3.6. *Let $F : M_1 \longrightarrow M_2$ be a map between smooth manifolds. For any smooth functions f_1 and f_2 on M_2, and any j-forms φ_1 and φ_2 on M_2,*

$$F^*(f_1\varphi_1 + f_2\varphi_2) = F^*(f_1)F^*(\varphi_1) + F^*(f_2)F^*(\varphi_2).$$

Now we consider some other constructions. Again we first consider them in a special case and then in general.

DEFINITION 4.3.7. Let M be a smooth manifold and let φ_1 and φ_2 be differential forms on M. In the notation of Definition 4.2.10, let $\psi_1 = k^*(\varphi_1|_\mathcal{U})$ and $\psi_2 = k^*(\varphi_2|_\mathcal{U})$. The *exterior product* $\varphi_1\varphi_2$ of φ_1 and φ_2 is the differential form defined by $\psi_1\psi_2 = k^*(\varphi_1\varphi_2|_\mathcal{U})$. ◇

LEMMA 4.3.8. *Let $k : \mathcal{O} \longrightarrow \mathcal{U}$ be a coordinate patch on M. Then for any differential forms φ_1 and φ_2 on $\mathcal{U}$, $k^*(\varphi_1)k^*(\varphi_2) = k^*(\varphi_1\varphi_2)$, i.e., exterior product commutes with pull-back by coordinate maps.*

Proof. This is true because we defined the exterior product precisely by this commutativity requirement. By Definition 4.3.7: $k^*(\varphi_1)\, k^*(\varphi_2) = \psi_1 \psi_2 = k^*(\varphi_1 \varphi_2)$. $\qquad\qquad\qquad\qquad\qquad\square$

THEOREM 4.3.9. *Let $F : M_1 \longrightarrow M_2$ be a smooth map between smooth manifolds. Let φ_1 be an i-form and let φ_2 be a j-form on M_2. Then*

$$F^*(\varphi_1 \varphi_2) = F^*(\varphi_1) F^*(\varphi_2),$$

i.e.,

$$\text{Pull-back commutes with exterior product.}$$

Equivalently, the following diagram commutes:

$$
\begin{array}{ccc}
\Omega^i(M_1) \times \Omega^j(M_1) & \xleftarrow{\;F^* \times F^*\;} & \Omega^i(M_2) \times \Omega^j(M_2) \\
\downarrow & & \downarrow \\
\Omega^{i+j}(M_1) & \xleftarrow{\;\;F^*\;\;} & \Omega^{i+j}(M_2)
\end{array}
$$

where the vertical maps are products.

Proof. This follows directly from Lemma 4.3.8, which tells us that pull-back of coordinate maps commutes with exterior product, and Theorem 3.4.1, which tells us that pull-back of smooth maps between open subsets of Euclidean space commutes with exterior product. $\quad\square$

DEFINITION 4.3.10. Let M be a smooth manifold and let φ be a differential form on M. In the notation of Definition 4.2.10, let $\psi = k^*(\varphi|_\mathcal{U})$. The *exterior derivative $d\varphi$ of φ* is the differential form defined by $d\psi = k^*(d\varphi|_\mathcal{U})$. $\qquad\qquad\qquad\diamond$

LEMMA 4.3.11. *Let $k : \mathcal{O} \longrightarrow \mathcal{U}$ be a coordinate patch on M. Then for any differential form φ on $\mathcal{U}$, $d(k^*(\varphi)) = k^*(d\varphi)$, i.e., exterior differentiation commutes with pull-back by coordinate maps.*

Proof. This is true because we defined the exterior derivative precisely by this commutativity requirement. By Definition 4.3.10: $d(k^*(\varphi)) = d\psi = k^*(d\varphi)$. $\qquad\qquad\qquad\qquad\qquad\square$

THEOREM 4.3.12. *Let $F : M_1 \longrightarrow M_2$ be a smooth map between smooth manifolds. Let φ be a j-form on M_2. Then*

$$dF^*(\varphi) = F^*(d\varphi),$$

i.e.,

Pull-back commutes with exterior differentiation.

Equivalently, the following diagram commutes:

$$
\begin{array}{ccc}
\Omega^j(M_1) & \xleftarrow{\ F^*\ } & \Omega^j(M_2) \\
d \downarrow & & \downarrow d \\
\Omega^{j+1}(M_1) & \xleftarrow{\ F^*\ } & \Omega^{j+1}(M_2)
\end{array}
$$

Proof. This follows directly from Lemma 4.3.11, which tells us that pull-back of coordinate maps commutes with exterior differentiation, and Theorem 3.4.8, which tells us that pull-back of smooth maps between open subsets of Euclidean space commutes with exterior differentiation. □

4.4 Orientations of manifolds—intuitive discussion

Our constructions so far have applied to all smooth manifolds. Thus we see we can do algebra (i.e., exterior algebra) and differential calculus (i.e., exterior differentiation) on arbitrary smooth manifolds. However, in the sequel we will want to do integral calculus, i.e., to integrate differential forms. We will see that it is not possible to do this in complete generality. Instead, we can only do this on oriented manifolds. In preparation for this work, we consider the question of orientability, and develop the notion of orientation, here.

We devote this section to a discussion of orientations of manifolds at an intuitive level. For this discussion, rather than consider abstract manifolds, we consider submanifolds of $\mathbb{R}^3$. In the next section we will give careful definitions, in general, and show how they relate to our intuitive discussion.

Recall we discussed orientations of vector spaces in general, and $\mathbf{R}^n$ in particular, in Section 2.3, and we will refer back to that discussion.

The notion of orientation applies to manifolds with boundary (and in fact, Stokes's theorem applies precisely to this situation), and so in this section we will use the word "manifold" to mean "manifold, possibly with boundary."

Manifolds are oriented, if possible, one component at a time, so in this section, unless stated otherwise, we assume all manifolds are connected.

First remember that a connected 0-manifold is just a point.

DEFINITION 4.4.1. Let p be a point in $\mathbb{R}^3$. An *orientation* of p is an assignment of sign (either $+$ or $-$) to p. $\diamond$

Thus, a typical oriented 0-manifold looks like this:

Note that a point has two possible orientations.

Now we come to 1-manifolds, or curves.

DEFINITION 4.4.2. Let C be a smooth curve in $\mathbb{R}^3$. An *orientation* of C is an assignment of a direction in which C is traversed. $\diamond$

We indicate a direction by an arrowhead. Thus, for example, a typical oriented 1-manifold looks like this:

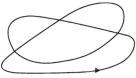

Note that a connected curve has two possible orientations.

In order to see how to generalize this to higher dimensions, we must examine this notion a little more carefully. At each point p of a smooth curve C, we have the tangent space of C at p, i.e., a line to tangent to C at p, depicted as follows:

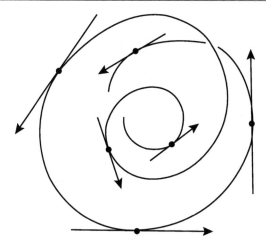

The tangent line is, of course, a line, i.e., a copy of $\mathbf{R}^1$, and if we remove the zero tangent vector, we are left with two half-lines. Then an orientation is a choice of one of the half-lines (indicated here by an arrowhead). Note that this is exactly what we have meant by an orientation of $\mathbf{R}^1$ in Section 2.3, where we choose the half-line in which the single basis vector is pointing. Further, and crucially, this is a continuous choice. It should be pretty clear geometrically that such a choice gives a direction in which the curve is traversed, and vice versa.

With this in mind, we come to 2-manifolds.

DEFINITION 4.4.3. Let S be a smooth surface in $\mathbb{R}^3$. An *orientation* of S is a continuous assignment of a direction of rotation to the tangent space at each point of S. ◇

This definition obviously needs considerable elaboration.

Let us begin by considering S to be a disk in the plane $\mathbb{R}^2$. Then we can consider the tangent space to S at each point of S to be the plane itself. We now pick a basis for the tangent space at each point. Our requirement is that when we consider the direction of rotation between the first and second vectors in our basis, this direction varies continuously. Note that a direction of rotation is exactly what we have meant by an orientation of $\mathbf{R}^2$ in Section 2.3.

Here is an example of such a choice, where we have indicated the first element in the basis by a solid arrow, the second by a dashed arrow, and our orientation (the direction between the two) by a curved arrow. (Here our choice is "counterclockwise.")

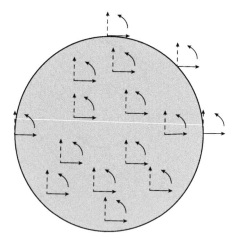

This discussion was particularly simple because we were considering a 2-disk S in $\mathbb{R}^2$. If we consider a 2-disk S in $\mathbb{R}^3$, the situation becomes more complicated because the tangent planes vary. But again an orientation is a continuous assignment of direction of rotation of the tangent space to S at each point of S.

Let us now consider another example, S = the 2-sphere.

Here we again have a tangent plane to S at each point. (This time, once again, these planes vary.) Again, we have chosen a pair of vectors forming a basis for the tangent space to S at each point, and a direction of rotation from the first to the second. (Again, we have chosen counterclockwise, viewing the 2-sphere from the front.)

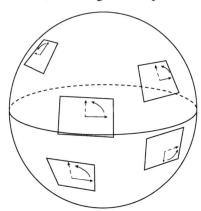

We should, however, notice a difference between the case of the disk and the sphere. In the case of the disk we can continuously move each of the two elements of the basis of the tangent space at a single point to be a basis of the tangent space to every point in the disk. This is particularly easy to see when the disk is in $\mathbb{R}^2$, as in this case the solid vector always points to the right and the dotted vector always points up, but can also be done for a disk in $\mathbb{R}^3$.

In the case of the 2-sphere this is impossible. We cannot even translate a single tangent vector around the sphere continuously, keeping it nonzero (much less translate a pair of vectors around, keeping them a basis for the tangent space at each point). For example, if we try to define a tangent vector everywhere so that it points east (along every circle of latitude), we find we cannot define it at the north or south poles; similarly, if we try to define it so that it points north (along every circle of longitude), we also find we cannot define it at the north or south poles. It is a theorem that no matter how we try to do this, there will always be at least one point where the vector is zero or undefined. Nevertheless, while we cannot get the individual vectors in the basis to move consistently, we can get the direction of rotation between them to do so, and so give the sphere an orientation.

By the way, it is an interesting geometric question to ask for which n-manifolds we can consistently translate a "frame" of n tangent vectors at each point. Such manifolds are called *parallelizable*, and the *only* parallelizable closed surface is the torus. You might try to visualize for yourself how such a frame would move around the torus.

We can always give an orientation to a point—just pick a sign—and to a curve—just pick a direction to traverse it. But an orientation of a surface is something more complicated, and you may justifiably ask whether every surface can be given an orientation. The answer is No! The first example of such a surface was given by August F. Möbius (1790–1868). It is the famous Möbius strip:

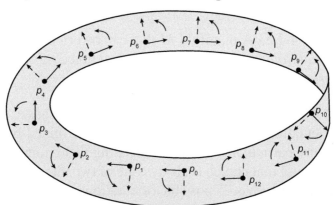

Suppose we start at p_0 with the illustrated pair of tangent vectors and the illustrated direction of rotation. We now translate these along, first to p_1, then to p_2, then to p_3, etc. Carefully follow this frame along (noting what happens to the dashed vector as we pass through the "half-twist" between p_9 and p_{10}). In particular, you see that at the nearby points p_0 and p_{12} the directions of rotation are *opposite* (at p_0 it is counterclockwise and at p_{12} it is clockwise), and thus we *cannot* orient the Möbius strip.

Now we come to 3-manifolds.

DEFINITION 4.4.4. Let B be a smooth 3-manifold in $\mathbb{R}^3$. An *orientation* of B is a continuous choice of handedness to the tangent space at each point of B. ◇

This definition obviously also needs considerable explanation.

First note that (just as in the case of a disk in the plane) we can consider the tangent space to B at each point of B to be $\mathbb{R}^3$ itself. We now pick a basis for the tangent space to B at each point of B. Our requirement is that the handedness of the basis varies continuously. Note that the handedness of a basis is exactly what we have meant by an orientation of $\mathbf{R}^3$ in Section 2.3.

Our discussion of orientations for 3-manifolds in $\mathbb{R}^3$ is relatively simple because we are considering 3-manifolds *in* $\mathbb{R}^3$; if we were considering 3-manifolds in $\mathbb{R}^n$ for $n > 3$, we would again have to consider different tangent 3-spaces at different points. (Compare our relatively simple discussion of the orientation of a 2-dimensional disk in $\mathbb{R}^2$ with that of a 2-dimensional disk in $\mathbb{R}^3$, or of a 2-sphere in $\mathbb{R}^3$.)

We now introduce some terminology.

DEFINITION 4.4.5. A manifold M that can be oriented is called an *orientable manifold*; and M together with a choice of orientation is called an *oriented manifold*. ◇

REMARK 4.4.6. If M is orientable and connected, it has exactly two possible orientations (since its orientation is determined by making a choice at any particular point and translating that choice around to the rest of M, and that choice is always one of two alternatives); if M has k components, it has 2^k possible orientations. ◇

The boundary ∂M of an n-manifold M is an $(n-1)$-manifold (or is empty). The boundary of an orientable manifold is always orientable, and indeed in a natural way we may obtain an orientation on ∂M from one on M. This orientation is known as the induced orientation on the boundary and is defined as follows:

DEFINITION 4.4.7. Let M be an oriented manifold with boundary ∂M. The *induced orientation* on ∂M is defined by the condition that the outward-pointing normal to M at ∂M, followed by the orientation of ∂M, agrees with the orientation of M.   ◇

Again this is a definition that requires elaboration. Since ∂M is an $(n-1)$-manifold when M is an n-manifold, the tangent space to ∂M at any point x of ∂M is $(n-1)$-dimensional, while the tangent space to M there is n-dimensional; thus there is one dimension left over, and we may choose an orientation for that 1-dimensional space either by pointing out of M or by pointing into M. We make the former choice. (We think of this dimension as being perpendicular to M, so we have phrased this as "choosing the outward-pointing normal," normal being a synonym for perpendicular.)

This gives us one vector, and an orientation for ∂M lets us choose a basis of $n-1$ vectors for the tangent space to ∂M at the point x. Putting these vectors together gives us a basis of n vectors at x, and hence an orientation for M. Our condition requires that the orientation for M so obtained is the same as the one M originally possessed.

Let us see some examples. First let us consider $n = 1$, so we have an oriented curve C as shown.

At the point q, the outward-pointing normal *agrees* with the direction the curve is traversed (the arrows point in the *same* direction) so we give q the orientation $+$. On the other hand, at the point p, the outward-pointing normal *disagrees* with the direction the curve is traversed (the arrows point in *opposite* directions) and so we give p the orientation $-$.

In particular, the boundary of C consists of the two points q and p, the first with orientation $+$ and the second with orientation $-$, and we write that as $\partial C = \{+q\} \cup \{-p\}$.

Next let us consider a disk D in the plane, and suppose we orient it counterclockwise, as we have done before.

We have drawn the outward-pointing normal at each of the points on the boundary with a solid line. Note that if we draw the vector tangent to the boundary as the dotted line indicates, then rotation from the first to the second vector is counterclockwise, agreeing with our orientation of the disk. (If we had drawn this vector pointing in the opposite direction, the direction of rotation from the first to the second would have been clockwise, disagreeing with our orientation of the disk.) This second, dotted vector gives the direction in which we traverse the boundary circle, i.e., it orients the boundary circle. In this case the orientation is to traverse the boundary circle counterclockwise. (We write $\partial D = C$, an equation of oriented manifolds.)

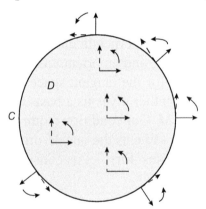

By the way, this example again shows that even when we can choose "constant" vectors to orient D, as we did in our first example of the orientation of a disc in $\mathbb{R}^2$, this would not suffice for us to be able to define the induced orientation on ∂D. For note that in that example, the second vector is perpendicular to the boundary circle at the "top" of the disk, and so does not give a direction to the circle there. Thus we must allow the individual vectors in our basis to vary.

Here is another example—the region R in the plane inside a big circle and outside two smaller ones. We again begin by orienting this region counterclockwise. Its boundary consists of three circles, and we ask what their induced orientations are.

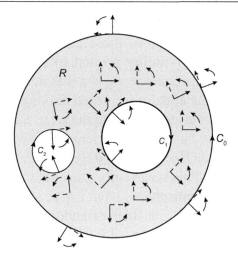

In each case, we have drawn the outward normals on the boundary circles. This is clear for the outermost circle, and a moment's reflection makes this clear for the two inner circles as well. (When we say "outer," we mean outer from the point of view of our manifold, i.e., pointing out of the region, as these two vectors do.) Then the induced orientation on C_0 is to traverse this curve counterclockwise, while the induced orientations on C_1 and C_2 are to traverse them clockwise. We again have an equation $\partial R = C_0 \cup C_1 \cup C_2$ between oriented manifolds.

Now we turn to an example in 3-space. Consider the 2-sphere S, again orienting it counterclockwise when viewed from the front as shown.

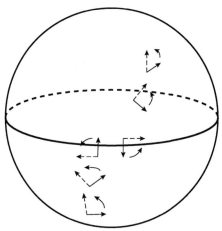

Now S is the union of the northern and southern hemispheres H_N and H_S respectively, and each of these gets an orientation from S. Noting that the outward-pointing normal to H_N (resp. H_S) points south (resp. north) on the equator, we see that $\partial H_N = E$ and $\partial H_S = E'$, where E denotes the equator traversed moving to the east and E' denotes the equator traversed moving to the west. Since E' is the same manifold as E but has the opposite orientation, we write $E' = -E$ (and $E = -E'$). Thus $\partial H_N = E$ and $\partial H_S = -E$.

Now $S = H_N \cup H_S$. This is certainly true as manifolds—the sphere is the union of its two hemispheres—but is true as oriented manifolds as well, since H_N and H_S get their orientations from S. Now S is a manifold without boundary, i.e., $\partial S = \emptyset$. Thus when we take the boundary of each side of the above equation, we get $\emptyset = \partial S = \partial(H_N \cup H_S) = \partial H_N \cup \partial H_S = E \cup -E$. This equation is correct, as we establish an arithmetic of oriented manifolds in which a manifold and its negative cancel out.

Now we consider the boundaries of oriented 3-manifolds in $\mathbb{R}^3$, which will be oriented surfaces.

Consider, for example, the solid unit ball B, oriented by the right-hand rule, as pictured below. Its boundary is the unit sphere, and our question is how to orient it.

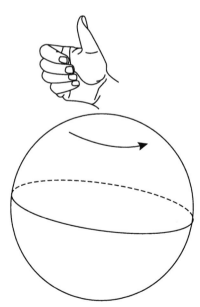

Let us position ourselves at the north pole. We take as our first vector the outward-pointing normal, which points straight up here. We stick up our right thumb in that direction, and see how the fingers of our right hand curl. Looking down from the top, i.e., from a position outside the sphere, we see they curl counterclockwise. The same holds true, as it must, when we look at the sphere from any outside vantage point. Thus $\partial B = S$, where S is the sphere oriented counterclockwise when viewed from outside.

We could also consider a solid body Q bounded by several surfaces—one, S_0, on the outside and several, say for example two, S_1 and S_2, on the inside. (Compare our consideration of the region R in the plane above.) We would then find that $\partial Q = S_0 \cup S_1 \cup S_2$, where S_0 is oriented counterclockwise when viewed from outside and S_1 and S_2 are oriented clockwise when viewed from outside.

In general, there is *no* preferred orientation on a manifold M. But there is one situation where we do have a preferred orientation. That is when $M = \mathbb{R}^n$, or when M is an n-manifold that is a submanifold of $\mathbb{R}^n$. We have given $\mathbb{R}^n$ a standard orientation (for $n \leq 3$), and that orientation restricts to an orientation, which we also call the standard orientation, of an n-dimensional submanifold of $\mathbb{R}^n$.

We will observe that if M is an n-manifold with boundary *in* $\mathbb{R}^n$, and we give M the standard orientation, then we can give $N = \partial M$ the induced orientation. In this way N gets a standard orientation as the boundary of M. But we should again explicitly observe that this is the standard orientation of N *as the boundary of M*, not a "standard" orientation of N as an abstract manifold–compare the situation $\partial R = C_0 \cup C_1 \cup C_2$ above where C_0 was oriented counterclockwise and C_1 and C_2 were oriented clockwise.

REMARK 4.4.8. We have chosen to follow a consistent philosophy in defining orientations by always considering tangent vectors to our manifold. (In technical terms, we have chosen to always orient the tangent bundle.) This philosophy also has the advantage that it is intrinsic to our manifold, i.e., that we can define it in terms of the manifold itself. Since we are considering abstract manifolds, this is essential for our purposes.

However, suppose we are considering manifolds that are submanifolds of Euclidean space. Then we may consider normal (i.e., perpendicular) vectors to our manifold (in technical terms, we consider the normal bundle). Then, *once we have decided how to orient Euclidean space*, we may pass from an orientation of the tangent bundle to an orientation of the normal bundle, and vice versa.

In general, the rule is: Suppose we have a connected orientable n-manifold M in $\mathbb{R}^N$. Choose an orientation of $\mathbb{R}^N$, given by a basis $\mathcal{B}$. Consider a point x of M. An orientation of the tangent (resp. normal) space to M at x is given by a basis $\mathcal{B}_{tangent}$ of the n-dimensional subspace of the tangent space of $\mathbb{R}^N$ consisting of vectors tangent to M at x (resp. by a basis $\mathcal{B}_{normal}$ of the $(N-n)$-dimensional subspace of the tangent space of $\mathbb{R}^N$ consisting of vectors normal to M at x), and the condition is that $\mathcal{B}_{normal}$ followed by $\mathcal{B}_{tangent}$ gives the same orientation of $\mathbb{R}^N$ as $\mathcal{B}$ does. $\diamond$

For example, suppose we give $\mathbb{R}^3$ the standard orientation, by the right-hand rule, and consider the issue of orienting a surface in $\mathbb{R}^3$. Suppose first we have oriented the surface in our sense, i.e., provided a direction of rotation. We now wish to consider the normal line to the surface at any given point and decide how to orient it, i.e., to choose in which direction to traverse it. We choose the direction by requiring that if our thumb points in that direction and our fingers curl in the chosen direction of rotation on the surface, the result is a right-hand orientation of $\mathbb{R}^3$. On the other hand, (so to speak) suppose we begin by considering the normal line at each point, and suppose we have continuously oriented these normal lines by providing a direction in which we traverse each of them. We then get an orientation of the surface in our sense, i.e., a direction of rotation, by pointing the thumb of our right hand in the chosen normal direction and then letting the direction of rotation be that in which the fingers of our right hand curl. Check this by rephrasing it as follows: Whatever choice we make for the direction in which we traverse the normal line (a choice we make by putting an arrowhead on the line), the arrow points out of the surface on one side while the arrow points into it on the other side. Call the side on which the arrow points out the *good* side, and the other side the *bad* side. Then our rule reads: The surface gets oriented by the direction of rotation that is counterclockwise when seen from the good side.

This discussion of orienting the normal bundle of a surface presupposes that our surface has two sides (one "good" and the other, "bad"). Does every surface have two sides? No. The Möbius strip is the most famous example of a 1-sided surface. Beginning at the point p_0 of the Möbius strip, think of the front of the Möbius strip there as the good side (and the back as the bad side). Then if you move around to $p_1, p_2, \ldots, p_{12}$ and back to p_0, you wind up on the back (or bad side). There is no distinction between good and bad—the Möbius strip has a single side. (Of course, we expected that we couldn't orient the normal bundle of the Möbius strip because we realized that we couldn't orient its tangent bundle. And, according to the discussion we just had, if we could do the one, we could do the other.)

If you think about this analysis for a minute, you will see that a surface in $\mathbb{R}^3$ is orientable if and only if it is two-sided. This generalizes to higher dimensions: An $(n - 1)$-dimensional submanifold of $\mathbb{R}^n$ is orientable if and only if it is two-sided.

Indeed, popular discussions of the Möbius strip say that it is nonorientable because it is "one-sided" rather than "two-sided." This is not the correct definition. But as we have just seen, for a surface in $\mathbb{R}^3$ "two-sidedness" is equivalent to orientability and "one-sidedness" is equivalent to nonorientability. Since popular discussions are aimed at a popular audience, they just talk about "sidedness" rather than introduce the correct and precise notions.

Not only is this the case for popular discussions, but calculus texts often also take the easy way out. They don't bother to orient points; they orient curves by selecting a direction in which to traverse them (as we do); they orient surfaces (when possible) by selecting a normal direction, with $\mathbb{R}^3$ always being given the right-hand orientation (usually without mentioning the word orientation).

REMARK 4.4.9. We have observed that the empty set $\emptyset$ is an n-manifold for every n. We may ask how to orient it. The answer is that the notion of orientation for $\emptyset$ is vacuous: $\emptyset$ is orientable, and has the single orientation of nothing.   ◇

REMARK 4.4.10. We conclude this section with an obvious bit of notation: If M is an oriented manifold, $-M$ denotes the same manifold but with the opposite orientation.

Note that $-(-M) = M$, as it should be. Also, the empty manifold acts very much like the number zero, in two ways: $-\emptyset = \emptyset$ and $M \cup \emptyset = M$, for any M. ◇

4.5   Orientations of manifolds—careful development

In this section we carefully develop the notion of an orientation of a manifold (possibly with boundary).

Again we recall that we defined the notion of an orientation for a vector space in Section 2.3, and again we refer the reader back to that section for the basic ideas as we develop this notion further.

We begin by citing some standard constructions and results from linear algebra. In this section all vector spaces are assumed to be finite-dimensional.

DEFINITION 4.5.1. Let V and W be vector spaces, and let $L : V \longrightarrow W$ be a linear transformation. Let $\mathcal{B}$ be a basis of V and $\mathcal{C}$ be a basis of W. Then the *matrix of L with respect to the bases $\mathcal{B}$ and $\mathcal{C}$* is the matrix

$$[L]_{\mathcal{C}}^{\mathcal{B}} = \big[[v_1]_{\mathcal{C}}| \ldots |[v_n]_{\mathcal{C}} \big],$$

where $\mathcal{B} = \{v_1, \ldots, v_n\}$. In case $W = V$ and $\mathcal{C} = \mathcal{B}$, we write $[L]_{\mathcal{B}}$ for $[L]_{\mathcal{C}}^{\mathcal{B}}$. ◇

THEOREM 4.5.2. *Let V and W be vector spaces, and let $L : V \longrightarrow W$ be a linear transformation. Let $\mathcal{B}$ be a basis of V and $\mathcal{C}$ be a basis of W. Then for any $v \in V$,*

$$[L(v)]_{\mathcal{C}} = [L]_{\mathcal{C}}^{\mathcal{B}}[v]_{\mathcal{B}}.$$

We make the following observation: Let V and W be two vector spaces and let $L : V \longrightarrow W$ be an isomorphism. If $\mathcal{B} = \{v_1, \ldots, v_n\}$ is a basis of V, then $L(\mathcal{B}) = \{L(v_1), \ldots, L(v_n)\}$ is a basis of W.

LEMMA 4.5.3. *Let V be a vector space and let $L : V \longrightarrow V$ be a linear transformation. Let $\mathcal{B}$ and $\mathcal{C}$ be bases of V. Then*

$$[L]_{\mathcal{C}}^{\mathcal{B}} = P_{\mathcal{C} \leftarrow L(\mathcal{B})}.$$

In particular,

$$[L]_{\mathcal{B}} = P_{\mathcal{B} \leftarrow L(\mathcal{B})}$$

and, if $I : V \longrightarrow V$ is the identity linear transformation,

$$[I]_{\mathcal{C}}^{\mathcal{B}} = P_{\mathcal{C} \leftarrow \mathcal{B}}.$$

Proof.   Compare Definitions 2.3.4 and 4.5.1.   □

THEOREM 4.5.4. *Let V be a vector space and let $L : V \longrightarrow V$ be a linear transformation. Let $\mathcal{B}$ and $\mathcal{C}$ be any two bases of V. Then*

$$[L(v)]_{\mathcal{C}} = P_{\mathcal{C} \leftarrow \mathcal{B}}[L(v)]_{\mathcal{B}} P_{\mathcal{B} \leftarrow \mathcal{C}}.$$

COROLLARY 4.5.5. *Let V be a vector space and let $L : V \longrightarrow V$ be a linear transformation. Let $\mathcal{B}$ and $\mathcal{C}$ be any two bases of V. Then*

$$\det\big([L(v)]_{\mathcal{C}}\big) = \det\big([L(v)]_{\mathcal{B}}\big).$$

Proof.   By Theorems 2.3.6 and 4.5.4,

$$[L(v)]_{\mathcal{C}} = P_{\mathcal{C} \leftarrow \mathcal{B}}[L(v)]_{\mathcal{B}} P_{\mathcal{B} \leftarrow \mathcal{C}} = \big(P_{\mathcal{B} \leftarrow \mathcal{C}}\big)^{-1} [L(v)]_{\mathcal{B}} P_{\mathcal{B} \leftarrow \mathcal{C}}$$

and the corollary follows from properties of determinants.   □

DEFINITION 4.5.6. Let V be a vector space and let $L : V \longrightarrow V$ be a linear transformation. The *determinant* $\det(L)$ is defined by $\det(L) = \det([L]_{\mathcal{B}})$ where $\mathcal{B}$ is any basis of V.   ◇

REMARK 4.5.7.   Definition 4.5.6 makes sense. For, while a priori it depends on the choice of a basis, by Corollary 4.5.5 all choices of bases give the same answer, so in fact it does not depend on the choice of basis.   ◇

With this material in hand, we now take up orientations.

DEFINITION 4.5.8. An *oriented vector space* V is a vector space with a choice of orientation.   ◇

DEFINITION 4.5.9. Let V and W be two oriented vector spaces, with orientations given by bases $\mathcal{B}$ and $\mathcal{C}$ respectively. Let $L : V \longrightarrow W$ be an isomorphism of vector spaces. Then L is *orientation-preserving* if $L(\mathcal{B})$ defines the same orientation on W as $\mathcal{C}$ does, and is *orientation-reversing* if $L(\mathcal{B})$ defines the opposite orientation on W as $\mathcal{C}$ does. ◇

LEMMA 4.5.10. *Let V be a vector space and $L : V \longrightarrow V$ be an isomorphism. Let $\mathcal{B}$ and $\mathcal{C}$ be any two bases of V. Then L is orientation-preserving (resp. orientation-reversing) on V with the orientation given by $\mathcal{B}$ if and only if L is orientation-reversing on V with the orientation given by $\mathcal{C}$.*

Proof. We are claiming that the determinants of the two change of basis matrices $P_{\mathcal{B} \leftarrow L(\mathcal{B})}$ and $P_{\mathcal{C} \leftarrow L(\mathcal{C})}$ have the same sign. In fact the determinants of these two matrices are equal as we now see:
By Lemma 4.5.3,

$$P_{\mathcal{B} \leftarrow L(\mathcal{B})} = [L]_{\mathcal{B}} \text{ and } P_{\mathcal{C} \leftarrow L(\mathcal{C})} = [L]_{\mathcal{C}},$$

and by Corollary 4.5.5 these matrices have the same determinant.   □

DEFINITION 4.5.11. Let V be a vector space and $L : V \longrightarrow V$ be an isomorphism. Then L is orientation-preserving (resp. orientation-reversing) if it is orientation-preserving (resp. orientation-reversing) on V with the orientation given by any basis $\mathcal{B}$.   ◇

REMARK 4.5.12. Again this is well defined, i.e., independent of the choice of $\mathcal{B}$, by Lemma 4.5.10. (Compare Remark 4.5.7.)   ◇

For our purposes, it is key to note that if we have two vector spaces, only one of which is oriented, any isomorphism between them allows us to transfer the orientation to the other one (with the result, of course, depending on the chosen isomorphism).

DEFINITION 4.5.13. Let $L : V \longrightarrow W$ be an isomorphism, and let V have the orientation given by a basis $\mathcal{B} = \{v_1, \ldots, v_n\}$. Then the orientation *induced* on W by L is the orientation of W given by the basis $\mathcal{C} = L(\mathcal{B}) = \{L(v_1), \ldots, (v_n)\}$.   ◇

REMARK 4.5.14.

(1) In the situation of Definition 4.5.13, if V has the orientation given by $\mathcal{B}$ and W has the orientation given by $\mathcal{C}$, then $L : V \longrightarrow W$ is orientation-preserving.

(2) In this situation, if W has the orientation given by $\mathcal{C}$ and we consider the inverse isomorphism $L^{-1} : W \longrightarrow V$, then the orientation induced on V by L^{-1} is just the original orientation given by $\mathcal{B}$.   ◇

LEMMA 4.5.15. *Let V be a vector space and let $\mathcal{B}$ be a basis of V. Let $L_1 : V \longrightarrow W$ and $L_2 : V \longrightarrow W$ be isomorphisms. Then $L_1(\mathcal{B})$ and $L_2(\mathcal{B})$ give the same orientation of W if and only if $L_2^{-1}L_1 : V \longrightarrow V$ is orientation-preserving.*

Proof. Let $\mathcal{B}' = L_2^{-1}L_1(\mathcal{B})$.
Let $\mathcal{C}_1 = L_1(\mathcal{B})$ and $\mathcal{C}_2 = L_2(\mathcal{B})$. Note that both matrices $[L_1]_{\mathcal{C}_1}^{\mathcal{B}}$ and $[L_2]_{\mathcal{C}_2}^{\mathcal{B}}$ are the identity matrix. Let I denote the identity linear transformation.
Then

$$
\begin{aligned}
[L_2^{-1}L_1]_{\mathcal{B}} &= [L_2^{-1}]_{\mathcal{B}}^{\mathcal{C}_1}[L_1]_{\mathcal{C}_1}^{\mathcal{B}} \\
&= ([L_2]_{\mathcal{C}_1}^{\mathcal{B}})^{-1}[L_1]_{\mathcal{C}_1}^{\mathcal{B}} = ([I]_{\mathcal{C}_1}^{\mathcal{C}_2}[L_2]_{\mathcal{C}_2}^{\mathcal{B}})^{-1}[L_1]_{\mathcal{C}_1}^{\mathcal{B}} \\
&= ([L_2]_{\mathcal{C}_2}^{\mathcal{B}})^{-1}[I]_{\mathcal{C}_2}^{\mathcal{C}_1}[L_1]_{\mathcal{C}_1}^{\mathcal{B}} = [I]_{\mathcal{C}_2}^{\mathcal{C}_1}.
\end{aligned}
$$

But

$$
[L_2^{-1}L_1]_{\mathcal{B}} = P_{\mathcal{B} \leftarrow \mathcal{B}'} \text{ and } [I]_{\mathcal{C}_2}^{\mathcal{C}_1} = P_{\mathcal{C}_1 \leftarrow \mathcal{C}_2},
$$

so these matrices, and hence their determinants, are equal (and hence have the same sign).   □

We now need to consider the special case of 0-dimensional vector spaces.

DEFINITION 4.5.16. A 0-dimensional *oriented vector space* V is a 0-dimensional vector space V with a choice of orientation (cf. Definition 2.3.11).
   If V is an oriented 0-dimensional vector space, with orientation given by $\varepsilon = \pm 1$, W is a 0-dimensional vector space, and $L : V \longrightarrow W$ is the unique linear transformation (given by $L(0) = 0$), the *induced orientation* on W is $\varepsilon = L_*(\varepsilon)$.

If V and W are oriented 0-dimensional vector spaces with orientations given by ε_V and ε_W, respectively, and $L : V \longrightarrow W$ is the unique linear transformation, then L is orientation-preserving (resp. orientation-reversing) if $\varepsilon_W = \varepsilon_V$ (resp. $\varepsilon_W = -\varepsilon_V$). $\diamond$

With this background out of the way, we now wish to orient (i.e., give an orientation to) smooth manifolds, when possible. Let us begin with the case of an open submanifold of $\mathbb{R}^n$, where this is always possible (and easy).

DEFINITION 4.5.17. Let $\mathcal{O}$ be an open subset of $\mathbb{R}^n$. An *orientation* of $\mathcal{O}$ is a choice of orientation on $T_p\mathcal{O}$ for every $p \in \mathcal{O}$ with the following property: For each point $p \in \mathcal{O}$, let $L_p : \mathbf{R}^n \longrightarrow T_p\mathcal{O}$ by

$$
L_p \left(\begin{bmatrix} a_1 \\ \vdots \\ a_n \end{bmatrix} \right) = \begin{bmatrix} a_1 \\ \vdots \\ a_n \end{bmatrix}_p .
$$

Then, for some orientation of $\mathbf{R}^n$, given by a basis $\mathcal{B}$, the chosen orientation of $T_p\mathcal{O}$ agrees with the orientation induced on $T_p\mathcal{O}$ by L_p, i.e., with the orientation on $T_p\mathcal{O}$ given by $\{L_p(\mathcal{B})\}$, for every $p \in \mathcal{O}$.

If $\mathcal{B}$ gives the standard orientation of $\mathbf{R}^n$, this is the standard orientation of $\mathcal{O}$, while if $\mathcal{B}$ gives the nonstandard orientation of $\mathbf{R}^n$, this is the nonstandard orientation of $\mathcal{O}$. $\diamond$

REMARK 4.5.18. Obviously a connected open subset $\mathcal{O}$ of $\mathbb{R}^n$ has exactly two orientations. Furthermore, in this case, an orientation of $T_p\mathcal{O}$ for any given point $p \in \mathcal{O}$ determines an orientation of $\mathcal{O}$. More precisely, this orientation on $\mathcal{O}$ is given by the requirement that the composition $L_q L_p^{-1} : T_p\mathcal{O} \longrightarrow T_q\mathcal{O}$ be orientation-preserving, for every point $q \in \mathcal{O}$. And, trivially, an orientation of $\mathcal{O}$ determines an orientation of $T_p\mathcal{O}$ for every point $p \in \mathcal{O}$.

If $\mathcal{O}$ is not connected, we may define an orientation on $\mathcal{O}$ by using Definition 4.5.17 to define an orientation on each component of $\mathcal{O}$. In this situation, some components may receive the standard orientation and some may receive the nonstandard orientation. $\diamond$

Now we wish to orient a general smooth manifold (if possible). The basic definition is almost the same, but it is much more subtle to see when it works.

DEFINITION 4.5.19. Let M be a smooth n-manifold. An *orientation* on M is a choice of orientation on T_pM for every $p \in M$ with the following property: There is an atlas $\mathcal{A} = \{k_i : \mathcal{O}_i \longrightarrow \mathcal{U}_i\}_{i \in I}$ on M and an orientation on each open subset $\mathcal{O}_i$ of $\mathbb{R}^n$ such that, if $q \in \mathcal{O}_i$ with $k_i(q) = p$, the map

$$(k_i)_* : T_q\mathcal{O}_i \longrightarrow T_pM$$

is orientation-preserving.

If M has an orientation, then M is *orientable*.   ◇

Of course, M has an atlas. But the potential difficulty here is that the orientations may not be consistent. That is, it is potentially the case that we might have a point $p \in M$ that is in two coordinate charts $\mathcal{U}_i$ and $\mathcal{U}_j$, say $p = k_i(q_i)$ and $p = k_j(q_j)$, and the orientation induced on T_pM from the orientation on $T_{q_j}\mathcal{O}_j$ by $(k_j)_*$ may differ from the orientation induced on T_pM from the orientation on $T_{q_i}\mathcal{O}_i$ by $(k_i)_*$. Thus M is orientable precisely when this difficulty does not arise. (Note that we are requiring that M has *some* atlas that we can use to orient it, not that *every* atlas works; indeed, we can always add to any atlas a new chart that gives an induced orientation at a point that disagrees with that given by the original atlas.)

LEMMA 4.5.20. *Let M be a connected orientable smooth manifold. Then the orientation of M is determined by the orientation of $T_{p_0}M$ for any point p_0 of M. In particular, M has exactly two orientations.*

Proof. Let $p_0 \in \mathcal{U}_0$ and let $q_0 \in \mathcal{O}_0$ with $k_0(q_0) = p_0$. Then the orientation of $T_{p_0}M$ determines an orientation of $T_{q_0}\mathcal{O}_0$, by Definition 4.5.19, and this determines an orientation of $T_q\mathcal{O}_0$ for every $q \in \mathcal{O}_0$, by Definition 4.5.17, which in turn determines an orientation of $T_p\mathcal{U}_0$ for every $p \in \mathcal{U}_0$, again by Definition 4.5.19. Now let $\mathcal{U}_1$ be any coordinate patch with $\mathcal{U}_0 \cap \mathcal{U}_1 \neq \emptyset$, and let $p_1 \in \mathcal{U}_0 \cap \mathcal{U}_1$. Since $p_1 \in \mathcal{U}_0$, the orientation on $T_{p_1}M$ is determined by the orientation of $T_{p_0}M$, and since $p_1 \in \mathcal{U}_1$, the orientation of $T_{p_1}M$ determines the orientation of T_pM for any $p \in \mathcal{U}_1$.

Now let p' be any point of M. It is easy to see that there is a sequence of coordinate charts $\mathcal{U}_0, \mathcal{U}_1, \ldots, \mathcal{U}_k$ with $\mathcal{U}_i \cap \mathcal{U}_{i+1} \neq \emptyset$ for $i = 0, 1, \ldots, k - 1$ and with $p' \in \mathcal{U}_k$. Then by successively applying the above argument, the orientation of $T_{p_0} M$ determines the orientation of $T_p M$.

The second conclusion then follows because $T_{p_0} M$ has exactly two orientations. $\qquad\square$

Since, strictly speaking, a smooth manifold is just the union of coordinate charts modulo an equivalence relation, we ought to be able to formulate the notion of orientability purely in terms of coordinate charts. We do that now. The condition in this theorem that each open subset $\mathcal{O}_i$ of $\mathbb{R}^n$ have the standard orientation is purely for convenience and simplicity. The theorem can be reformulated more generally without this condition, but this formulation is the most useful one.

THEOREM 4.5.21. *Let M be a smooth n-manifold. Then M is orientable if and only if M has an atlas $\mathcal{A} = \{k_i : \mathcal{O}_i \longrightarrow \mathcal{U}_i\}_{i \in I}$, with every $\mathcal{O}_i$ having the standard orientation of an open subset of $\mathbb{R}^n$, such that: If $\mathcal{U}_i \cap \mathcal{U}_j \neq \emptyset$, then for every $p \in \mathcal{U}_i \cap \mathcal{U}_j$, $p = k_i(q_i) = k_j(q_j)$,*

$$\left((k_j)_* L_{q_j}\right)^{-1} \left((k_i)_* L_{q_i}\right) : \mathbf{R}^n \longrightarrow \mathbf{R}^n$$

is orientation-preserving.

Proof. This follows directly from Lemma 4.5.15. $\qquad\square$

REMARK 4.5.22.

(1) Note that, for fixed i and j, the map in the statement of the theorem is a linear transformation that depends continuously on the point p; write it as $\ell_p : \mathbf{R}^n \longrightarrow \mathbf{R}^n$. Then the condition of the theorem can equally well be stated as

$$\det(\ell_p) > 0.$$

(2) In case $\mathcal{U}_i \cap \mathcal{U}_j$ is connected, if this condition is satisfied for any $p \in \mathcal{U}_i \cap \mathcal{U}_j$, it is satisfied for every $p \in \mathcal{U}_i \cap \mathcal{U}_j$, as ℓ_p is an isomorphism for every p, so its determinant can never be 0.

(3) In the standard basis of $\mathbf{R}^n$, the matrix of ℓ_p is just the Jacobian matrix $\left((k_j)^{-1} k_i\right)'(q_i)$ and its determinant is the Jacobian determinant. $\qquad\diamond$

EXAMPLE 4.5.23. In this example we construct the open Möbius strip M as an abstract smooth manifold. We let $\mathcal{O}_1 = (-1, 1) \times (-1, 1)$ and $\mathcal{O}_2 = (-1, 1) \times (-1, 1)$, both subsets of $\mathbb{R}^2$. We let M be the identification space obtained as follows: For $1/2 < x < 1$ the point $(x, y) \in \mathcal{O}_1$ is identified with the point $(-1 + x, y) \in \mathcal{O}_2$, and for $-1 < x < -1/2$ the point $(x, y) \in \mathcal{O}_1$ is identified with the point $(1 + x, -y) \in \mathcal{O}_2$.

If $p \in M$ is in the image of the first identification, then, in the standard basis of $\mathbf{R}^2$,

$$\ell_p = \begin{bmatrix} 1 & 0 \\ 0 & 1 \end{bmatrix} \text{ with determinant } 1 > 0,$$

while if $p \in M$ is in the image of the second identification, then, in the standard basis of $\mathbf{R}^2$,

$$\ell_p = \begin{bmatrix} 1 & 0 \\ 0 & -1 \end{bmatrix} \text{ with determinant } -1 < 0.$$

(This does not show by itself that M is nonorientable, as we may have merely made a bad choice of atlas. But it is in fact true that M is nonorientable.)   ◇

REMARK 4.5.24. The above discussion goes through virtually unchanged for the case of a 0-manifold, but is much too elaborate for that case. A connected 0-manifold M is just a point p, and an orientation of M is once again an orientation of $T_p M$. But $T_p M$ is a 0-dimensional vector space, so this orientation is just a choice of sign $\varepsilon_p = \pm 1$. Thus an oriented 0-manifold is a pair (p, ε_p), which we shall (almost always) abbreviate to $+p$ or $-p$ according as $\varepsilon_p = 1$ or $\varepsilon_p = -1$.   ◇

Our development of orientability and orientations for manifolds goes through essentially unchanged for manifolds with boundary, except that instead of coordinate charts that are the images of open sets in $\mathbb{R}^n$, we need to consider coordinate charts that are the images of open sets in $\mathbb{R}^n_-$ (or in $\mathbb{R}^1_+$).

Suppose we have an oriented manifold with boundary M. We wish to see how to obtain an orientation on its boundary ∂M from the given orientation on M. In order to do so, we again need to fist develop a bit more linear algebra.

DEFINITION 4.5.25. Let the vector space V be the direct sum of the vector spaces U_1 and U_2, $V = U_1 \oplus U_2$.

First suppose that both U_1 and U_2 are positive-dimensional. If U_1 has an orientation given by a basis $\mathcal{B}_1$, U_2 has an orientation given by a basis $\mathcal{B}_2$, and V has an orientation given by a basis $\mathcal{C}$, then these orientations are *compatible* if $\mathcal{B}_1 \cup \mathcal{B}_2$ gives the same orientation on V as $\mathcal{C}$ does.

Now suppose that U_1 is positive-dimensional and U_2 is 0-dimensional. If U_1 has an orientation given by a basis $\mathcal{B}_1$, U_2 has an orientation given by a sign ε_2, and V has an orientation given by a basis $\mathcal{C}$, then these orientations are *compatible* if $\mathcal{B}_1$ gives the same orientation on V as $\mathcal{C}$ does and $\varepsilon_2 = 1$, or $\mathcal{B}_1$ gives the opposite orientation on V as $\mathcal{C}$ does and $\varepsilon_2 = -1$.

In any case, if any two of U_1, U_2, and V have orientations, the orientation on the third that makes all three compatible is the *induced* orientation on that vector space. ◇

(Note the word "induced" has two different but related meanings when used with regard to orientation. In fact, this word is heavily used in mathematics, with the common meaning that something induced on an object Y somehow comes from something on a different object X.)

EXAMPLE 4.5.26. For $n > 1$, let $V = \mathbf{R}^n$, as in Definition 4.6.17, let $U_1 = \{v \in V \mid a_2 = \ldots = a_n = 0\}$, and let $U_2 = \{v \in V \mid a_1 = 0\}$. Let $\mathcal{B}_1 = \{e_1\}$ be the standard basis of U_1 and $\mathcal{B}_2$ be the basis of U_2 induced from the standard basis of $\mathbf{R}^{n-1}$ by the inclusion of $\mathbf{R}^{n-1}$ as the last (rather than the first) $n - 1$ coordinates of $\mathbf{R}^n$; concretely $\mathcal{B}_2 = \{e_2, \ldots, e_n\}$. Let us call this the standard basis of U_2. Let V have basis $\mathcal{C} = \mathcal{E}$, the standard basis. Orient U_1, U_2, and V by these bases. Then these orientations are compatible, as $\mathcal{B}_1 \cup \mathcal{B}_2 = \mathcal{C}$.

For $n = 1$, orient $U_1 = V = \mathbf{R}^1$ by the standard basis $\mathcal{B}_1 = \mathcal{C} = \mathcal{E}$, and let U_2 have the orientation induced from the standard orientation of $\mathbf{R}^0$ by the unique linear map of $\mathbf{R}^0$ into $\mathbf{R}^1$. Then again these orientations are compatible, as $\mathcal{B}_1 = \mathcal{C}$ and U_2 has the orientation $+1$.

Otherwise stated, the induced orientation on U_2 from the standard orientations on U_1 and V is the standard orientation on U_2. ◇

EXAMPLE 4.5.27.

(1) Let $n > 1$. Consider the vector space $V = T_0\mathbb{R}^n$ with its standard basis $\mathcal{C} = \mathcal{E}$, which gives V an orientation. (Here 0 denotes the origin.) Consider the subset $\mathbb{R}^n_-$ of $\mathbb{R}^n$. Let U_1 be the subspace of V with basis $\mathcal{B}_1 = \{e_0^1\}$, with orientation given by that basis. Note that e_0^1 is an outward-pointing tangent vector to $\mathbb{R}^n_-$ at 0. Let $T_0\mathbb{R}^{n-1}$ have its standard basis, and recall we have decided to identify $\mathbb{R}^{n-1}$ with the boundary $\partial\mathbb{R}^n_-$ by $(x_2, \ldots, x_n) \mapsto (0, x_2, \ldots, x_n)$. Let $\mathcal{B}_2$ be the image of the standard basis of $T_0\mathbb{R}^{n-1}$ under this identification, which we will call the standard basis of $T_0\mathbb{R}^{n-1}$, so that $\mathcal{B}_2$ gives an orientation of $U_2 = T_0\partial\mathbb{R}^n_-$. (This orientation is just the induced orientation from the standard orientation of $T_0\mathbb{R}^{n-1}$ under this identification.) Then the orientations of U_1, U_2, and V are compatible.

This compatibility is the reason why we chose to identify $\mathbb{R}^{n-1}$ with this subset of $\mathbb{R}^n$ rather than with the "usual" one.

(2) Let $n = 1$. Proceed exactly as above, except let U_2 have the orientation induced by the standard orientation of $T_0\mathbb{R}^0$ by the (unique) inclusion of $T_0\mathbb{R}^0$ into $T_0\mathbb{R}^1$. This orientation is just $+1$. Then $\mathcal{B}_1 = \mathcal{C} = \mathcal{E}$ so again the orientations of U_1, U_2, and V are compatible.

Otherwise stated, the (basis consisting of) an outward-pointed normal to $\mathbb{R}^n_-$ at 0 and the standard basis of $T_0\mathbb{R}^n$ induce the standard orientation on $T_0\partial\mathbb{R}^n_-$.

Of course, this remains true if we replace 0 by any point in $\partial\mathbb{R}^n_-$.

(3) Let $n = 1$ and consider instead $\mathbb{R}^n_+$. Note that now e_0^1 is an inward-pointing tangent vector to $\mathbb{R}^n_-$ at 0. To obtain an outward-pointing tangent vector to $\mathbb{R}^n_-$ at 0 we instead choose $-e_0^1$. Thus in this case $\mathcal{B}_1 = \{-e_0^1\}$ and $\mathcal{C} = \{e_0^1\}$, so that if we give U_2 the orientation -1 then the orientations of U_1, U_2, and V are compatible.

Otherwise stated, the (basis consisting of) an outward-pointed normal to $\mathbb{R}^n_+$ at 0 and the standard basis of $T_0\mathbb{R}^n$ induce the nonstandard orientation on $T_0\partial\mathbb{R}^n_+$. (It is precisely for this reason that it is convenient to have $\mathbb{R}^1_+$ available.)   ◇

With this example in hand it is almost immediate how to give the boundary of an n-manifold with boundary in $\mathbb{R}^n_-$ an induced orientation, given an orientation of $\mathbb{R}^n_-$. Then it is easy to give the boundary ∂M of a general n-manifold an induced orientation, given an orientation of M. (This applies, with obvious modifications, to $\mathbb{R}^1_+$.)

DEFINITION 4.5.28. Let M be an oriented n-dimensional manifold with boundary in $\mathbb{R}^n_-$, and suppose that $\partial M = M \cap \partial \mathbb{R}^n_-$. (Thus $T_p M$ is oriented for every point $p \in M$.) The *induced* orientation on ∂M is the orientation obtained as follows: Let p be any point in ∂M. Let $V = T_p M$. Let $U_1 \subseteq V$ be the subspace with basis $\{e^1_p\}$, oriented by that basis. (Note that e^1_p is an *outward-pointing* normal to M at p.) Then the orientation of $U_2 = T_p \partial M$ is the orientation of U_2 induced from the orientations of U_1 and V.   ◇

REMARK 4.5.29. Identifying $\mathbb{R}^{n-1}$ with $\partial \mathbb{R}^n_-$ as above, it is easy to check that the orientation of $\mathbb{R}^{n-1}$ induced by the standard orientation of $\mathbb{R}^n_-$ is the standard orientation of $\mathbb{R}^{n-1}$.

In particular, in low dimensions: The orientation "right-hand" on $\mathbb{R}^3_-$ induces the orientation "counterclockwise" on $\mathbb{R}^2$; the orientation "counterclockwise" on $\mathbb{R}^2_-$ induces the orientation "increasing" on $\mathbb{R}^1$; and the orientation "increasing" on $\mathbb{R}^1_-$ induces the orientation "plus" on $\mathbb{R}^0$.   ◇

DEFINITION 4.5.30. Let M be an oriented n-manifold with boundary, with orientation given as in Definition 4.5.19, where now each set $\mathcal{O}_i$ is an oriented open subset of $\mathbb{R}^n_-$. The *induced* orientation on ∂M is the orientation given as follows: Let $p \in \partial M$. Let $p \in \mathcal{U}_i$, $p = k_i(q)$ for $q \in \partial \mathcal{O}_i$. Let $T_q \partial \mathbb{R}^n_-$ be given the induced orientation as in Definition 4.5.28. Then the orientation of $T_p \partial M$ is the orientation induced from the orientation on $T_q \partial \mathbb{R}^n_-$ by the restriction of the isomorphism $(k_i)_*$ to this subspace.   ◇

EXAMPLE 4.5.31. Let M be the 1-manifold $M = [-1, 1]$, the 1-disk in $\mathbb{R}^1$. Let M have the induced orientation as a subset of $\mathbb{R}^1$. Then, as an oriented 0-manifold, $\partial M = \{1\} \cup \{-1\}$, with $\{1\}$ having the orientation $+1$ and $\{-1\}$ having the orientation -1.

More generally, let C be a parameterized smooth curve in $\mathbb{R}^n$, given by $f : I \longrightarrow \mathbb{R}^n$, where $I = [a, b]$ is a closed interval in $\mathbb{R}^1$. Let I have its standard orientation and give C the induced orientation. Let

$p = f(a)$ and $q = f(b)$. (Thus, C is a curve going from p to q.) Then, as an oriented manifold,

$$\partial C = \{q\} \cup -\{p\}. \qquad \diamond$$

For an n-manifold in $\mathbb{R}^n$, we have the notion of a standard orientation. We now consider n-manifolds with boundary in $\mathbb{R}^n$.

DEFINITION 4.5.32. Let M be an n-manifold with boundary in $\mathbb{R}^n$. The *standard orientation* of M is the standard orientation of M as defined in Definition 4.5.17 for all points in $\text{int}(M)$, noting that this definition extends to points on ∂M as well. The *standard orientation* on ∂M is the orientation induced on ∂M from the standard orientation on M as in Definition 4.5.30. (Note that this comes from the outward-pointing normal to M at every point of ∂M.) $\diamond$

Finally, we record what it means for a diffeomorphism between two oriented smooth manifolds to be orientation-preserving. The definition is the obvious one.

DEFINITION 4.5.33. Let $n > 0$. Let $k : M \longrightarrow N$ be a diffeomorphism from the connected oriented smooth n-manifold M to the connected oriented smooth n-manifold N. Then k is *orientation-preserving* (resp. *orientation-reversing*) if for some (and hence, by Lemma 4.5.20, any) point $p \in M$, $k' : T_p M \longrightarrow T_{k(p)} N$ is orientation-preserving (resp. orientation-reversing). $\diamond$

For $n = 0$ we need a separate definition as there is no derivative map k'.

DEFINITION 4.5.34. Let $M = (p, \varepsilon_p)$ and $N = (q, \varepsilon_q)$ be oriented 0-manifolds. An oriented map $(k, \varepsilon_k) : M \longrightarrow N$ is a pair (k, σ_k) where k is the (unique) function $k(p) = q$ and $\sigma_k = \pm 1$. The oriented map (k, ε_k) is *orientation-preserving* (resp. *orientation-reversing*) if $\varepsilon_q = \sigma_k \varepsilon_p$ (resp. $\varepsilon_q = -\sigma_k \varepsilon_p$). $\diamond$

4.6 Partitions of unity

This section is devoted to proving a purely technical result, the existence of partitions of unity (Theorem 4.6.3). This result will be very useful to us.

The proof divides into two parts. The first part, which is the key, is a clever construction of an appropriate function $f : \mathbb{R} \longrightarrow \mathbb{R}$. The second part applies some topological arguments to get from f to what we need.

LEMMA 4.6.1.

(1) There is a smooth function $f_1 : \mathbb{R} \longrightarrow \mathbb{R}$ such that $f_1(x) = 0$ for $x \leq 0$ and $f_1(x)$ is strictly increasing for $x > 0$.

(2) There is a smooth function $f_2 : \mathbb{R} \longrightarrow \mathbb{R}$ such that $f_2(x) = 0$ for $x \leq -1$, $f_2(x)$ is positive and strictly increasing for $-1 < x \leq 0$, $f_2(x)$ is positive and strictly decreasing for $0 \leq x < 1$, and $f_2(x) = 0$ for $x \geq 1$.

(3) There is a smooth function $f_3 : \mathbb{R} \longrightarrow \mathbb{R}$ such that $f_3(x) = 0$ for $x \leq -1$, $f_3(x)$ is positive and strictly increasing for $-1 < x < 0$, and $f_3(x) = 1$ for $x \geq 0$.

(4) There is a smooth function $f_4 : \mathbb{R} \longrightarrow \mathbb{R}$ such that $f_4(x) = 0$ for $x \leq -2$, $f_4(x)$ is positive and strictly increasing for $-2 < x < -1$, $f_4(x) = 1$ for $-1 \leq x \leq 1$, $f_4(x)$ is positive and strictly decreasing for $1 < x < 2$, and $f_4(x) = 0$ for $x \geq 2$.

Proof.

(1) For $x > 0$, let $e(x)$ be the function $e(x) = \exp(-1/x)$. The function $e(x)$ is positive and strictly increasing for $x > 0$.

Also, $\lim_{x \downarrow 0} e(x) = 0$, so $e(x)$ extends to a continuous function for $x \geq 0$ by defining $e(0) = 0$, and then, by considering difference quotients, an inductive argument shows that the right-hand derivatives of all orders of $e(x)$ at 0 are equal to 0. Thus $f_1(x)$ defined by

$$f_1(x) = e(x) \text{ for } x > 0, \quad f_1(x) = 0 \text{ for } x \leq 0$$

is a function with the desired properties.

(2) Let $f_2(x) = f_1(1 - x^2)$.

(3) Let $f_3(x) = 0$ for $x \leq -2$, $f_3(x) = f_2(x)/f_2(1 + x)$ for $-2 < x < 0$, and $f_3(x) = 1$ for $x \geq 0$.

(4) Let $f_4(x) = f_3(1+x) f_3(1-x)$. $\qquad\qquad\qquad\qquad\qquad$ $\square$

Recall that, for a subspace Y of a topological space X, $\overline{Y}$ denotes the closure of Y.

DEFINITION 4.6.2. Let M be a smooth manifold with boundary and let $\mathcal{A} = \{\mathcal{U}_i\}$ be an open cover of M. A smooth *partition of unity* subordinate to $\mathcal{A}$ is a set of smooth functions $\{\varepsilon_j : M \longrightarrow \mathbb{R}\}$ such that:

(1) For each j, $\mathrm{supp}(\varepsilon_j) = \overline{\{p \in M | \varepsilon_j(p) \neq 0\}} \subset \mathcal{U}_i$ for some i.

(2) For each j, $\varepsilon_j(p) \geq 0$ for every $p \in M$.

(3) For every $p \in M$, $\{j | \varepsilon_j(p) \neq 0\}$ is finite.

(4) $\sum_j \varepsilon_j = 1$, i.e., for every $p \in M$, $\sum_j \varepsilon_j(p) = 1$. $\qquad$ $\diamond$

(Note that by (3), for any point p the sum in (4) only has finitely many nonzero terms, so it makes sense.)

THEOREM 4.6.3. *Let M be a smooth manifold with boundary and let $\mathcal{A} = \{\mathcal{U}_i\}$ be any open cover of M. Then there exists a smooth partition of unity subordinate to $\mathcal{A}$.*

Proof. For simplicity of notation, we first prove this in the case that M is a manifold. We then briefly the describe the minor modifications that need to be made to handle the case that M is a manifold with boundary.

Note that if $\mathcal{A}'$ is a refinement of $\mathcal{A}$, then a partition of unity subordinate to $\mathcal{A}'$ is subordinate to $\mathcal{A}$. Thus in the course of this proof we will freely replace open covers by refinements.

First let us suppose M is compact.

Choose an arbitrary atlas on M. Given any open cover $\mathcal{A}$, by taking refinements we may suppose that every set in $\mathcal{A}$ is contained in some set in that atlas. Now each set in the atlas is the image of an open set in $\mathbb{R}^n$. Any open set $\mathcal{O}$ in $\mathcal{R}^n$ may be written as the union of open sets with compact closures contained in $\mathcal{O}$. For our purposes it is convenient to use open "bricks" rather than the usual open balls, i.e., we use open sets in $\mathbb{R}^n$ of the form

$$\{(x_1, \ldots, x_n) \mid |x_i - x_i^0| < \delta_i, i = 1, \ldots, n\}$$

for some fixed point $(x_i^0, \ldots, x_n^0)$ and some fixed positive real numbers $\delta_1, \ldots, \delta_n$. (In fact, we only need countably many such sets, as we may choose all the x_i^0 and δ_i rational.)

Thus, again by taking refinements, we may assume that $\mathcal{A}$ consists of sets of this form. Now M is compact, and $\mathcal{A}$ is an open cover, so there is a finite subcover $\{\mathcal{U}_1, \ldots, \mathcal{U}_t\}$.

We now refine this subcover to get a slightly "smaller" subcover $\{\mathcal{V}_1, \ldots, \mathcal{V}_t\}$. We proceed inductively. Suppose $\mathcal{V}_1, \ldots, \mathcal{V}_{s-1}$ have been defined so that $\mathcal{A}_{j-1} = \{\mathcal{V}_1, \ldots, \mathcal{V}_{j-1}, \mathcal{U}_j, \ldots, \mathcal{U}_t\}$ is a cover. Then $\mathcal{U}_j$ is the image of an open brick B in $\mathbb{R}^n$ under a smooth map $k_j : B \longrightarrow M$, and furthermore k_j extends to its closure $\overline{B}$. Consider the subset $W = k_j(\overline{B} - B) \subset M$. This subset is disjoint from $\mathcal{U}_j$, and $\mathcal{A}_{j-1}$ is a cover of M, so must be contained in the union of the other elements of the cover. Consider $k^{-1}(W \cap \mathcal{V}_1) \cup \ldots \cup k^{-1}(W \cap \mathcal{U}_t)$. This is a relatively open subset of the closed brick $\overline{B}$, a compact subset of $\mathbb{R}^n$, so its complement in $\overline{B}$ is a closed (in fact, compact) subset of $\mathbb{R}^n$, and hence these two sets are a positive distance apart. Thus there is a $\gamma_j > 0$ so that, if B is the brick

$$B = \{(x_1, \ldots, x_n) \mid |x_i - x_i^0| < \delta_i, i = 1, \ldots, n\}$$

and $\mathcal{V}_j$ is the image of the (slightly) smaller brick

$$B' = \{(x_1, \ldots, x_n) \mid |x_i - x_i^0| < \delta_i - \gamma_j, i = 1, \ldots, n\}$$

under k_j, then $\mathcal{A}_j = \{\mathcal{V}_1, \ldots, \mathcal{V}_{j-1}, \mathcal{V}_j, \mathcal{U}_{j+1}, \ldots, \mathcal{U}_t\}$ is still a cover of M.

Now that we have the cover $\mathcal{A}_t$, we construct the partition of unity. Suppose that $B = B_j$ is as above, and consider the function

$$g_j(x_1, \ldots, x_n) = \prod_{i=1}^{n} f_2((x_i - x_i^0)/(\delta_i - \gamma_j))$$

where $f_2(x)$ is the function in Lemma 4.6.1(2).

Let $k_j : B_j \longrightarrow M$ be as above, and consider the function $\zeta_j :$ $M \longrightarrow \mathbb{R}$ defined by

$$\begin{aligned}
\zeta_j(p) &= g_j(x_1, \ldots, x_n) \text{ if } p = k_j(x_1, \ldots, x_n) \in \mathcal{U}_j \\
&= 0 \text{ if } p \notin \overline{\mathcal{V}_j}.
\end{aligned}$$

Note that each ζ_j is a smooth function on M as $g_j(p) = 0$ for $p \in \mathcal{U}_j - \overline{\mathcal{V}_j}$ (so that the two halves of the definition of $\zeta_j(p)$ agree on this open set, which is the overlap of the domains in the two halves of the definition). Also note that $g_j(p) \neq 0$ if $p \in \mathcal{V}_j$. Finally note that p is in at least one set $\mathcal{V}_j$, since $\mathcal{A}_t$ is a cover, so $\zeta_j(p) \neq 0$ for at least one value of j.

Thus we obtain a partition of unity by defining

$$\varepsilon_j(p) = \frac{\zeta_j(p)}{\sum_i \zeta_i(p)}.$$

Next consider the case that M is not compact. By Theorem 4.1.6, there is a countable collection $\{\mathcal{D}_1, \mathcal{D}_2, \ldots\}$ of compact subsets of M with $\mathcal{D}_i \subseteq \operatorname{int}(\mathcal{D}_{i+1})$ for each i, with $M = \cup_i \mathcal{D}_i$. Set $\mathcal{D}_0 = \emptyset$ for convenience.

By taking refinements, if necessary, we may assume that each element of our cover $\mathcal{A}$ is contained in the open set $\operatorname{int}(\mathcal{D}_{i+1}) - \mathcal{D}_{i-1}$ for some $i \geq 1$. Now apply the above argument to successively obtain finite covers $\mathcal{A}^i = \{\mathcal{U}_1^i, \ldots, \mathcal{U}_{s_i}^i\}$ of the compact sets $\mathcal{D}_i - \operatorname{int}(\mathcal{D}_{i-1})$. Let $\mathcal{A} = \{\mathcal{U}_1^1, \ldots, \mathcal{U}_{s_1}^1, \mathcal{U}_1^2, \ldots, \mathcal{U}_{s_2}^2 \ldots\} = \{\mathcal{U}_1, \mathcal{U}_2, \ldots\}$, inductively perform the "shrinking" operation to get a cover $\{\mathcal{V}_1, \mathcal{V}_2, \ldots\}$, and inductively define the smooth functions $\{\zeta_1, \zeta_2, \ldots\}$. As before, it is the case that every point p of M is contained in at least one element of this cover, but it is also the case that every point p of M is contained in only finitely many elements of the cover.

Thus in this case as well we obtain a partition of unity by defining

$$\varepsilon_j(p) = \frac{\zeta_j(p)}{\sum_i \zeta_i(p)}.$$

In case M has a boundary, we merely need to modify our open covers to handle ∂M. Thus, in addition to the images of open sets in $\mathbb{R}^n$, we also begin with the images of relatively open sets in $\mathbb{R}^n_-$ of the form

$$B_- = \{(x_1, \ldots, x_n) | - \delta_1 < x_1 \leq 0, \ |x_i - x_i^0| < \delta_i, i = 2, \ldots, n\},$$

which contain the images of the smaller relatively open sets

$$
\begin{aligned}
B'_- = \{(x_1, \ldots, x_n) | &- (\delta_1 - \gamma_j) < x_1 \leq 0, \\
&|x_i - x_i^0| < \delta_i - \gamma_j, i = 2, \ldots, n\},
\end{aligned}
$$

which also form a cover, and proceed as above. $\square$

REMARK 4.6.4. For future reference, we note that, in the notation of the proof of Theorem 4.6.3, if $\mathcal{V}_j = k_j(B')$ then $\varepsilon_j = 0$ on $\partial(k_j(B'))$, while if $\mathcal{V}_j = k_j(B'_-)$ then $\varepsilon_j = 0$ on all but one "face" of $\partial(k_j(B'))$, but ε_j is not 0 on $k_j(B' \cap \mathbb{R}^{n-1})$. $\diamond$

4.7 Smooth homotopies and the Converse of Poincaré's Lemma in general

In this section we generalize the results of Section 3.5. In Chapter 3, we only considered open sets of $\mathbb{R}^n$, so our results were restricted to that case. But in this chapter we have developed the notions of smooth manifolds in general, so we can consider the general situation.

As we will see, the results generalize almost word-for-word. Also, given the results of Section 3.5, the proofs here are almost trivial, except that in one crucial place at the start we must use partitions of unity.

LEMMA 4.7.1. *Let N be a smooth n-manifold with boundary. Then there is a linear map, for every $\ell \geq 0$,*

$$\tilde{\xi} : \Omega^\ell(N \times I) \longrightarrow \Omega^{\ell-1}(N)$$

with the property that, for any ℓ-form φ on $N \times I$,

$$i_1^*(\varphi) - i_0^*(\varphi) = \tilde{\xi}(d\varphi) + d(\tilde{\xi}(\varphi)).$$

Proof. Again, for simplicity, we prove this in the case that N is a manifold. The case that N is a manifold with boundary is almost exactly the same.

We proved this as Lemma 3.5.3 in the case that N was an open subset of $\mathbb{R}^n$. Clearly if it is true for N it is true for any manifold diffeomorphic to N, so Lemma 3.5.3 holds for any manifold N diffeomorphic to an open subset of $\mathbb{R}^n$.

In particular, let $\mathcal{A} = \{k_i : \mathcal{O}_i \longrightarrow \mathcal{U}_i\}$ be a locally finite atlas on N. Then the lemma is true for every $\mathcal{U}_i$. Denote the map so constructed on $\mathcal{U}_i$ by $\tilde{\xi}_i$. Observe from Corollary 3.5.4 that if φ has compact support on $\mathcal{U}_i \times I$, then $\tilde{\xi}_i(\varphi)$ has compact support on $\mathcal{U}_i$, an open subset of N, so $\tilde{\xi}_i(\varphi)$ extends to a smooth ℓ-form on N by defining it to be 0 outside of $\mathcal{U}_i$. To avoid introducing new notation, we will also denote this extension by $\tilde{\xi}_i(\varphi)$.

Choose once and for all a partition of unity subordinate to $\mathcal{A}$, $\{\varepsilon_j : N \longrightarrow \mathbb{R}\}$, and for each j, let $\tilde{\varepsilon}_j : N \times I \longrightarrow \mathbb{R}$ by $\tilde{\varepsilon}_j(p,t) = \varepsilon_j(p)$. Then $\{\tilde{\varepsilon}_j\}$ is a partition of unity on $N \times I$, with $\operatorname{supp}(\tilde{\varepsilon}_j) = \operatorname{supp}(\varepsilon_j) \times I$.

Now let φ be an ℓ-form on $N \times I$. Let $\varphi_j = \tilde{\varepsilon}_j\varphi$ for each j. Then $\varphi = \sum_j \varphi_j$, and for each j, $\operatorname{supp}(\varphi_j) \subset \mathcal{U}_i \times I$ for some $i = i(j)$. Thus, by the observation at the beginning of this proof we may construct $\tilde{\xi}_i(\varphi_j)$, and then

$$\tilde{\xi}(\varphi) = \tilde{\xi}\left(\sum_j \varphi_j\right) = \sum_j \tilde{\xi}_{i(j)}(\varphi_j)$$

(a sum of ℓ-forms defined on N, as each φ_j has compact support) and we are done.   □

DEFINITION 4.7.2. Let M be a smooth m-manifold with boundary and let N be a smooth n-manifold with boundary. Let $f_0 : M \longrightarrow N$ and $f_1 : M \longrightarrow N$ be smooth maps. A *smooth homotopy* between f_0 and f_1 is a smooth map $F : M \times I \longrightarrow N$ with

$$F(x_1, \ldots, x_m, 0) = f_0(x_1, \ldots, x_m) \quad \text{and}$$
$$F(x_1, \ldots, x_m, 1) = f_1(x_1, \ldots, x_m)$$

for every $(x_1, \ldots, x_m) \in M$.

If there exists a smooth homotopy between f_0 and f_1, then f_0 and f_1 are said to be *smoothly homotopic*.

In particular, if there exists a smooth homotopy between $f : M \longrightarrow N$ and a constant map $c : M \longrightarrow N$ (i.e., for some point $q_0 \in N$, $c(p) = q_0$ for every point $p \in M$) then f is *smoothly null-homotopic*.   ◇

THEOREM 4.7.3. *Let M be a smooth m-manifold with boundary and let N be a smooth n-manifold with boundary. Let $f_0 : M \longrightarrow N$ and $f_1 : M \longrightarrow N$ be smoothly homotopic smooth maps. Then there is a linear map, defined for every $j \geq 0$,*

$$\Xi : \Omega^j(N) \longrightarrow \Omega^{j-1}(M)$$

such that, for any k-form φ on N,

$$f_1^*(\varphi) - f_0^*(\varphi) = d(\Xi(\varphi)) + \Xi(d\varphi).$$

Proof. We define this map just as before: Let $F : M \times I \longrightarrow N$ be a homotopy. Then

$$\Xi(\varphi) = \tilde{\tilde{\xi}}(F^*(\varphi))$$

where $\tilde{\tilde{\xi}}$ is the map in Lemma 4.7.1.   □

COROLLARY 4.7.4. *Let M be a smooth m-manifold with boundary and let N be a smooth n-manifold with boundary. Let $f_0 : M \longrightarrow N$ and $f_1 : M \longrightarrow N$ be smoothly homotopic smooth maps. Let φ be a closed k-form on N. Then $f_1^*(\varphi) - f_0^*(\varphi)$ is exact.*

Proof. Exactly the same as before: We simply let $\psi = \Xi(\varphi)$ as in the conclusion of Theorem 4.7.3 (noting that $\Xi(d\varphi) = 0$ as $d\varphi = 0$, since φ is closed, by hypothesis).   □

DEFINITION 4.7.5. Let M be a smooth m-manifold with boundary. Then M is *smoothly contractible* if the identity map from M to itself is smoothly null-homotopic.   ◇

Now we have our final generalization of the converse of Poincaré's Lemma.

COROLLARY 4.7.6 (*Converse of Poincaré's Lemma*). *Let M be a smoothly contractible m-manifold with boundary. Let φ be a closed k-form defined on $M, k > 0$. Then φ is exact.*

Proof. Exactly as before: Let $f_1 : M \longrightarrow M$ be the identity map and $f_0 : M \longrightarrow M$ be the map taking M to a point. Then $f_1^*(\varphi) = \varphi$ for any φ. Also, f_0 is a constant map, so for $k > 0$, $f_0^*(\varphi) = 0$ for any φ. Thus, by Corollary 4.7.4,

$$\varphi = f_1^*(\varphi) = f_0^*(\varphi) + d(\xi(\varphi)) = 0 + d(\xi(\varphi) = d(\xi(\varphi))$$

is exact. $\qquad\square$

4.8 Exercises

1. Fill in the details of the proof of Lemma 4.6.1.

2. Let M be a smooth m-manifold and let N be a smooth n-manifold.

   (a) Show that $M \times N$ is a smooth $(m + n)$-manifold.

   (b) If M and N are each orientable, show that $M \times N$ is orientable.

3. Let M and N be smooth manifolds and let $f : M \longrightarrow N$ be a smooth map.

   (a) Let A and B be smooth submanifolds of M and N respectively. If $f(A) \subseteq B$, show that the restriction $f|A : A \longrightarrow B$ is a smooth map.

   (b) Suppose that f is a diffeomorphism. Let A be a smooth submanifold of M and set $B = f(A)$. Show that B is a smooth submanifold of N and furthermore that $f|A : A \longrightarrow B$ is a diffeomorphism.

4. Show that the Möbius strip is nonorientable, as claimed.

5. Let M be a connected smooth manifold. Show that for any two points p and q of M, there is a diffeomorphism $f : M \longrightarrow M$, smoothly homotopic to the identity map of M, with $f(p) = q$.

6. Let M be a smooth n-manifold with boundary. A *volume form* on M is an n-form φ that is nonzero at every point of M.

   (a) Show that M has a volume form if and only if M is orientable.

   (b) Show how a choice of volume form on M determines an orientation of M.

7. Let $T = \{(x, y, z) \in \mathbb{R}^3 | (x^2 + y^2 - 4)^2 + z^2 = 1\}$.

   (a) Show that T is a smooth surface.

   (b) Show that T is diffeomorphic to $S^1 \times S^1$.

8. Let $\mathbb{F} = \mathbb{R}$ or $\mathbb{C}$. Let $N = \mathbb{F}^{n+1} - \{(0, \ldots, 0)\}$ and define an equivalence relation $\sim$ on N by $(x_1, \ldots, x_{n+1}) \sim (y_1, \ldots, y_{n+1})$ if there is a nonzero element $z \in \mathbb{F}$ with $y_i = zx_i, i = 1, \ldots, n+1$. Let M be the quotient space $N/ \sim$. Show that M is a smooth manifold of dimension dn, where $d = 1$ if $\mathbb{F} = \mathbb{R}$ and $d = 2$ if $\mathbb{F} = \mathbb{C}$, with the quotient map $q : N \longrightarrow M$ a smooth map. This quotient space is usually denoted $\mathbb{RP}^n$ or $\mathbb{CP}^n$ and is called real or complex *projective space*, and the image of the point $(x_1, \ldots, x_{n+1})$ is usually denoted by $[x_1, \ldots, x_{n+1}]$.

9. Let N be a smooth n-manifold. Let G be a group acting smoothly on N, i.e., for each element g of G there is a smooth map $f_g : N \longrightarrow N$, such that $f_{g_1 g_2} = f_{g_1} f_{g_2}$ for any g_1 and g_2 in G, and f_e is the identity map on N, where e is the identity element of G. (Note that each f_g is then necessarily a diffeomorphism.) Suppose that G acts freely and properly discontinuously on N, i.e., that any two points p and q of N have neighborhoods U_p and U_q with $f_g(U_p) \cap U_q = \emptyset$ unless $g = e$. (In case G is finite, this condition is equivalent to G acting freely on N, i.e., for any point p of N, $f_g(p) \neq p$ unless $g = e$.)

(a) Show that the quotient space N/G is a smooth manifold. (This quotient space is often called the *orbit space* of the action.)

(b) Let $N = \mathbb{R}$ and $G = \mathbb{Z}$ and let G act on N by $f_n(x) = x + n$. Show that the quotient space $M = N/G$ is diffeomorphic to S^1.

(c) Let $M = S^n$ and $G = \mathbb{Z}/2\mathbb{Z}$ (the group of order 2) and let the nontrivial element of G act on N by the antipodal map, $a(x_1, \ldots, x_{n+1}) = (-x_1, \ldots, -x_{n+1})$. Show that the quotient space $M = N/G$ is diffeomorphic to $\mathbb{RP}^n$.

10. A smooth map $f : M_1 \longrightarrow M_2$ from the smooth manifold M_1 to the smooth manifold M_2 is an *embedding* if the map $f : M_1 \longrightarrow f(M_1)$ is a diffeomorphism.

(a) Let $B = \{p = (x_1, \ldots, x_n) \in \mathbb{R}^n | |x_i| < 2, i = 1, \ldots, n\}$. Construct a smooth map $f : \mathbb{R}^n \longrightarrow \mathbb{R}^{n+1}$ such that $f|B$, the restriction of f to B, is an embedding and such that $f(p) = (0, \ldots, 0)$ (the origin in $\mathbb{R}^{n+1}$) for $p \notin B$.

(b) Let M be an arbitrary compact smooth n-dimensional manifold. Show that there is an embedding $f : M \longrightarrow \mathbb{R}^N$ for some sufficiently large N. (The value of N may depend on the specific manifold M.)

11. A *Lie group* G is a smooth manifold with a group structure such that multiplication $m : G \times G \longrightarrow G$ by $m(g_1, g_2) = g_1 g_2$ and inversion $i : G \longrightarrow G$ by $i(g) = g^{-1}$ are both smooth maps.

(a) Show that $G = GL_n(\mathbb{R})$ is a Lie group.

(b) The *orthogonal group* O_n is the subgroup of $GL_n(\mathbb{R})$ defined by

$$O_n = \{A \in GL_n(\mathbb{R}) | {}^t A A = I\},$$

where I is the identity matrix and ${}^t A$ denotes the transpose of the matrix A. Show that O_n is a compact smooth submanifold of $GL(n, \mathbb{R})$ of dimension $n(n-1)/2$.

(c) Show that O_n is a Lie group.

(d) Show that the identity map on $GL_n(\mathbb{R})$ is smoothly homotopic to a smooth map $r : GL_n(\mathbb{R}) \longrightarrow O_n$, with the restriction $r|O_n$ the identity map.

5 Vector Bundles and the Global Point of View

We have introduced vector fields and differential forms as "local" objects, i.e., objects that are defined on coordinate charts, and then we have obtained vector fields and differential forms on a smooth manifold M by "piecing together" locally defined vector fields and differential forms.

In this chapter we show how to regard vector fields and differential forms as "global" objects, i.e., as objects defined on all of M. These objects will be sections of smooth vector bundles, which we proceed to define.

As we will see, vector bundles are themselves defined by a piecing together construction. Thus our point here is not to eliminate this sort of construction—it is unavoidable—but rather to introduce a new point of view, in which we think of vector fields and differential forms globally and hence as something intrinsic to a smooth manifold.

We will only be dealing with finite-dimensional vector spaces and we shall implicitly assume that any vector space V has its canonical topology and smooth structure, making it into a smooth manifold (see Example 4.1.21(c)). We then also observe that the set of linear transformations from V to itself has a canonical topology and smooth structure as well (see Example 4.1.21(d)).

The material in this chapter provides an illuminating viewpoint, and one which is essential for more advanced work. But it is not strictly necessary for the study of integration on manifolds, the principal objective of this book. As such, we will take the liberty of being rather sketchy in some our arguments here.

Differential Forms, Second Edition. http://dx.doi.org/10.1016/B978-0-12-394403-0.00005-0

5.1 The definition of a vector bundle

In this section we define vector bundles in general, and make some important constructions related to vector bundles.

First we will go global to local, i.e., start with the notion of a vector bundle and obtain coordinate charts. Then we will go local to global, i.e., start with coordinate charts and obtain a vector bundle. This parallels our approach in Section 4.1.

DEFINITION 5.1.1. Let V be a vector space. A *smooth vector bundle* with *base space* B, *total space* E, *fiber* V, and *projection map* π consists of the following:

(i) Smooth manifolds E and B and a smooth map $\pi : E \longrightarrow B$ from E onto B such that for every point $b \in B$, the inverse image $V_b = \pi^{-1}(b)$ has the structure of a vector space, and as a vector space is isomorphic to V.

(ii) A pair of atlases $\mathcal{A} = \{k_i : \mathcal{O}_i \longrightarrow \mathcal{U}_i\}$ of B and $\tilde{\mathcal{A}} = \{\tilde{k}_i : \tilde{\mathcal{O}}_i \longrightarrow \tilde{\mathcal{U}}_i\}$ of E that are related as follows:

$$\tilde{\mathcal{O}}_i = \mathcal{O}_i \times V, \quad \tilde{\mathcal{U}}_i = \pi^{-1}(\mathcal{U}_i)$$

and for every $a \in \mathcal{O}_i$ and $v \in V$,

$$\pi(\tilde{k}_i(a, v)) = k_i(a),$$

i.e., if $b = k_i(a)$, then for any fixed a,

$$\tilde{k}_i(\{a\} \times V) = V_b.$$

Equivalently, the following diagram commutes (where π_1 denotes projection on the first factor):

$$
\begin{array}{ccc}
\mathcal{O}_i \times V & \xrightarrow{\tilde{k}_i} & \tilde{\mathcal{U}}_i \\
\pi_1 \downarrow & & \downarrow \pi \\
\mathcal{O}_i & \xrightarrow{k_i} & \mathcal{U}_i
\end{array}
$$

(iii) For any fixed a, if $b = k_i(a)$, then the map

$$V \longrightarrow V_b \quad \text{defined by} \quad v \mapsto k_i(a, v)$$

is a vector space isomorphism. ◇

REMARK 5.1.2. Here is the key point of Definition 5.1.1. Note that by part (i) of this definition, as a set E is just the disjoint union of the fibers $\{V_b | b \in B\}$. But we want E not just to be a set, but rather a smooth manifold. Thus we must have a smooth structure on this set, but we want a smooth structure that fits together with the smooth structure on B and the projection $\pi : E \longrightarrow B$, and that is what we give in part (ii) of the definition. Furthermore, we not only want each fiber V_b to be abstractly isomorphic to the vector space V, but we want to have this isomorphism given by our coordinate maps, and that is what we require in part (iii) of the definition. ◇

We will often abbreviate smooth vector bundle to vector bundle. Also, we will often abbreviate Definition 5.1.1 by saying that $\pi : E \longrightarrow B$ is a vector bundle with fiber V. Sometimes we will abbreviate this even further, and say that E is a vector bundle over B, when it is clear what the projection map π and the fiber V are.

We have a notion of equivalence of pairs of atlases for smooth vector bundles that exactly parallels our notion of equivalence of atlases for smooth manifolds. We will say that two smooth vector bundles over the same base whose definitions differ only in the choice of equivalent atlases for the base are strictly equivalent, and we will regard them as being the same. This parallels our work for smooth manifolds, as we may simply take a maximal atlas satisfying the conditions of Definition 5.1.1, and then every smooth vector bundle is strictly equivalent to a unique smooth vector bundle whose atlas is maximal in this sense.

However, in the case of bundles we also want to consider a more general sort of equivalence.

DEFINITION 5.1.3. Let $\pi : E \longrightarrow B$ and $\pi' : E' \longrightarrow B$ be two smooth vector bundles over the same bases B with the same fiber V. Then $F : E \longrightarrow E'$ is an *equivalence* of bundles, and the bundles E and E' are *equivalent* if for every point $b \in B$, the restriction of F to the fiber V_b is a vector space isomorphism from V_b to V_b. ◇

REMARK 5.1.4. Note that in Definition 5.1.3, the map F must take the fiber in E over every point of B to the fiber in E' over the *same* point.   ◇

Here is a simple, but basic, example of a vector bundle.

EXAMPLE 5.1.5. Let B be a smooth manifold, let V be a vector space. Set $E = B \times V$, and let $\pi : E \longrightarrow V$ be projection on the first factor. Then E is a smooth vector bundle over B with fiber V. E is called a *trivial* bundle. Note that if $\mathcal{A} = \{k_i : \mathcal{O}_i \longrightarrow \mathcal{U}_i\}$ is an atlas for B, then $\tilde{\mathcal{A}} = \{\tilde{k}_i : \tilde{\mathcal{O}}_i \longrightarrow \tilde{\mathcal{U}}_i\}$ is an atlas for E, where $\tilde{\mathcal{O}}_i = \mathcal{O}_i \times V, \tilde{\mathcal{U}}_i = \mathcal{U}_i \times V$, and $\tilde{k}_i = k_i \times \mathrm{id}$ (i.e., $\tilde{k}_i(a, v) = (k_i(a), v)$).◇

EXAMPLE 5.1.6.

(1) $T\mathbb{R}^n = \mathbb{R}^n \times \mathbf{R}^n$, with $\pi : T\mathbb{R}^n \longrightarrow \mathbb{R}^n$ the projection on the first factor, is a trivial smooth vector bundle. (Note we already gave $T\mathbb{R}^n$ a smooth structure in Example 4.1.21(e).) Furthermore, for any open subset $\mathcal{O}$ of $\mathbb{R}^n$, if $T\mathcal{O} = \mathcal{O} \times \mathbf{R}^n = \pi^{-1}(\mathcal{O})$, then $\pi : T\mathcal{O} \longrightarrow \mathcal{O}$ is a smooth vector bundle. (An open subset of a smooth manifold is a smooth manifold.)

(2) More generally, let $\pi : E \longrightarrow B$ be a smooth vector bundle. Let $\mathcal{A}$ be an atlas as in Definition 5.1.1, and let $\mathcal{U} = \mathcal{U}_i$ be any open set in that atlas. Let $E_{\mathcal{U}} = \pi^{-1}(\mathcal{U})$. Then $\pi : E_{\mathcal{U}} \longrightarrow \mathcal{U}$ is a trivial bundle.   ◇

REMARK 5.1.7. If $\pi : E \longrightarrow B$ is equivalent to a trivial vector bundle, we will say that $\pi : E \longrightarrow B$ is *trivial* as well. We will call a particular choice of equivalence $F : E \longrightarrow B \times V$ a *trivialization* of $\pi : E \longrightarrow B$.   ◇

REMARK 5.1.8. We see from Example 5.1.6(2) that any vector bundle is a union of trivial bundles. But in general these trivial bundles fit together in nontrivial ways. In order to see how they fit together, we must investigate what happens when we change coordinate charts.

(i) First suppose we have a pair of coordinate charts $k_{i_1} : \mathcal{O}_{i_1} \longrightarrow \mathcal{U}_{i_1}$ and $k_{i_2} : \mathcal{O}_{i_2} \longrightarrow \mathcal{U}_{i_2}$ on B. Let $\mathcal{O}_{i_1 i_2} = k_{i_1}^{-1}(\mathcal{U}_{i_1} \cap \mathcal{U}_{i_2}) \subseteq \mathcal{O}_{i_1}$ and

$\mathcal{O}_{i_2 i_1} = k_{i_2}^{-1}(\mathcal{U}_{i_1} \cap \mathcal{U}_{i_2}) \subseteq \mathcal{O}_{i_2}$. Then we can consider $\ell_{i_2 i_1} = k_{i_2}^{-1} \circ k_{i_1}$: $\mathcal{O}_{i_1 i_2} \longrightarrow \mathcal{O}_{i_2 i_1}$, and $\tilde{\ell}_{i_2 i_1} = \tilde{k}_{i_2}^{-1} \circ \tilde{k}_{i_1} : \mathcal{O}_{i_1 i_2} \times V \longrightarrow \mathcal{O}_{i_2 i_1} \times V$.

For any point $b \in \mathcal{U}_{i_1} \cap \mathcal{U}_{i_2}$, let $b = k_{i_1}(a)$ and $b = k_{i_2}(a')$, so that $\ell_{i_2 i_1}(a) = a'$. In this case the restriction of $\tilde{\ell}_{i_2 i_1}$ maps $\{a\} \times V$ to $\{a'\} \times V$, i.e.,

$$\tilde{\ell}_{i_2 i_1}(a, v) = (a', v').$$

Restricting our attention to the second coordinate we see that, for each a, we have a map

$$L_{i_2 i_1}(a) : V \longrightarrow V \quad \text{defined by} \quad L_{i_2 i_1}(a)(v) = v'$$

and so we can rewrite $\tilde{\ell}_{i_2 i_1}(a, v)$ as

$$\tilde{\ell}_{i_2 i_1}(a, v) = (\ell_{i_2 i_1}(a), L_{i_2 i_1}(a)(v)).$$

On the one hand we may fix the point a. Then condition (iii) of Definition 5.1.1 implies that each map $L_{i_2 i_1}(a) : V \longrightarrow V$ is a vector space isomorphism.

On the other hand we may let a vary. Then we obtain a map

$$\mathcal{U}_{i_1} \cap \mathcal{U}_{i_2} \longrightarrow GL(V) \quad \text{defined by} \quad a \mapsto L_{i_2 i_1}(a)$$

and the condition that E be a smooth manifold implies that this is a smooth map. Here $GL(V)$ denotes the group of vector space isomorphisms from V to V, which is a smooth manifold. (By choosing a basis of V, we may identify $GL(V)$ with the group of invertible n-by-n matrices, where n is the dimension of V. This identification depends on the basis, but the smooth structure does not.)

(ii) Next suppose we have a triple of coordinate charts $k_{i_1} : \mathcal{O}_{i_1} \longrightarrow \mathcal{U}_{i_1}$, $k_{i_2} : \mathcal{O}_{i_2} \longrightarrow \mathcal{U}_{i_3}$, and $k_{i_3} : \mathcal{O}_{i_3} \longrightarrow \mathcal{U}_{i_3}$ on B. We have already seen in Corollary 4.1.29 that we may form $\ell_{i_2 i_1}$, $\ell_{i_3 i_2}$, and $\ell_{i_3 i_1}$ and that we have the compatibility conditions

$$\ell_{i_3 i_1} = \ell_{i_3 i_2} \circ \ell_{i_2 i_1}.$$

In a similar fashion we may form $\tilde{\ell}_{i_2 i_1}$, etc., $L_{i_2 i_1}$, etc., and we similarly have compatibility conditions

$$\tilde{\ell}_{i_3 i_1} = \tilde{\ell}_{i_3 i_2} \circ \tilde{\ell}_{i_2 i_1}, \quad L_{i_3 i_1} = L_{i_3 i_2} \circ L_{i_2 i_1}.$$

(Since, as maps, $\tilde{\ell}_{i_2 i_1} = (\ell_{i_2 i_1}, L_{i_2 i_1})$, etc., the compatibility condition for $\tilde{\ell}$ is equivalent to the pair of compatibility conditions for ℓ and L.) ◇

Definition 5.1.1 was "global" to "local," that is, we started with the global notion of a smooth vector bundle and obtained a collection of local objects, trivial bundles over open sets, that fit together well. We now give the "local" to "global construction," obtaining a smooth vector bundle from a compatible collection of trivial bundles. (Compare the two directions of Corollary 4.1.30 and Theorem 4.1.31.)

THEOREM 5.1.9. *Let V be a fixed vector space. Let $\{\mathcal{O}_i\}$ be a collection of open sets in $\mathbb{R}^n$. Suppose that for each ordered pair (i_1, i_2) there is an open subset (which may be empty) $\mathcal{O}_{i_1 i_2}$ of $\mathcal{O}_{i_1}$ and a smooth map $\tilde{\ell}_{i_2 i_1} : \mathcal{O}_{i_1 i_2} \times V \longrightarrow \mathcal{O}_{i_2 i_1} \times V$, where $\tilde{\ell}_{i_2 i_1} = (\ell_{i_2 i_1}, L_{i_2 i_1})$, with $\ell_{i_2 i_1}$ a diffeomorphism and $L_{i_2 i_1}$ a vector space isomorphism. Suppose further that the collection of maps $\{\tilde{\ell}_{i_2 i_1}\}$ satisfies the compatibility relations $\tilde{\ell}_{i_3 i_1} = \tilde{\ell}_{i_3 i_2} \circ \tilde{\ell}_{i_2 i_1}$, or equivalently that $\ell_{i_3 i_1} = \ell_{i_3 i_2} \circ \ell_{i_2 i_1}$ and $L_{i_3 i_1} = L_{i_3 i_2} \circ L_{i_2 i_1}$, whenever the composition is defined, for every triple (i_1, i_2, i_3).*

Let B be the space obtained by identifying points in $\{\mathcal{O}_i\}$ where $q_1 \in \mathcal{O}_1$ is identified with $q_2 \in \mathcal{O}_2$ if $q_2 = \ell_{i_2 i_1}(q_1)$, and let E be the space obtained by identifying points in $\{\mathcal{O}_i \times V\}$ where $(q_1, v_1) \in \mathcal{O}_1 \times V$ is identified with $(q_2, v_2) \in \mathcal{O}_2$ if $(q_2, v_2) = \tilde{\ell}_{i_2 i_1}(q_1, v_1)$.

Then there is a well-defined map $\pi : E \longrightarrow B$ defined by $\pi(p, v) = p$ where (p, v) is the image of any point (q_1, v_1) under this identification.

Suppose that B is a separable Hausdorff space. If $\mathcal{U}_i$ is the image of $\mathcal{O}_i$ under this identification, and $k_i : \mathcal{O}_i \longrightarrow \mathcal{U}_i$ is the identification map, then $\mathcal{A} = \{k_i : \mathcal{O}_i \longrightarrow \mathcal{U}_i\}$ is a smooth atlas on B, giving B the structure of a smooth manifold, $\tilde{\mathcal{A}} = \{\tilde{k}_i = k_i \times \mathrm{id} : \mathcal{O}_i \times V \longrightarrow \mathcal{U}_i \times V\}$ is a smooth atlas on E, giving E the structure of *a smooth manifold, and $\pi : E \longrightarrow B$ is a smooth vector bundle with fiber V. Also, $\tilde{\ell}_{i_2 i_1} = \tilde{k}_{i_2}^{-1} \circ \tilde{k}_{i_1}$ for every (i_1, i_2).*

Proof.   Analogous to the proof of Theorem 4.1.31. $\square$

REMARK 5.1.10.   We will be applying Theorem 5.1.9 in the following manner: We will start with a smooth manifold B. Then we will constructing vector bundles beginning with an atlas on B. Since B is a smooth manifold to begin with, the atlas will automatically have the property that when we perform the required identifications, the space we get will be separable Hausdorff. Furthermore, because of the topology on $\mathbf{R}^n$, once B is separable Hausdorff, E is automatically separable Hausdorff as well.   ◇

DEFINITION 5.1.11.   A smooth *section* of the smooth vector bundle $\pi : E \longrightarrow B$ is a smooth map $s : B \longrightarrow E$ with $\pi \circ s = \mathrm{id} : B \longrightarrow B$.

Equivalently, a smooth *section* of the smooth vector bundle $\pi : E \longrightarrow B$ is a smooth map $s : B \longrightarrow E$ with $s(b) \in V_b = \pi^{-1}(b)$ for every $b \in B$.   ◇

DEFINITION 5.1.12.   Let $\pi : E \longrightarrow B$ be a vector bundle. A section s of E is *nonzero* if for every $b \in B$, $s(b)$ is a nonzero element of the fiber $V_b = \pi^{-1}(b)$.

Sections $s_1, \ldots, s_k$ of E are *linearly independent* if for every $b \in B$, $\{s_1(b), \ldots, s_k(b)\}$ is a linearly independent subset of the fiber $V_b = \pi^{-1}(b)$.   ◇

Note that a single section s is linearly independent if and only if s is nonzero.

THEOREM 5.1.13.   *Let $\pi : E \longrightarrow B$ be a vector bundle with k-dimensional fiber V. Then $\pi : E \longrightarrow B$ is trivial if and only if there exist k linearly independent sections $s_1, \ldots, s_k$ of E. In particular, if $\pi : E \longrightarrow B$ is a vector bundle with 1-dimensional fiber V, then $\pi : E \longrightarrow B$ is trivial if and only if there exists a nonzero section s.*

Proof.   On the one hand, suppose E is trivial, so that we have an equivalence $F : E \longrightarrow B \times V$. Choose any basis $\{v_1, \ldots, v_k\}$

of V. If $s_i(b) = F^{-1}(b, v_i)$, $i = 1, \ldots, k$, then $s_1, \ldots, s_k$ are linearly independent sections of E.

On the other hand, suppose we have linearly independent sections $s_1, \ldots, s_k$ of E. Choose a basis $\{v_1, \ldots, v_k\}$ of V. Then we may define an equivalence $F : E \longrightarrow B \times V$ as follows: For $e \in E$ let $b = \pi(e)$, so that $e \in V_b$. Then there are unique $c_1, \ldots, c_k$ so that $e = c_1 s_1(b) + \cdots + c_k s_k(b)$. Let $F(e) = (b, c_1 v_1 + \cdots + c_k v_k)$. $\square$

Finally, we consider the question of orientation.

DEFINITION 5.1.14. Let $\pi : E \longrightarrow B$ be a smooth vector bundle with fiber V, as in Definition 5.1.1. An *orientation* of this vector bundle is obtained as follows: Choose an orientation of V. Then for each $b \in B$, the fiber V_b is given the orientation induced by the isomorphism $V \longrightarrow V_b$ of Definition 5.1.1(iii). An *orientable* bundle is one that has an orientation. $\diamond$

Note that there was a choice in Definition 5.1.14. The map $V \longrightarrow V_b$ is the map $v \mapsto k_i(a, v)$ and that depends on the choice of coordinate patch $k_i : \mathcal{O}_i \longrightarrow \mathcal{U}_i$. Thus, this definition makes sense, and the bundle is orientable, if and only if the orientation of V_b is independent of this choice, i.e., if whenever b is also in the image of some other coordinate patch $k_j : \mathcal{O}_j \longrightarrow \mathcal{U}_j$, then the map $v \mapsto k_j(a, v)$ induces the same orientation on V_b. We may rephrase this condition as follows.

THEOREM 5.1.15. *Let $\pi : E \longrightarrow B$ be a smooth vector bundle with fiber V, as in Definition 5.1.1. This bundle is orientable if and only if, in the language and notation of Remark 5.1.8(ii), the map*

$$\mathcal{U}_{i_1} \cap \mathcal{U}_{i_2} \longrightarrow GL(V) \; defined \; by \; a \mapsto L_{i_2 i_1}(a)$$

has its image in $GL_+(V)$, the subgroup of orientation-preserving vector space isomorphisms from V to V.

Proof. Choose an orientation of V.

Let b be any point in B, and suppose that $b \in \mathcal{U}_i \cap \mathcal{U}_j$. Let $k_i : \mathcal{O}_i \longrightarrow \mathcal{U}_i$ and $k_j : \mathcal{O}_j \longrightarrow \mathcal{U}_j$, and suppose that $b = k_i(a) = k_j(a')$. Then it is easy to check that the maps $v \mapsto k_i(a, v)$ and $v \mapsto k_j(a', v)$

induce the same orientation on V_b if and only if the map

$$L_{i_2 i_1}(a) : V \longrightarrow V$$

is orientation-preserving.   □

REMARK 5.1.16.

(1) We have observed that, by choosing a basis of V, we may identify $GL(V)$ with the group of invertible n-by-n matrices, where n is the dimension of V. Under this identification, $GL_+(V)$ is the subgroup of matrices with positive determinant. Again, this identification depends on the basis, but the question of whether the image is in $GL_+(V)$ does not.

(2) Note that $GL(V)$ is a smooth manifold (indeed, we may identify it with an open subset of $\mathbb{R}^{n^2}$, where n is the dimension of V). This manifold is disconnected. More precisely, it has two components, and $GL_+(V)$ is one of those components (the component containing the identity linear transformation).   ◊

REMARK 5.1.17. In the situation of Theorem 5.1.15, if B is connected and the bundle is orientable, an orientation is obtained by choosing an orientation of any fiber V_{b_0}. For this orientation determines an orientation of V, and then this orientation of V determines an orientation of V_b for every b. More generally, if B has more than one component, the bundle is orientable if its restriction to each component of B is orientable, and then an orientation is determined by a choice of orientation of the fiber over a single point in each component of B. Thus, if B has k components, an orientable bundle over B has 2^k orientations.   ◊

5.2 The dual bundle, and related bundles

Having defined a vector bundle, we now wish to define its dual. In order to do so, we must first consider the notion of the dual of a vector space.

DEFINITION 5.2.1. Let V be a vector space. Its *dual space* V^* is the vector space

$$V^* = \{\text{linear functions } f : V \longrightarrow \mathbb{R}\}. \qquad \diamond$$

REMARK 5.2.2. In the situation of Definition 5.2.1, the value $f(v) \in \mathbb{R}$. In this situation, we will often denote $f(v)$ by $\langle f, v \rangle$. $\quad \diamond$

If V is a finite-dimensional vector space, then V^* is a vector space of the same dimension as V. (Thus V^* is isomorphic to V, but it is important *not* to identify V^* with V as there is no natural way to do so.)

DEFINITION 5.2.3. Let V and W be vector spaces and let V^* and W^* be their dual spaces. Let $L : V \longrightarrow W$ be a linear transformation. Then the *dual* or *adjoint* of L is the linear transformation $L^* : W^* \longrightarrow V^*$ defined by

$$\langle L^*(g), v \rangle = \langle g, L(v) \rangle. \qquad \diamond$$

LEMMA 5.2.4.

(1) The adjoint id* *of the identity linear transformation* id $: V \longrightarrow V$ *is the identity linear transformation* id $: V^* \longrightarrow V^*$.

(2) If $L : V \longrightarrow W$ *and* $M : W \longrightarrow X$ *are linear transformations, then* $(M \circ L)^* = L^* \circ M^*$.

Proof. Routine (by now). $\qquad\qquad\qquad\qquad\qquad\qquad \square$

COROLLARY 5.2.5.

(1) If $L : V \longrightarrow W$ *is an isomorphism, then* $(L^{-1})^* = (L^*)^{-1}$.

(2) If $L : V \longrightarrow W$ *and* $M : W \longrightarrow X$ *are isomorphisms, then* $((M \circ L)^{-1})^* = (M^{-1})^* \circ (L^{-1})^*$.

Proof. For (1), note by Lemma 5.2.4 that

$$\text{id} = \text{id}^* = (L \circ L^{-1})^* = (L^{-1})^* \circ L^* \quad \text{and}$$
$$\text{id} = \text{id}^* = (L^{-1} \circ L)^* = L^* \circ (L^{-1})^*.$$

For (2), note by Lemma 5.2.4 and (1) that

$$((M \circ L)^{-1})^* = ((M \circ L)^*)^{-1} = (L^* \circ M^*)^{-1}$$
$$= (M^*)^{-1} \circ (L^*)^{-1} = (M^*)^{-1} \circ (L^*)^{-1}. \qquad \square$$

Suppose we have constructed a vector bundle with fiber V as in Theorem 5.1.9. It is very tempting to try to simply construct its dual bundle, with fiber V^*, the same way, simply replacing the map $L_{i_2 i_1}$ on each fiber by its dual $L^*_{i_2 i_1}$. But that won't work, because the dual map goes the wrong way. That is, suppose we replace the family of maps $\{\tilde{\ell}_{i_2 i_1} = (\ell_{i_2 i_1}, L_{i_2 i_1})\}$ by the family of maps $\{\tilde{m}_{i_2 i_1} = (\ell_{i_2 i_1}, L^*_{i_2 i_1})\}$. Then the compatibility condition $L_{i_3 i_1} = L_{i_3 i_2} \circ L_{i_2 i_1}$ for the original bundle does *not* translate into the compatibility condition $L^*_{i_3 i_1} = L^*_{i_3 i_2} \circ L^*_{i_2 i_1}$ for the dual bundle, but rather, from Lemma 5.2.4 (2), to $L^*_{i_3 i_1} = L^*_{i_2 i_1} \circ L^*_{i_3 i_2}$, which is wrong.

The fix for that is very simple, though. Namely, we note that each $L_{i_2 i_1}$ is an isomorphism, so we may instead consider its inverse $L^{-1}_{i_2 i_1}$. Then, by Corollary 5.2.5 (2), we *do* have the correct compatibility condition $\left(L^{-1}_{i_3 i_1}\right)^* = \left(L^{-1}_{i_3 i_2}\right)^* \circ \left(L^{-1}_{i_2 i_1}\right)^*$.

DEFINITION 5.2.6. Let $\{\tilde{\ell}_{i_2 i_1} = (\ell_{i_2 i_1}, L_{i_2 i_1})\}$ be as in Theorem 5.1.9, defining a smooth vector bundle $\pi : E \longrightarrow B$ with fiber V. Consider the new set of data $\{\tilde{\ell}^*_{i_2 i_1} = (\ell_{i_2 i_1}, (L^{-1}_{i_2 i_1})^*)\}$. Then the smooth vector bundle $\pi^* : E^* \longrightarrow B$ with fiber V^* defined by applying the construction in Theorem 4.4.9 to this new set of data is the *dual bundle* to $\pi : E \longrightarrow B$. $\diamond$

DEFINITION 5.2.7. Let V be a vector space. For a fixed positive integer k, let $\Lambda^k(V^*)$ be the vector space

$$\Lambda^k(V^*) = \{\text{alternating multilinear functions}$$
$$f : V \times \cdots \times V \longrightarrow \mathbb{R}\},$$

where there are k factors.

Let V and W be vector spaces and let $L : V \longrightarrow W$ be a linear transformation. Then $\Lambda^k(L^*) : \Lambda^k(W^*) \longrightarrow \Lambda^k(V^*)$ is the linear transformation defined by

$$(\Lambda^k(L^*)(f))(v_1, \ldots, v_k) = f(L(v_1), \ldots, L(v_k)). \qquad \diamond$$

DEFINITION 5.2.8. Let $\{\tilde{\ell}_{i_2 i_1} = (\ell_{i_2 i_1}, L_{i_2 i_1})\}$ be as in Theorem 5.1.9, defining a smooth vector bundle $\pi : E \longrightarrow B$ with fiber V. Consider the new set of data $\{\tilde{\ell}^*_{i_2 i_1} = (\ell_{i_2 i_1}, \Lambda^k((L^{-1}_{i_2 i_1})^*))\}$. Then for any positive integer k, the smooth vector bundle $\pi^* : E^* \longrightarrow B$ with fiber $\Lambda^k(V^*)$ defined by applying the construction in Theorem 5.1.9 to this new set of data is the kth *exterior power* of the dual bundle to $\pi : E \longrightarrow B$.   ◇

We extend this definition to $k = 0$ as follows.

DEFINITION 5.2.9. The 0th exterior power of any bundle $\pi :$ $E \longrightarrow B$ is the trivial bundle $\pi : B \times \mathbf{R} \longrightarrow B$.   ◇

This discussion has been rather abstract. But now let us be more concrete.

Note that in the following we use the notation $^t A$ rather than A^t for the transpose of a matrix A.

Also, we let ent be the function that picks out the entry of a 1-by-1 matrix, i.e., $\text{ent}([a]) = a$.

EXAMPLE 5.2.10. Let

$$V = \mathbf{R}^n = \left\{ v = \begin{bmatrix} a_1 \\ \vdots \\ a_n \end{bmatrix} \mid a_i \in \mathbb{R} \right\}.$$

Then

$$V^* = (\mathbf{R}^n)^* = \left\{ f = \begin{bmatrix} \alpha_1 \\ \vdots \\ \alpha_n \end{bmatrix} \mid \alpha_i \in \mathbb{R} \right\},$$

where

$$\langle f, v \rangle = f(v) = \text{ent}(^t f v) = \alpha_1 a_1 + \cdots + \alpha_n a_n.$$ ◇

LEMMA 5.2.11. *Let V and V^* be as in Example 5.2.10, and similarly for W and W^*. Let $L : V \longrightarrow W$ be the linear transformation with matrix A. Then $L^* : W^* \longrightarrow V^*$ is the linear transformation with matrix $^t A$.*

Proof. Let L^* have matrix B. Then on the one hand, for $v \in V$ and $g \in W^*$,

$$\langle g, L(v) \rangle = \langle g, Av \rangle = \text{ent}(^t g \, Av)$$

and on the other hand

$$\langle L^*(g), v \rangle = \langle Bg, v \rangle = \text{ent}(^t(Bg)v) = {}^t g \, {}^t B v.$$

By the definition of the adjoint, these must be equal for every g and v. Hence we must have

$$^t B = A \text{ and hence } B = {}^t A. \qquad \square$$

Recall that for any invertible matrix A, $(^t A)^{-1} = {}^t(A^{-1})$ (compare Corollary 5.2.5 (1)). We denote this common value by $^t A^{-1}$.

DEFINITION 5.2.12. For each of the $m = \binom{n}{k}$ k-element subsets $(j_1, \ldots, j_k)$ of $\{1, \ldots, n\}$, order $j_1, \ldots, j_k$ in lexicographic (i.e., increasing) order, and then order these subsets in lexicographic order. For example:

For $k = 0$ we have the single 0-tuple ().

For $k = 1$ we have the n 1-tuples $(1), \ldots, (n)$.

For $k = 2$ we have the $\binom{n}{2}$ 2-tuples $(1, 1), \ldots, (1, n)$, $(2, 1), \ldots, (2, n), \ldots, (n, n)$.

For $k = n$ we have the single n-tuple $(1, 2, \ldots, n)$.

Call these k-tuples $t_1, \ldots, t_m$.

Given an n-by-n matrix A, let $B = \Lambda^k (^t A^{-1})$ be the m-by-m matrix whose entry in position (r, s) is the determinant of the k-by-k submatrix of $^t A^{-1}$ formed by taking rows t_r and columns t_s (in that order). (In case $k = 0$, the determinant of the empty matrix is defined to be 1.)

In particular:

For $k = 0$, B is the 1-by-1 matrix $B = [1]$.

For $k = 1$, B is the n-by-n matrix $B = {}^t A^{-1}$.

For $k = n$, B is the 1-by-1 matrix $B = [\det({}^t A^{-1})]$.   ◇

DEFINITION 5.2.13. Let $\{\tilde{\ell}_{i_2 i_1} = (\ell_{i_2 i_1}, L_{i_2 i_1})\}$ be as in Theorem 5.1.9, defining a smooth vector bundle $\pi : E \longrightarrow B$ with fiber $\mathbf{R}^n$. Suppose that, for each point $p \in \mathcal{O}_{i_1 i_2}$ the linear transformation $L_{i_2 i_1}(p)$ is given by multiplication by the matrix $A_{i_2 i_1}(p)$.

Let $\pi^* : E^* \longrightarrow B$ be the bundle with fiber $\mathbf{R}^m$ defined by $\{\tilde{\ell}_{i_2 i_1} = (\ell_{i_2 i_1}, L^*_{i_2 i_1})\}$ where the linear transformation $L_{i_2 i_1}(p)^*$ is given by multiplication by the matrix $\Lambda^k({}^t A_{i_2 i_1}(p)^{-1})$ as in Definition 5.2.12.

Then $\pi^* : E^* \longrightarrow B$ is the kth exterior power of the dual bundle to $\pi : E \longrightarrow B$.   ◇

REMARK 5.2.14. There was a certain amount of choice involved in Definition 5.2.13. But if we had made a different choice, we would have obtained an equivalent bundle.

Similarly, if we had started with $V = \mathbf{R}^n$ and chosen a different basis of V, we would have obtained different matrices, but again an equivalent bundle.

More generally, we could apply this construction to a vector bundle with fiber an arbitrary vector space V of dimension n. The first step would be to choose a basis of V. By taking the coordinates of $v \in V$ in that basis, we obtain an isomorphism from V to $\mathbf{R}^n$, and then from any linear transformation $L : V \longrightarrow V$ we obtain the n-by-n matrix that is the matrix of L in that basis. Then we proceed as before. Again, the construction depends on our choice of basis, but different bases yield equivalent bundles.   ◇

We apply this to the question of orientability of bundles.

THEOREM 5.2.15. *Let M be a smooth manifold, and let $\pi : E \longrightarrow M$ be a vector bundle with fiber V of dimension n. Then $\pi : E \longrightarrow M$ is orientable if and only if the bundle $\pi^* : E^* \longrightarrow M$ with fiber $\Lambda^n(V^*)$ (of dimension 1) is trivial.*

Proof.   We may assume M is connected as otherwise we just work one component at a time.

First suppose that π^* is trivial. By Theorem 5.1.13, π^* has a nonzero section s. This determines an orientation of π as follows: Let $q \in M$. A choice of basis for V_q gives an orientation of V_q. Choose a basis $\mathcal{B} = \{v_1, \ldots, v_n\}$ with $s(v_1, \ldots, v_n) > 0$.

Now for the converse. Suppose that π is orientable. By Theorem 5.1.13, in order to show that π^* is trivial we need only show that π^* has a nonzero section. Choose an orientation of π.

We may assume (taking refinements as necessary) that we have an atlas on M satisfying the conditions of both Theorem 4.1.31 and Theorem 5.1.9 for the bundle π and hence for the bundle π^* as well. Choose a partition of unity $\{\varepsilon_j : M \longrightarrow \mathbb{R}\}$ subordinate to this atlas.

Since the restriction of π to each $\mathcal{U}_i$ is trivial, we may certainly choose a section s_i of π^* on $\mathcal{U}_i$ so that for each point $q \in \mathcal{U}_i$, if $\mathcal{B} = \{v^1, \ldots, v^n\}$ is a basis of V_q giving the chosen orientation of V_q, $s_i(v^1, \ldots, v^n) > 0$. Then, for each j, the section $\varepsilon_j s_i$ extends to a smooth section of π^* over all of M, this section being identically 0 outside of some coordinate patch. Thus if we let

$$s = \sum_{j,i} \varepsilon_j s_i,$$

then s is a section of π^* on M. It remains to show that s is a nonzero section. But, since π is orientable, if $q \in \mathcal{U}_{i_1} \cap \mathcal{U}_{i_2}$, and $\mathcal{B}$ is as above, then $s_{i_1}(v^1, \ldots, v^n)$ and $s_{i_2}(v^1, \ldots, v^n)$ are both positive; hence each term $\varepsilon_j s_i(v^1, \ldots, v^n)$ is nonnegative, and, since they are not all 0, $s(v^1, \ldots, v^n)$ is always positive. $\square$

COROLLARY 5.2.16.   *Let M be a smooth manifold, and let π : $E \longrightarrow M$ be an orientable vector bundle with fiber V of dimension n. Then, by Theorem 5.2.15, an orientation of π is given by a choice of a nonzero section s of $\Lambda^n(V^*)$. Two nonzero sections s_1 and s_2 of $\Lambda^n(V^*)$ give the same orientation (resp. opposite orientations) of π if $s_2 = t s_1$ for some function $t : M \longrightarrow \{positive\ real\ numbers\}$ (resp. $t : M \longrightarrow \{negative\ real\ numbers\})$.*

Proof.   Immediate from the proof of Theorem 5.2.15.   □

5.3 The tangent bundle of a smooth manifold, and related bundles

Now we specialize our constructions to the situation we are interested in, that of the tangent bundle. We have done all the work, so we can now simply reap the fruits of our labor.

DEFINITION 5.3.1.   Let $V = \mathbf{R}^n$. Let M be a smooth n-manifold and let $\mathcal{A} = \{k_i : \mathcal{O}_i \longrightarrow \mathcal{U}_i\}$ be a smooth atlas on M. For any ordered pair (i_1, i_2) let $\mathcal{O}_{i_1 i_2} = k_{i_1}^{-1}(\mathcal{U}_{i_1} \cap \mathcal{U}_{i_2}) \subseteq \mathcal{O}_{i_1}$, and let $\ell_{i_2 i_1} = k_{i_2}^{-1} \circ k_{i_1} : \mathcal{O}_{i_1 i_2} \longrightarrow \mathcal{O}_{i_2 i_1}$. Define

$$\tilde{\ell}_{i_2 i_1} : T\mathcal{O}_{i_1 i_2} = \mathcal{O}_{i_1 i_2} \times V \longrightarrow T\mathcal{O}_{i_2 i_1} = \mathcal{O}_{i_2 i_1} \times V$$

by

$$\tilde{\ell}_{i_2 i_1} = (\ell_{i_2 i_1}, L_{i_2 i_1}), \quad \text{where } L_{i_2 i_1} = \ell'_{i_2 i_1}, \text{ the derivative of } \ell_{i_2 i_1}.$$

Then $\pi : TM \longrightarrow M$, the bundle obtained by the construction of Theorem 5.1.9, is the *tangent bundle* of M.

(Note that $\{\ell_{i_2 i_1}\}$ satisfies the compatibility condition in Theorem 5.1.9 as $\mathcal{A}$ is an atlas, and that $\{L_{i_2 i_1}\}$ satisfies the compatibility condition by the chain rule.)   ◇

REMARK 5.3.2.   Note that this agrees with our previous constructions, in that we have already seen that the map induced on tangent vectors by a smooth map f is the linear map f'.   ◇

Recall we defined a section of a vector bundle in the last section.

EXAMPLE 5.3.3.   A smooth vector field on a smooth manifold M is a smooth section of the tangent bundle TM of M.   ◇

DEFINITION 5.3.4.   Let M be a smooth manifold. The *cotangent bundle* T^*M is the dual bundle to the tangent bundle TM.   ◇

DEFINITION 5.3.5. A 1-form on M is a smooth section of the cotangent bundle T^*M.   ◇

Having obtained 1-forms in this manner, we now obtain k-forms by a straightforward generalization.

DEFINITION 5.3.6. Let M be a smooth manifold. For any nonnegative integer k, the bundle $\Lambda^k(T^*M)$ is the kth *exterior power of the cotangent bundle T^*M*.   ◇

We can (and shall) be quite specific about this definition, since it is so important for us.

COROLLARY 5.3.7. *Let M be a smooth n-manifold. The bundle defined by $\{\tilde{\ell}_{i_2 i_1} = (\ell_{i_2 i_1}, L^*_{i_2 i_1})\}$ as in Definition 5.2.8, where $L_{i_2 i_1}(p) = \ell'_{i_2 i_1}(p)$, is $\Lambda^k(T^*M)$, the kth exterior power of the cotangent bundle of M.*

DEFINITION 5.3.8. A k-form on M is a smooth section of the bundle $\Lambda^k(T^*M)$.   ◇

REMARK 5.3.9. We have previously defined a 0-form on M to be a smooth function on M. That agrees with Definition 5.2.9, where we identify the smooth function $f : M \longrightarrow \mathbb{R}$ with the section $s : M \longrightarrow M \times \mathbf{R}$ given by $s(x) = (x, [f(x)])$.

(We are being very precise in this remark by writing $[f(x)]$ rather than $f(x)$, remembering that $[f(x)]$ is really a vector with a single entry.)   ◇

Finally, we come to the notion of an orientation of a smooth manifold.

DEFINITION 5.3.10. Let M be a smooth manifold. M is *orientable* if its tangent bundle TM is orientable. In this case, an *orientation* of M is an orientation of TM.   ◇

COROLLARY 5.3.11. *Let M be a smooth n-manifold. Then M is orientable if and only if the bundle $\Lambda^n(T^*M)$ is trivial.*

Proof.   Immediate from Definition 5.3.10 and Theorem 5.2.15.   □

5.4 Exercises

1. Carry out the proof of Theorem 5.1.9.

2. Let E be a smooth vector bundle over the smooth manifold M. Let A be a smooth submanifold of M. Show that $E|A$, the restriction of E to A, is a smooth vector bundle over A.

3. Let $\pi : E \longrightarrow N$ be a smooth vector bundle over the smooth manifold N with fiber V. Let M be a smooth manifold and let $f : M \longrightarrow N$ be a smooth map. Let

$$f^*(E) = \{(m, e) \in M \times E \mid f(m) = \pi(e)\}.$$

Let $\pi_1 : f^*(E) \longrightarrow M$ be projection onto the first coordinate, $\pi_1(m, e) = m$. Show that $\pi_1 : f^*(E) \longrightarrow M$ is a smooth vector bundle over the smooth manifold N with fiber V. $f^*(E)$ is the *pull-back* of E via f.

4. Show that a vector bundle is orientable if and only if its dual bundle is orientable.

5. Let M be an n-manifold. Show that its tangent bundle TM is an orientable $2n$-manifold.

6. Let S^1 be the unit sphere in $\mathbb{R}^2$. Let

$$T = \{(x, y) \in S^1 \mid y > -1/2\} \text{ and } B = \{(x, y) \in S^1 \mid y < 1/2\}.$$

For $j = 0, 1$ define $\ell^j : T \cap B \longrightarrow \{[\pm 1] \subset GL_1(\mathbb{R})\}$ by

$$\ell^j((x, y)) = 1 \text{ if } x > 0 \text{ and } \ell^j((x, y)) = (-1)^j \text{ if } x < 0.$$

These define 1-dimensional vector bundles ξ^j over S^1 as in Theorem 5.1.9.

(a) Show that ξ^0 is a trivial bundle but ξ^1 is not a trivial bundle.

(b) Show that every 1-dimensional vector bundle over S^1 is equivalent to ξ^0 or ξ^1.

7. Suppose we are given two sets of data as in Theorem 5.1.9, one for a vector space V' and one for a vector space V'', where the maps for V' are denoted by primes and the maps for V'' are denoted by double primes, defining vector bundles E' and E'' over B respectively. Let $V = V' \oplus V''$ and, in the notation of that theorem, $L_{i_2 i_1} = \left(L'_{i_2 i_1}, L''_{i_2 i_1} \right)$.

(a) Show that $\{\tilde{\ell}_{i_2 i_1}\}$ defines a vector bundle E over B. This vector bundle $E = E' \oplus E''$ is called the *direct sum* of the vector bundles E' and E''.

(b) Show that if any two of E', E'', and E are orientable, so is the third.

8. Show that any n-dimensional vector bundle over S^1 is equivalent to $\xi^0 \oplus \tau^{n-1} = \tau^n$ or to $\xi^1 \oplus \tau^{n-1}$, but that these two vector bundles are not equivalent to each other. Here τ denotes a trivial bundle and τ^k is $\tau \oplus \cdots \oplus \tau$ where there are k summands.

9. Let $S^k \subset \mathbb{R}^{k+1} \subseteq \mathbb{R}^n$. Show that $T\mathbb{R}^n | S^k$ is equivalent to $TS^k \oplus \tau^{n-k}$ where $T\mathbb{R}^n | S^k$ denotes the restriction of $T\mathbb{R}^n$ to S^k and τ denotes a trivial bundle.

10. Let $\pi_V : E_V \longrightarrow B$ be a smooth vector bundle with fiber V. Suppose that there is a subspace W of V such that, in the notation of Remark 5.1.8(ii), every linear transformation $L_{i_2 i_1}(a) : V \longrightarrow V$ preserves the subspace W. Let $L^0_{i_2 i_1}(a)$ denote the restriction of this map to W. Show that $\{L^0_{i_2 i_1}(a)\}$ define a smooth vector bundle $\pi_W : E_W \longrightarrow B$ with fiber W, as in Theorem 5.1.9. The bundle E_W is called a *subbundle* of E_V.

Let $\pi_V : E_V \longrightarrow B$ be a smooth vector bundle with fiber $V = \mathbb{R}^n$. Suppose that, in the notation of Remark 5.1.8(ii), every linear transformation $L_{i_2 i_1}(a) : V \longrightarrow V$ has its image in $O_n \subset GL_n(\mathbb{R})$. (Recall that O_n is the subgroup of $GL_n(\mathbb{R})$ that preserves the length of vectors and the angle between nonzero vectors.) It is a theorem, which we will not prove here, that in fact every vector bundle is equivalent to such a bundle. Observe that in this situation, the length of a vector in any fiber, and the angle between two nonzero vectors in any fiber, is well-defined.

11. In this situation, let $D = \{\mathbf{v} \in E \mid s\|\mathbf{v}\| \leq 1\}$ and let $S = \{\mathbf{v} \in E \mid \|\mathbf{v}\| = 1\}$. Show that D is a smooth manifold with boundary S. D (resp. S) is called the *unit disk bundle* (resp. *unit sphere bundle*) of E.

12. In this situation, let $\pi_W : E_W \longrightarrow B$ be a smooth subbundle of $\pi_V : E_V \longrightarrow B$. Let $W^\perp$ be the orthogonal complement of W in V.

(a) Show that there is a smooth subbundle $\pi_{E^\perp} : E_{W^\perp} \longrightarrow B$ of $\pi_V : E_V \longrightarrow B$.

(b) Show that $E_V = E_W \oplus E_{W^\perp}$.

6 Integration of Differential Forms

In this chapter we will see how to find

$$\int_M \varphi,$$

the integral over an arbitrary oriented n-manifold M with boundary of the n-form φ defined on M.

We will take for granted the theory of ordinary definite integrals over subsets of $\mathbb{R}^n$, and we will be reducing our integrals to these. But we begin by reviewing this theory, especially because there are some fine points we need to attend to.

While our theory of integration of forms works for any oriented manifold, we will illustrate this theory by looking at submanifolds of Euclidean space. This is not a restriction of the theory, but rather a way of making it concrete.

Let us begin by pointing out a potential notational confusion, and how we will go about avoiding it. Let M be an oriented n-manifold with boundary in $\mathbb{R}^n$, and consider

$$\int_M f\,dx_1 \cdots dx_n.$$

There are two things we might mean by this:

(1) The integral of the n-form $f\,dx_1 \cdots dx_n$ over the n-manifold with boundary M.

(2) The ordinary definite integral of the function f over the subset M of $\mathbb{R}^n$.

In order to keep these two meanings straight we will use two distinct notations. The above notation will always have meaning (1). For

Differential Forms, Second Edition. http://dx.doi.org/10.1016/B978-0-12-394403-0.00006-2

meaning (2), we will write

$$\int_M f \, dV,$$

where the symbol dV denotes us that we are integrating the function f with respect to the volume form in $\mathbb{R}^n$. It will often be the case that we need to keep track of the parameterization of $\mathbb{R}^n$—indeed, it will often be crucial to do so—and in that case we will write $dV_{x_1 \ldots x_n}$ when $\mathbb{R}^n = \{(x_1, \ldots, x_n)\}$.

"Volume" is a word we often associate with $\mathbb{R}^3$, with "length" and "area" being the corresponding concepts in $\mathbb{R}^1$ and $\mathbb{R}^2$. To help with these intuitive associations, in doing concrete examples we will often use $d\ell_x$, dA_{xy}, and dV_{xyz} to denote the length, area, and volume forms in $\mathbb{R}^1, \mathbb{R}^2$, and $\mathbb{R}^3$ respectively, when we use the usual parameterizations (and different subscripts when we use different parameterizations).

6.1   Definite integrals in $\mathbb{R}^n$

In this section we consider the theory of definite integrals over subsets of $\mathbb{R}^n$. We assume $n \geq 1$. (As usual, $n = 0$ is a special case that, on the one hand, is easier, but, on the other hand, must be handled separately.) Let M be a smooth n-manifold with boundary in $\mathbb{R}^n$ and let f be a smooth function defined on M. We assume without further ado that the reader is familiar with the integral

$$\int_M f \, dV.$$

(Of course, this integral is defined more generally, but this is the only case we will need.) In particular, we recall that $\mathrm{Vol}_n(M)$, the n-dimensional volume of M, is defined by

$$\mathrm{Vol}_n(M) = \int_M dV.$$

However, in this section we review some of the fine points associated with these integrals.

To begin with, consider a brick B in $\mathbb{R}^n$, as in Definition 4.1.46. Of course, $\text{Vol}_n(B) = (b_1 - a_1) \cdots (b_n - a_n)$.

We recall Fubini's theorem, which tells us how to evaluate definite integrals over bricks. We simply state this theorem for a brick in $\mathbb{R}^2$, for notational convenience, but it applies without change to a brick in $\mathbb{R}^n$ for general n.

THEOREM 6.1.1. *(Fubini's theorem). Let $f(x_1, x_2)$ be a smooth function on the brick B in $\mathbb{R}^2$ as in Definition 4.1.46. Let $I_1 = [a_1, b_1]$ and $I_2 = [a_2, b_2]$, intervals in $\mathbb{R}$. Then*

$$\int_B f(x_1, x_2) dA_{x_1, x_2} = \int_{I_2} \left[\int_{I_1} f(x_1, x_2) d\ell_{x_1} \right] d\ell_{x_2}$$
$$= \int_{I_1} \left[\int_{I_2} f(x_1, x_2) d\ell_{x_2} \right] d\ell_{x_1}.$$

REMARK 6.1.2. Thus Fubini's theorem states that we may evaluate a multiple integral as an iterated integral, and furthermore that the order in which we do the iteration is immaterial.   ◇

REMARK 6.1.3. To make the reader feel more comfortable, we recall a very familiar notation, which we shall use. As usual,

$$\int_a^b f(x) dx$$

denotes the ordinary definite integral of the function $f(x)$ from $x = a$ to $x = b$. (We can distinguish this from the integral of the 1-form $f(x) dx$ as we have written both upper and lower limits for the integral.)   ◇

Next we recall the change-of-variable formula for multiple integrals, a formula that is absolutely *crucial* for us.

THEOREM 6.1.4. *Let N be a compact n-dimensional manifold with boundary in $\mathbb{R}^n$, with $\mathbb{R}^n = \{(x_1, \ldots, x_n)\}$. Let M be a compact n-dimensional manifold with boundary in $\mathbb{R}^n = \{(u_1, \ldots, u_n)\}$. Let $k \colon M \longrightarrow N$ be a diffeomorphism. Let $f(x_1, \ldots, x_n)$ be a smooth* *function on* N. *Then*

$$\int_N f(x_1, \ldots, x_n) dV_{x_1, \ldots, x_n}$$

$$= \int_M f(k(u_1, \ldots, u_n)) |\det(k'(u_1, \ldots, u_n))| dV_{u_1, \ldots, u_n}.$$

REMARK 6.1.5. We recall that k' is simply the derivative of k, which is an n-by-n matrix, as in Remark 3.2.4. Classically, this matrix is called the Jacobian matrix of k and its determinant is called the Jacobian determinant of k.   ◇

REMARK 6.1.6. Note carefully that the absolute value $|\det(k')|$, and not just $\det(k')$, appears in Theorem 6.1.4.   ◇

The appearance of the absolute value sign in Theorem 6.1.4 makes this result *very* unnatural. Thus we introduce a new definition in order for us to obtain a natural formula.

DEFINITION 6.1.7. Let M be a compact oriented n-dimensional manifold with boundary in $\mathbb{R}^n$. Set

$\varepsilon_M = 1$ if M has the standard orientation of a region in $\mathbb{R}^n$,

$ = -1$ if M has the nonstandard orientation of a region in $\mathbb{R}^n$.

Let $f(x_1, \ldots, x_n)$ be a smooth function on M. The *signed definite integral* of f over M is defined by

$$\int_{s M} f dV = \varepsilon_M \int_M f dV.$$ ◇

With this definition we have the following natural version of Theorem 6.1.4.

THEOREM 6.1.8. *Let* N *be a compact oriented* n-*dimensional manifold with boundary in* $\mathbb{R}^n$, *with* $\mathbb{R}^n = \{(x_1, \ldots, x_n)\}$. *Let* M *be a compact oriented* n-*dimensional manifold with boundary in* $\mathbb{R}^n = \{(u_1, \ldots, u_n)\}$. *Let* $k \colon M \longrightarrow N$ *be a diffeomorphism. Set*

$$\varepsilon_k = 1 \text{ if } k \text{ is orientation-preserving,}$$
$$= -1 \text{ if } k \text{ is orientation-reversing.}$$

Let $f(x_1, \ldots, x_n)$ be a smooth function on N. Then

$$\int_s \int_N f(x_1, \ldots, x_n) dV_{x_1, \ldots, x_n}$$

$$= \varepsilon_k \int_s \int_M f(k(u_1, \ldots, u_n)) \det(k'(u_1, \ldots, u_n)) dV_{u_1, \ldots, u_n}.$$

Proof. Given Theorem 6.1.4, we merely have to check that the signs work. We do this in the case that k is orientation-preserving.

If M and N both have the standard orientations, then k orientation-preserving implies that $\det(k')$ is positive, and this reduces to Theorem 6.1.4.

If M and N both have the nonstandard orientations, then k orientation-preserving again implies that $\det(k')$ is positive, and this again reduces to Theorem 6.1.4, as in both cases the signed definite integral is the negative of the ordinary definite integral.

If one of M and N has the standard orientation and the other has the nonstandard orientation, then k orientation-preserving implies that $\det(k')$ is negative, and for one of M and N the signed definite integral is equal to the ordinary definite integral while for the other it is equal to the negative of the ordinary definite integral, so this again reduces to Theorem 6.1.4.

In the case that k is orientation-reversing, the signs (of both ε and $\det(k')$) reverse. $\square$

REMARK 6.1.9. In fact, in $\mathbb{R}^1$, the definite integral with its usual notation *is* a signed definite integral. Let $I = [a, b]$ be an interval in $\mathbb{R}^1$. Recall that the standard orientation on I is "increasing" and the nonstandard orientation on I is "decreasing." Thus if I has the standard orientation, $x \in I$ "goes from" a to b, while if it has the nonstandard orientation $x \in I$ "goes from" b to a. But note that, for any smooth function $f(x)$ on I,

$$\int_a^b f(x) d\ell_x = \int_s \int_I f(x) d\ell_x \text{ if } I \text{ has the standard orientation,}$$

$$\int_b^a f(x) d\ell_x = -\int_a^b f(x) d\ell_x = \int_s \int_I f(x) d\ell_x \text{ if } I \text{ has the}$$
$$\text{nonstandard orientation.} \qquad \diamond$$

REMARK 6.1.10. We should also remind the reader of the change-of-variable formula for integrals in $\mathbb{R}$. Let $J = [c, d]$ be a closed interval in $\mathbb{R} = \{x\}$, and let $I = [a, b]$ be a closed interval in $\mathbb{R} = \{u\}$. Let $k: I \longrightarrow J$ be a diffeomorphism. Then for any smooth function $f(x)$ on J,

$$\int_c^d f(x)dx = \int_{k^{-1}(c)}^{k^{-1}(d)} f(k(u))k'(u)du.$$

The key point to note is that in this formula we have the natural $k'(u)$ rather than the unnatural $|k'(u)|$. Why is that? Remark 6.1.9 gives us the answer. It is because in this case the definite integral written with limits is a signed definite integral.

There are two cases to consider.

In the first case, $k: I \longrightarrow J$ is orientation-preserving, where I and J have their standard orientations ("increasing"). In this case, $k^{-1}(c) = a$ and $k^{-1}(d) = b$, and $k'(u)$ is always positive, so $|k'(u)| = k'(u)$. Then this formula immediately translates to

$$\int_{sJ} f(x)d\ell_x = \int_{sI} f(u)k'(u)d\ell_u.$$

In the second case, $k: I \longrightarrow J$ is orientation-reversing, where I and J have their standard orientations ("increasing"). In this case, $k^{-1}(c) = b$ and $k^{-1}(d) = a$, and $k'(u)$ is always negative, so $|k'(u)| = -k'(u)$. Now the definite integral on the right-hand side is from b to a rather than from a to b. Again, interchanging the limits of a definite integral changes the sign of the integral. Thus this formula again translates to

$$\int_{sJ} f(x)d\ell_x = \int_{sI} f(u)k'(u)d\ell_u. \qquad \diamond$$

REMARK 6.1.11. Of course, if M is a compact n-manifold with boundary in $\mathbb{R}^n$, and f is any smooth function on M,

$$\int_M fdV$$

is defined. But if not, the integral may not be defined. This problem already manifests itself in the case of the constant function $f(x_1, \ldots, x_n) = 1$. Then the above integral becomes

$$\int_M dV = \mathrm{Vol}_n(M)$$

so that if M has infinite volume (e.g., $M = \mathbb{R}^n$) then the integral is not defined.

But there is one situation where this problem cannot arise. Regardless of whether M is compact, if f is a function defined on M with compact support, i.e., $f(x_1, \ldots, x_n) = 0$ for $(x_1, \ldots, x_n)$ outside of a compact set C, then

$$\int_M f\, dV = \int_C f\, dV$$

is defined. ◇

REMARK 6.1.12. We will almost exclusively be concentrating on the case M compact, or at least f with compact support. But we will remark that even if f does not have compact support, it is sometimes, but not always, possible to define the integral. The definition here mimics the familiar definition of "improper integrals." Let M be as above and suppose we have a sequence of compact submanifolds with boundary $\{D_i\}$ with $\cup_i D_i = M$. Let $z_i = \int_{D_i} f\, dV$. If $\{z_i\}$ converge to a limit z for each $\{D_i\}$, and furthermore that limit is the same for all $\{D_i\}$, then we may define $\int_M f\, dV$ by

$$\int_M f\, dV = z.$$ ◇

6.2 Definition of the integral in general

Let M be an oriented manifold, possibly with boundary, and let φ be a smooth n-form on M. In this section we define, under the appropriate conditions,

$$\int_M \varphi.$$

We proceed step by step. As usual, the first step is to define the integral when M is a submanifold of $\mathbb{R}^n$, for some n. Then we define the integral for a smooth manifold M that is diffeomorphic to a submanifold of $\mathbb{R}^n$, using the diffeomorphism to pull back to this submanifold. Finally, we define the integral for general M, using an atlas on M and pulling back to submanifolds of $\mathbb{R}^n$. We assume throughout that $n \geq 1$.

DEFINITION 6.2.1. Let $n \geq 1$. Let M be a compact oriented n-manifold with boundary in $\mathbb{R}^n$ and let φ be an arbitrary smooth n-form on M, or, more generally, let M be an oriented n-manifold with boundary in $\mathbb{R}^n$ and let φ be a smooth n-form on M with compact support. Express φ as $\varphi = f(x_1, \ldots, x_n)dx_1 \ldots dx_n$.
Then the *integral* of φ over M is the signed definite integral

$$\int_M \varphi = {}_s\!\int_M \varphi(\mathbf{e}^1, \ldots, \mathbf{e}^n)dV_{x_1,\ldots,x_n}$$

as in Definition 6.1.7. ◇

The very first thing we do is to get a concrete formula for this integral.

LEMMA 6.2.2. *In the situation of Definition 6.2.1, express φ as $\varphi = f(x_1, \ldots, x_n)dx_1 \cdots dx_n$. Then*

$$\int_M \varphi = {}_s\!\int_M f(x_1, \ldots, x_n)dV_{x_1,\ldots,x_n}.$$

Proof. If $\varphi = f(x_1, \ldots, x_n)dx_1 \cdots dx_n$ then $\varphi(\mathbf{e}^1, \ldots, \mathbf{e}^n) = f(x_1, \ldots, x_n)$. □

Now we have a theorem about how the integral behaves under diffeomorphisms.

THEOREM 6.2.3. *Let M and N be oriented n-manifolds with boundary in $\mathbb{R}^n$ and let $k: M \longrightarrow N$ be a diffeomorphism. Set*

$$\varepsilon_k = 1 \text{ if } k \text{ is orientation-preserving,}$$
$$= -1 \text{ if } k \text{ is orientation-reversing.}$$

Let φ be a smooth n-form on N with compact support. Then

$$\int_M k^*(\varphi) = \varepsilon_k \int_N \varphi.$$

Proof. Let $M \subseteq \mathbb{R}^n = \{(u_1, \ldots, u_n)\}$ and $N \subseteq \mathbb{R}^n = \{(x_1, \ldots, x_n)\}$. Write $\varphi = f\,dx_1 \cdots dx_n$ where $f = f(x_1, \ldots, x_n)$. Then, by Corollary 3.3.21,

$$k^*(\varphi) = f(k)\det(k')du_1 \cdots du_n.$$

Now by Lemma 6.2.2,

$$\int_N \varphi = {}_s\!\int_N f(x_1, \ldots, x_n)dV_{x_1,\ldots,x_n}$$

and

$$\int_M k^*(\varphi) = {}_s\!\int_M f(k(u_1, \ldots, u_n))\det(k'(u_1, \ldots, u_n))dV_{u_1,\ldots,u_n}.$$

But, by Theorem 6.1.8, these are equal.   □

REMARK 6.2.4. On the one hand, the proof of Theorem 6.2.3 was almost trivial, because we carefully set up everything in advance. But, on the other hand, this theorem will be *crucial* for future developments.   ◇

DEFINITION 6.2.5. Let N be an oriented n-manifold with boundary and let φ be a smooth n-form on N with compact support. Suppose that there is an orientation preserving diffeomorphism $k\colon M \longrightarrow N$ from some smooth submanifold with boundary M of $\mathbb{R}^n$ to N. Then the *integral* of φ over N is

$$\int_N \varphi = \int_M k^*(\varphi).$$ ◇

Again, let us obtain a concrete expression (in fact, two concrete expressions) for this integral.

LEMMA 6.2.6. *In the situation of Definition 6.2.5,*

$$\int_N \varphi = \int_M \left(k^*(\varphi) \right)(\mathbf{e}^1, \ldots, \mathbf{e}^n) dV_{x_1, \ldots, x_n}$$

$$= \int_M \varphi((k_*(\mathbf{e}^1), \ldots, k_*(\mathbf{e}^n)) dV_{x_1, \ldots, x_n}.$$

Proof. The first equality is simply Definition 6.2.1, and the second equality is simply the definition of the pullback $k^*(\varphi)$.   □

As usual, we need to see that Definition 6.2.5 makes sense, i.e., that the integral only depends on N and not on the choice of M or k.

LEMMA 6.2.7. *The integral $\int_N \varphi$ as in Definition 6.2.5 is well defined.*

Proof. In the situation of Definition 6.2.5, let $k_1 \colon M_1 \longrightarrow N$ and $k_2 \colon M_2 \longrightarrow N$ be orientation preserving diffeomorphisms. Then we must show that

$$\int_{M_1} (k_1)^*(\varphi) = \int_{M_2} (k_2)^*(\varphi).$$

Let $\ell \colon (k_2)^{-1} \circ (k_1)$. Then $\ell \colon M_1 \longrightarrow M_2$ is an orientation preserving diffeomorphism. Let $\psi = (k_2)^*(\varphi)$. Observe that

$$\begin{aligned}
\ell^*(\psi) &= \ell^*((k_2)^*(\varphi)) \\
&= ((k_2)^{-1}(k_1))^*((k_2)^*(\varphi)) \\
&= ((k_1)^*((k_2)^{-1})^*)((k_2)^*(\varphi)) \\
&= ((k_1)^*((k_2)^*)^{-1})((k_2)^*(\varphi)) \\
&= ((k_1)^*)(((k_2)^*)^{-1}(k_2)^*)))(\varphi) \\
&= (k_1)^*(\varphi).
\end{aligned}$$

Then, from Theorem 6.2.3, we know that

$$\int_{M_1} \ell^*(\psi) = \int_{M_2} \psi,$$

i.e.,

$$\int_{M_1} (k_1)^*(\varphi) = \int_{M_2} (k_2)^*(\varphi),$$

as claimed.   □

Now we come to the general definition.

DEFINITION 6.2.8. Let M be a smooth oriented n-manifold with boundary. Let $\mathcal{A} = \{k_i : \mathcal{O}_i \longrightarrow \mathcal{U}_i\}$ be an atlas giving the orientation of M, as in Theorem 4.5.21. Let $\{\varepsilon_j : M \longrightarrow \mathbb{R}\}$ be a partition of unity subordinate to $\{\mathcal{U}_i\}$, as in Definition 4.6.2, and for each j choose $i = i(j)$ with $\text{supp}(\varepsilon_j) \subset \mathcal{U}_i$.

Let φ be an n-form on M. Let $\varphi_j = \varepsilon_j \varphi$ (so that $\varphi_j = 0$ outside of $\mathcal{U}_i$ and $\varphi = \sum_j \varphi_j$). Then the *integral* of φ over M is

$$\int_M \varphi = \sum_j \int_{\mathcal{O}_i} (k_i)^*(\varphi_j). \qquad \diamond$$

THEOREM 6.2.9. *The integral $\int_M \varphi$ as in Definition 6.2.8 is well defined.*

Proof. We shall sketch the proof, as writing it out is rather messy.

Again we must show the value of the integral is independent of the choice of atlas and partition of unity. Note that if we keep the same open sets in the atlas and vary the maps $\{k_i\}$, this is true by Lemma 6.2.7. If we keep the open sets in the atlas the same and vary the partition of unity, this is still true, as if we have partitions of unity $\{\varepsilon_j^1\}$ and $\{\varepsilon_{j'}^2\}$, then for any j', $\text{supp}(\varepsilon_j^1 \varepsilon_{j'}^2) \subseteq \text{supp}(\varepsilon_j^1) \subset \mathcal{U}_i$. If $\mathcal{A}'$ is a refinement of $\mathcal{A}$ (i.e., if every $\mathcal{U}_{i'}' \subseteq \mathcal{U}_i$ for some i), this is again true. But then if $\mathcal{A}'$ is any atlas, then $\mathcal{A}$ and $\mathcal{A}'$ have a common refinement $\mathcal{A}''$, and the result is still true.   □

REMARK 6.2.10. Note that in Definition 6.2.8 it was *absolutely necessary* that N be orientable. For if N were nonorientable, there would be some pair of charts such that $\ell = k_2^{-1} k_1$ was orientation reversing. Then, referring to the proof of Lemma 6.2.7, Theorem 6.2.3

would *not* give us what we need, namely that

$$\int_{M_1} \ell^*(\psi) = \int_{M_2} \psi,$$

but instead would give us

$$\int_{M_1} \ell^*(\psi) = -\int_{M_2} \psi. \qquad\qquad \diamond$$

We now have the following direct generalization of Theorem 6.2.3.

THEOREM 6.2.11. *Let M and N be oriented n-manifolds with boundary and let $k: M \longrightarrow N$ be a diffeomorphism. Set*

$$\varepsilon_k = 1 \ \textit{if k is orientation-preserving,}$$
$$= -1 \ \textit{if k is orientation-reversing.}$$

Let φ be a smooth n-form on N with compact support. Then

$$\int_M k^*(\varphi) = \varepsilon_k \int_N \varphi.$$

Proof. Clear from Theorem 6.2.3 and Definition 6.2.8.   □

COROLLARY 6.2.12. *Let M be an oriented n-manifold with boundary and let $-M$ denote M with the opposite orientation. Let φ be a smooth n-form on M with compact support. Then*

$$\int_{-M} \varphi = -\int_M \varphi.$$

Proof. This is the special case of Theorem 6.2.11 where $N = M$ and k is the identity map.   □

We also record the following result for the record.

THEOREM 6.2.13. *Let M be an oriented n-manifold with boundary that is a union of components $\{M_i\}$. Let φ be a smooth n-form on M*

with compact support. Then

$$\int_M \varphi = \sum_i \int_{M_i} \varphi.$$

Proof. Clear.   □

While Definition 6.2.8 is an excellent definition in theory, we never use it in practice. It is very cumbersome, as it requires us to construct partitions of unity whenever we wish to evaluate integrals. Instead, to practically compute integrals we use a "tiling" of our manifolds, as follows.

DEFINITION 6.2.14. Let N be an oriented smooth n-manifold with boundary. A *tiling* of N is $\mathcal{T} = \{k_i \colon M_i \longrightarrow N\}$ satisfying the following conditions: Let $N_i = k_i(M_i)$, let $M_i^\circ = \mathrm{int}(M_i)$ and let $N_i^\circ = k_i(M_i^\circ)$. Then:

(i) for each i, M_i is a smooth oriented compact n-dimensional manifold with boundary in $\mathbb{R}^n$, and $k_i \colon M_i \longrightarrow N_i$ is a smooth map;

(ii) for each i, $k_i \colon M_i^\circ \longrightarrow N_i^\circ$ is an orientation preserving diffeomorphism (where N_i° has the orientation induced by being a subset of N);

(iii) for each $i_1 \neq i_2$, $N_{i_1}^\circ \cap N_{i_2}^\circ = \emptyset$;

(iv) $N = \cup_i N_i$.

(v) Any compact subset of N intersects only finitely many N_i.   ◇

REMARK 6.2.15. The essence of this definition is that N is "tiled" by the $\{N_i\}$, i.e., it is covered by the $\{N_i\}$, which intersect only along $\{N_i - N_i^\circ\}$, with each N_i a compact subset of N, and each $\{N_i^\circ\}$ a submanifold of N.

Thus, as opposed to a partition of unity of N, which is a covering of N by open sets (which certainly overlap), a tiling of N is a covering of N by compact sets with no point in the "interior" N_i° of more than one tile.

It may seem more natural at first glance to require that each k_i be a diffeomorphism on all of M_i, not just its interior M_i°, but it is very convenient to have the additional flexibility obtained by not making this stronger requirement.

We impose condition (v) to ensure that the tiles do not "pile up" anywhere. Note in particular that this condition implies that if N is compact then there are only finitely many tiles in any tiling.   ◇

Here are several (important) examples of tilings.

EXAMPLE 6.2.16.

(a) Consider the unit circle $S^1 \subset \mathbb{R}^2$, $S^1 = \{(x, y)|x^2 + y^2 = 1\}$. Let $I = [0, 2\pi] \subset \mathbb{R}^1$. Then we have a tiling of S^1 by one tile, $k: I \longrightarrow S^1$ given by $k(\theta) = (\cos\theta, \sin\theta)$. (Here is a case where k is a diffeomorphism on $I°$ but not on I.)

(b) Consider the unit sphere $S^2 \subset \mathbb{R}^3$, $S^2 = \{(x, y, z)|x^2+y^2+z^2 = 1\}$. S^2 is covered by two coordinate patches $k_+: \mathbb{R}^2 \longrightarrow S^2$ and $k_-: \mathbb{R}^2 \longrightarrow S^2$, where the functions k_+ and k_- are given by

$$k_+(u, v) = \left(\frac{2u}{u^2 + v^2 + 1}, \frac{2v}{u^2 + v^2 + 1}, \frac{1 - (u^2 + v^2)}{u^2 + v^2 + 1} \right),$$

$$k_-(u, v) = \left(\frac{2u}{u^2 + v^2 + 1}, \frac{-2v}{u^2 + v^2 + 1}, \frac{(u^2 + v^2) - 1}{u^2 + v^2 + 1} \right).$$

Every point of the unit sphere except $(0, 0, -1)$ is in the image of k_+ (i.e., is $k_+(u, v)$ for some $(u, v) \in \mathbb{R}^2$) and every point of the unit sphere except $(0, 0, 1)$ is in the image of k_- (i.e., is $k_-(u, v)$ for some $(u, v) \in \mathbb{R}^2$). However, every point of S^2 except $(0, 0, 1)$ and $(0, 0, -1)$ is in both sets in the covering. Also, neither of the two sets in the cover is compact. Thus, this does not give us a tiling. But we can easily modify it to obtain one.

Let H_N (resp. H_S) denote the northern (resp. southern) hemisphere of S^2,

$$H_N = \{(x, y, z) \in S^2 | z \geq 0\}, \qquad H_S = \{(x, y, z) \in S^2 | z \leq 0\}.$$

Note that S^2 is covered by H_N and H_S, but H_N and H_S intersect only on their boundaries.

Let T be the unit disk in $\mathbb{R}^2$,

$$T = \{(u, v)|u^2 + v^2 \leq 1\}.$$

Then

$$\{k_+: T \longrightarrow H_N, \ k_-: T \longrightarrow H_S\}$$

is a tiling of S^2 (by two tiles).

(c) Let M be a smooth oriented compact n-dimensional manifold with boundary in $\mathbb{R}^n$ and let $f: M \longrightarrow \mathbb{R}^k$ be a smooth map. Let N be the *graph* of f, i.e.,

$$N = \{(m, p) \in M \times \mathbb{R}^k | p = f(m)\}.$$

Then $k: M \longrightarrow N$ by $k(m) = (m, f(m))$ is a tiling of N by a single tile. $\diamond$

Here is how we compute integrals in practice. The proof of this theorem is quite technical, and so we omit it.

THEOREM 6.2.17. *Let N be an oriented smooth n-manifold with boundary. Let T be a tiling of N as in Definition 6.2.14. Let φ be an n-form on N with compact support. Then*

$$\int_N \varphi = \sum_i \int_{M_i} (k_i)^*(\varphi).$$

6.3 The integral of a 0-form over a point

We have deferred the case of a 0-manifold, but in fact this case is very easy.

We begin with a preliminary definition.

DEFINITION 6.3.1. Let p be a point and let f be a function defined at p. The *evaluation map* Eval_p is defined by

$$\text{Eval}_p(f) = f(p).$$

If (p, ε_p) is an oriented point, the signed evaluation map $_s\mathrm{Eval}_p$ is defined by

$$_s\mathrm{Eval}_{(p,\varepsilon_p)}(f) = \varepsilon_p f(p). \qquad \diamond$$

Now an oriented 0-manifold M is just a union of oriented points, and a smooth 0-form on M is just a function on M. (The condition of smoothness on the function is vacuous for an isolated point–there are no derivatives to take.) Then we make the following definition.

DEFINITION 6.3.2. Let $M = (p, \varepsilon_p)$ be an oriented point, and let $\varphi = f$ be a smooth 0-form on M. Then

$$\int_M \varphi = {}_s\mathrm{Eval}_{(p,\varepsilon_p)}(\varphi). \qquad \diamond$$

REMARK 6.3.3. This definition can be phrased more simply: If $\pm p$ is an oriented point, and f is a function defined at p, then

$$\int_{+p} f = f(p) \text{ and } \int_{-p} f = -f(p).$$

We have phrased it as we did to make a (psychological) point. We think of integrating a form over an oriented manifold as performing an operation on that form, and in dimension 0, that operation is just evaluation. $\qquad \diamond$

EXAMPLE 6.3.4. Let $p \in \mathbb{R}^3$ be the point $p = (2, 3, 4)$ and let f be the function $f(x, y, z) = x^2 y^2 + z^3$. Then

$$\int_{+p} f = f(2, 3, 4) = 80 \text{ and } \int_{-p} f = -f(2, 3, 4) = -80. \quad \diamond$$

It is easy to check from Definition 6.3.2 that all of the constructions and results of the previous section go through (much more easily) for oriented 0-manifolds. (In that section, to find integrals over n-manifolds we had to start by using submanifolds of $\mathbb{R}^n$ as "references" because we only knew how to evaluate integrals there. But there is no need for that step here, as we know how to evaluate functions at points in general.) We simply summarize these results here. Before doing so we recall that pullback on 0-forms is given by composition, i.e., that $k^*(f) = f \circ k$.

THEOREM 6.3.5. *Let* $M = (p, \varepsilon_p)$ *and* $N = (q, \varepsilon_q)$ *be oriented* 0-*manifolds and let* $(k, \sigma_k) \colon M \longrightarrow N$ *be an oriented diffeomorphism. Set*

$$\varepsilon_k = 1 \ \textit{if } k \ \textit{is orientation-preserving},$$
$$= -1 \ \textit{if } k \ \textit{is orientation-reversing}.$$

Let φ *be a smooth* 0-*form on* N *with compact support. Then*

$$\int_M k^*(\varphi) = \varepsilon_k \int_N \varphi.$$

COROLLARY 6.3.6. *Let* M *be an oriented* 0-*manifold let* $-M$ *denote* M *with the opposite orientation. Let* φ *be a smooth* 0-*form on* M *with compact support. Then*

$$\int_{-M} \varphi = - \int_M \varphi.$$

THEOREM 6.3.7. *Let* M *be an oriented* 0-*manifold with boundary that is a union of components* $\{M_i\}$. *Let* φ *be a smooth* 0-*form on* M *with compact support. Then*

$$\int_M \varphi = \sum_i \int_{M_i} \varphi.$$

REMARK 6.3.8. We stated these results so as to parallel our previous results. But, more simply, a 0-manifold M is a union of points and a 0-form (i.e., a function) φ on a M has compact support if and only if it is nonzero on only finitely many of these points.   ◇

6.4 The integral of a 1-form over a curve

In this section we investigate integrals of 1-forms over oriented 1-manifolds, i.e., over oriented curves. In addition to developing the theory of these in general, we will see that in the special case of curves in $\mathbb{R}^n$, for any n, they correspond to classical "line integrals."

Let $\mathbb{R}^1 = \{t\}$, and first consider an oriented interval I in $\mathbb{R}^1$. I is the archetype of an oriented curve. Consider a 1-form φ on I and write $\varphi = f(t)dt$. Note that dt is the volume form on $\mathbb{R}^1$ in the standard orientation of $\mathbb{R}^1$. Then, directly from Lemma 6.2.2, we see that

$$\int_I \varphi = {}_s\!\!\int_I f(t)d\ell_t.$$

We observe that if I has the standard orientation, then the sign in the signed definite integral is positive.

Now we deal with curves in $\mathbb{R}^n$ for $n > 1$. The theory works for any value of n, but for simplicity we will choose our examples in $\mathbb{R}^3$. Recall our theory from Section 6.2: Let C be an oriented curve in $\mathbb{R}^n$, parameterized by $k(t)$ for $t \in I = [a, b]$. (Here $[a, b]$ denotes the closed interval $a \le t \le b$.) We assume that $k: I \longrightarrow C$ is orientation-preserving.

Let φ be a 1-form defined on C. Then

$$\int_C \varphi = \int_I k^*(\varphi)$$

and, furthermore:

(a) $\int_C \varphi$ depends only on the oriented curve C, but is independent of the parameterization k; and

(b) $\int_{-C} \varphi = -\int_C \varphi$ where $-C$ denotes C with the opposite orientation.

Also, we can concretely evaluate this integral as follows:

$$\int_C \varphi = {}_s\!\!\int_I k^*(\varphi)(\mathbf{e}^1)d\ell_t = {}_s\!\!\int_I (\varphi)(k_*(\mathbf{e}^1))d\ell_t$$
$$= {}_s\!\!\int_I \varphi(\mathbf{k}'(t))d\ell_t.$$

where we use the fact that $k_*(\mathbf{e}_t^1) = \mathbf{k}'(t)$. We also recall that we can evaluate $k^*(\varphi)$ by "substitution," and that if $k^*(\varphi) = A(t)dt$, then $k^*(\varphi)(\mathbf{e}^1) = A(t)$.

Before proceeding further we observe that in this situation we have tiled C with a single tile.

EXAMPLE 6.4.1.

(a) Let C be the oriented curve given by $(x, y, z) = k(t) = (2t + 1, t^2, t^3)$, $1 \le t \le 3$, and let φ be the 1-form $\varphi = (3x - 1)^2 dx + 5zdy + 2dz$. Then

$$k^*(\varphi) = (3(2t + 1) - 1)^2(2dt) + 5(t^3)(2tdt) + 2(3t^2 dt)$$
$$= (10t^4 + 78t^2 + 48t + 8)dt$$

and so

$$\int_C \varphi = \int_1^3 (10t^4 + 78t^2 + 48t + 8)dt$$
$$= 2t^5 + 26t^3 + 24t^2 + 8t \Big|_1^3 = 1428 - 60 = 1368.$$

(b) Let C be the oriented curve given by $(x, y, z) = k(t) = (1 - t^2, t^3 + t, 0)$, $0 \le t \le 2$. Let $\varphi = -21xydx + dz$. Then

$$k^*(\varphi) = (-21(1 - t^2)(t^3 + t))(-2t) + 0$$
$$= (-42t^6 + 42t^2)dt$$

and so

$$\int_C \varphi = \int_0^2 (-42t^6 + 42t^2)dt$$
$$= -6t^7 + 14t^3 \Big|_0^2 = -144 - 0 = -144.$$

(c) Let C be the curve in the plane that is the graph of the function $y = 1 + x^2$, $-1 \le x \le 1$, oriented in the direction of increasing x, and let $\varphi = ydx + xdy$. Then C is the parameterized curve $k(t) = (t, 1 + t^2)$, $-1 \le t \le 1$, and

$$k^*(\varphi) = ((1 + t^2)(dt) + t(2tdt))$$
$$= (1 + 3t^2)dt$$

and so

$$\int_C \varphi = \int_{-1}^{1} (1 + 3t^2)dt$$

$$= t + t^3 \Big|_{-1}^{1} = 2 - (-2) = 4.$$

(As a matter of practice, note that in this example it was not necessary to introduce the variable t. We could have parameterized the curve by $k(x) = (x, 1 + x^2)$ with the same result.)

(d) Let C_1 be the curve in the plane parameterized by $k_1(t) = (t - 1, 2t - 1)$, $1 \leq t \leq 2$, and let C_2 be the curve parameterized by $k_2(u) = (2 - u, 5 - 2u)$, $1 \leq u \leq 2$. Let $\varphi = ydx$. Then

$$\int_{C_1} \varphi = \int_{1}^{2} (2t - 1)(1)dt = t^2 - t \Big|_{1}^{2} = 2$$

and

$$\int_{C_2} \varphi = \int_{1}^{2} (5 - 2u)(-1)du = u^2 - 5u \Big|_{1}^{2} = -2.$$

This second answer should come as no surprise. Observe that C_1 is the straight line segment running from the point $(0, 1)$ to the point $(1, 3)$, while C_2 is the straight line segment running from the point $(1, 3)$ to the point $(0, 1)$. In other words, $C_2 = -C_1$, and we see $\int_{C_2} \varphi = -\int_{C_1} \varphi$, as it should be
    Observe further that C_1 is also the line segment given by $k_3(x) = (x, 2x + 1)$, $0 \leq x \leq 1$, and, using this parameterization, we find that

$$\int_{C_1} \varphi = \int_{0}^{1} (2x + 1)dx = 2,$$

verifying, in this very simple case, that the two different parameterizations k_1 and k_3 give the same value for the integral.

(e) Let C be the oriented curve parameterized by $k_1(t) = (t, \sqrt{1 + t^2})$, $-1 \leq t \leq 1$, and let $\varphi = y^2 dx$. Then $\mathbf{k}_1'(t) = \begin{bmatrix} 1 \\ t/\sqrt{1 + t^2} \end{bmatrix}$ and

$$\int_C \varphi = \int_{-1}^{1} (1 + t^2)dt = 8/3.$$

On the other hand, C is also parameterized by $k_2(\theta) = (\tan\theta,$ $\sec\theta)$, $-\pi/4 \le t \le \pi/4$. Then $\mathbf{k}_2'(\theta) = \begin{bmatrix} \sec^2\theta \\ \sec\theta\tan\theta \end{bmatrix}$ and

$$\int_C \varphi = \int_{-\pi/4}^{\pi/4} (\sec^2\theta)(\sec^2\theta)d\theta = \int_{-\pi/4}^{\pi/4} \sec^4\theta\, d\theta = 8/3.$$

(f) Let C be the oriented curve parameterized by $k_1(x) = (x, x^2)$, $-2 \le x \le 2$, and let $\varphi = 5xy\,dy$. Then

$$\int_C \varphi = \int_{-2}^{2} 5x^3(2x)dx = \int_{-2}^{2} 10x^4 dx = 2x^5 \Big|_{-2}^{2} = 128.$$

On the other hand, C is also parameterized by $k_2(t) = (t^3 + t, t^6 + 2t^4 + t^2)$, $-1 \le t \le 1$. This parameterization gives

$$\int_C \varphi = \int_{-1}^{1} 5(t^3 + t)(t^6 + 2t^4 + t^2)(6t^5 + 8t^3 + 2t)dt$$

$$= \int_{-1}^{1} (30t^{14} + 130t^{12} + 220t^{10} + 180t^8 + 70t^6 + 10t^4)dt$$

$$= 2t^{15} + 10t^{13} + 20t^{11} + 20t^9 + 10t^7 + 2t^{15} \Big|_{-1}^{1} = 128.$$

Another parameterization of C is $k_3(\theta) = (4\sin\theta, 16\sin^2\theta)$, $-\pi/6 \le \theta \le \pi/6$. With this parameterization

$$\int_C \varphi = \int_{-\pi/6}^{\pi/6} (320\sin^3\theta)(32\sin\theta\cos\theta)d\theta$$

$$= 2048\sin^5\theta \Big|_{-\pi/6}^{\pi/6} = 128.$$

We thus see that all three parameterizations give the same value, as we expect.   ◇

These were all chosen as illustrative cases, without any particular significance. But the next two cases are so important that we present them separately.

EXAMPLE 6.4.2. Let C be the unit circle in the (x, y)-plane, parameterized by $k(\theta) = (\cos\theta, \sin\theta)$, $0 \le \theta \le 2\pi$. (This is the tiling of Example 6.2.16(a).) Thus C is traversed once counterclockwise, beginning and ending at the point $(1, 0)$. Let φ^1 be the 1-form of Example 1.2.17,

$$\varphi^1 = \frac{-y}{x^2 + y^2} dx + \frac{x}{x^2 + y^2} dy.$$

Then

$$k^*(\varphi^1) = \frac{-\sin\theta}{\cos^2\theta + \sin^2\theta}(-\sin\theta\, d\theta) + \frac{\cos\theta}{\cos^2\theta + \sin^2\theta}(\cos\theta\, d\theta)$$
$$= d\theta$$

and so

$$\int_C \varphi^1 = \int_0^{2\pi} d\theta = 2\pi. \qquad\qquad \diamond$$

EXAMPLE 6.4.3. Let C be unit circle in the (x, y)-plane, as in Example 6.4.2, and let ω be the 1-form of Example 1.2.21(1),

$$\omega = \frac{x}{x^2 + y^2} dx + \frac{y}{x^2 + y^2} dy.$$

Then

$$k^*(\omega) = \frac{\cos\theta}{\cos^2\theta + \sin^2\theta}(-\sin\theta\, d\theta) + \frac{\cos\theta}{\cos^2\theta + \sin^2\theta}(\sin\theta\, d\theta)$$
$$= 0$$

and then

$$\int_C \omega = \int_0^{2\pi} 0\, d\theta = 0. \qquad\qquad \diamond$$

We now write down a theorem which simply codifies the procedure we have been using.

THEOREM 6.4.4. *Let C be the oriented curve in $\mathbb{R}^n$ given by*

$$(x_1, \ldots, x_n) = k(t) = (k_1(t), \ldots, k_n(t)) \text{ for } t \in I = [a, b]$$

with the orientation induced by the standard orientation on $\mathbb{R}^1$ (i.e., C traversed in the direction of increasing t).
Let φ be the 1-form on C given by

$$\varphi = A_1(x_1, \ldots, x_n)dx_1 + \cdots + A_n(x_1, \ldots, x_n)dx_n.$$

Then $\int_C \varphi$ is given by

$$\int_C \varphi = \int_a^b \big(A_1(k(t))k_1'(t) + \cdots + A_n(k(t))k_n'(t)\big)dt.$$

Proof. By definition,

$$\int_C \varphi = \int_a^b \varphi(\mathbf{k}'(t))dt.$$

But

$$\mathbf{k}'(t) = \begin{bmatrix} k_1'(t) \\ \vdots \\ k_n'(t) \end{bmatrix},$$

so $dx_i(\mathbf{k}'(t)) = k_i'(t)$, giving the stated value for $\varphi(k'(t))$. □

REMARK 6.4.5. Our theory tells us that a 1-form φ on a curve C is a function on tangent vectors *to C* at every point of M. In the above examples, and in Theorem 6.4.4, the form φ that we considered was a function on tangent vectors *to $\mathbb{R}^2$ or $\mathbb{R}^3$* at every point of C. But this was just because it was easy to write down φ if it was expressed in that way.

Consider the unit circle C in the (x, y)-plane, traversed counterclockwise. We can describe two 1-forms on C. The first, ρ_1, is defined by $\rho_1(\mathbf{v}_p) = $ where $\mathbf{v}_p$ is chosen to be the unit tangent vector to C at p.

Note this gives us a 1-form because this tangent vector varies smoothly as we traverse C. The second, ρ_2, is simply defined by $\rho_2(\mathbf{v}_p) = 0$. Then, as 1-forms on C, the forms φ^1 and ψ of Example 6.4.2 are just ρ_1 and ρ_2, respectively. $\diamond$

REMARK 6.4.6. We have developed our theory of integration for smooth curves C, i.e., for smooth 1-manifolds. In fact the theory goes through without change for 1-manifolds with corners. These are also known as piecewise smooth curves. They are exactly what the name suggests, a typical example being the following:

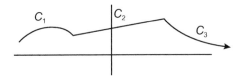

If this is C, then we define

$$\int_C \varphi = \int_{C_1} \varphi + \int_{C_2} \varphi + \int_{C_3} \varphi$$

where each of the integrals on the right is already defined, being an integral over a smooth curve.

Actually, we may extend our theory even further, to objects which are not manifolds, but rather only "immersed" manifolds, a typical one being:

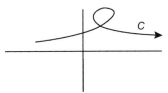

Here C is not a manifold because of the self-intersection point; nevertheless, we may parameterize C by $r(t) = (f(t), g(t))$, and then $\int_C \varphi$ is given by the exact same formula as before. $\diamond$

EXAMPLE 6.4.7. Let us now compute a few more integrals of 1-forms over oriented curves. Let us consider the following curves (where we stick to $\mathbb{R}^2$ for simplicity):

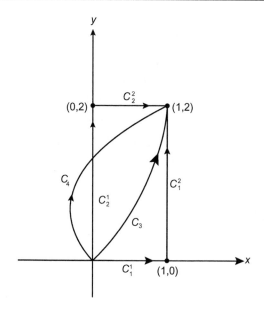

Here C_1 is the piecewise smooth curve $C_1 = C_1^1 \cup C_1^2$, where C_1^1 is a horizontal segment and C_1^2 is a vertical segment. Similarly, C_2 is the piecewise smooth curve $C_2 = C_2^1 \cup C_2^2$, where C_2^1 is a vertical segment and C_2^2 is a horizontal segment. C_3 is the smooth curve parameterized by $r_3(x) = (x, x + x^3)$, $0 \le x \le 1$, and C_4 is the smooth curve parameterized by $r_4(y) = (y^2 - 3y/2, y)$, $0 \le y \le 2$. All of these are oriented curves beginning at $(0, 0)$ and ending at $(1, 2)$.

Now let $\varphi_1 = 2xy\,dx$ and let $\varphi_2 = x^2\,dy$. Let $\varphi = \varphi_1 + \varphi_2$. Of course, $\int_{C_i} \varphi = \int_{C_i} \varphi_1 + \int_{C_i} \varphi_2$ for each i. We compute the following:

$$\int_{C_1} \varphi_1 = 0 \qquad \int_{C_1} \varphi_2 = 2 \qquad \int_{C_1} \varphi = 2$$

$$\int_{C_1} \varphi_1 = 2 \qquad \int_{C_1} \varphi_2 = 2 \qquad \int_{C_1} \varphi = 2$$

$$\int_{C_1} \varphi_1 = 7/6 \qquad \int_{C_1} \varphi_2 = 5/6 \qquad \int_{C_1} \varphi = 2$$

$$\int_{C_1} \varphi_1 = 8/5 \qquad \int_{C_1} \varphi_2 = 2/5 \qquad \int_{C_1} \varphi = 2$$

It is certainly no surprise that the values of the four integrals for φ_1 differ, and likewise for φ_2—in fact, this is only to be expected, as the integrals should depend on the paths.

What is surprising is that all four integrals for φ have the same value. That is, at least for these four curves, the integral *doesn't* depend on the path. Indeed, for *any* path C from $(0, 0)$ to $(1, 2)$, $\int_C \varphi = 2$.

What is so special about the 1-form φ that makes this true? The answer is that φ is exact. Here $\varphi = dF$ where F is the 0-form (i.e., function) $F(x, y) = x^2 y$.   ◇

We now wish to investigate the phenomenon illustrated by this example further. Let us begin by formalizing it.

DEFINITION 6.4.8.   Let φ be a 1-form defined on a smooth manifold M. Then $\int_C \varphi$ is *path-independent* if the value of this integral depends only on the boundary ∂C of the oriented curve C, for any oriented smooth curve C in M.   ◇

REMARK 6.4.9.   Recall that if C is an oriented curve in M from p to q, $\partial C = \{q\} \cup -\{p\}$. Thus this definition says that $\int_C \varphi$ is path-independent if it depends only on the points p and q (in order), not on the particular path from p to q in M.   ◇

It is convenient to reformulate this slightly:

LEMMA 6.4.10.   *Let φ be a smooth 1-form on a smooth manifold M. Then $\int_C \varphi$ is path-independent if and only if for every closed curve C_0 in M, $\int_{C_0} \varphi = 0$.*

Proof.   We need to show that this condition is equivalent to the condition in the definition.

Recall that a closed curve C_0 is one whose boundary is empty.

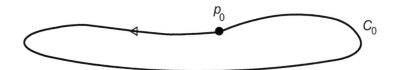

Thus, picking any point p_0 on C_0, we may regard the oriented curve C_0 as a path from p_0 to p_0.

First, suppose the condition in the definition holds. Then $\int_{C_0} \varphi = \int_{C_1} \varphi$, where C_1 is any other path from p_0 to p_0. In particular, we may take C_1 to be the constant path at p_0. But then it is clear that $\int_{C_1} \varphi = 0$. (If $k: I \longrightarrow C_1$ is a parameterization of the constant path, then $k^*(\varphi) = 0$.)

On the other hand, suppose the condition in the lemma holds, and pick any two points p_0 and p_1 in M.

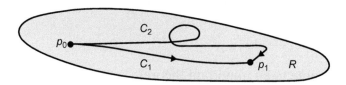

Let C_1 and C_2 be any two paths in M from p_0 to p_1. Then $C = C_2 \cup -C_1$ is an oriented closed curve in M, so by the condition in the lemma, $\int_C \varphi = 0$. But then

$$0 = \int_C \varphi = \int_{C_2 \cup -C_1} \varphi = \int_{C_2} \varphi + \int_{-C_1} \varphi = \int_{C_2} \varphi - \int_{C_1} \varphi,$$

so $\int_{C_2} \varphi = \int_{C_1} \varphi$, giving the condition in the definition.   □

REMARK 6.4.11. In the proof of Lemma 6.4.10, we picked a point p_0 on C_0, regarded C_0 as a path from p_0 to p_0, and used that to find $\int_{C_0} \varphi$. What if we had picked a different point p_1, regarded C_0 as a path from p_1 to p_1, and used that to find $\int_{C_0} \varphi$? Would we have obtained the same value for the integral?

The answer is Yes, which may be seen as follows. (Note this is true whether $\int_{C_0} \varphi$ is zero or not.) To see this consider the following situation:

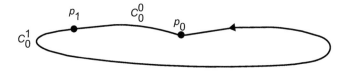

We have broken up C_0 into two pieces, C_0^0 and C_0^1. Regarding C_0 as beginning at p_0, we have

$$\int_{C_0} \varphi = \int_{C_0^0 \cup C_0^1} \varphi = \int_{C_0^0} \varphi + \int_{C_0^1} \varphi$$

while regarding C_0 as being at p_1, we have

$$\int_{C_0} \varphi = \int_{C_0^1 \cup C_0^0} \varphi = \int_{C_0^1} \varphi + \int_{C_0^0} \varphi$$

and these are equal.   ◇

We now come to an important theorem.

THEOREM 6.4.12. *Let φ be a 1-form on a smooth manifold M. Then φ is exact if and only if $\int_C \varphi$ is path-independent.*

Proof. First suppose that φ is exact, so that $\varphi = dF$ for some 0-form (i.e., function) F on M.

Let p and q be arbitrary points in M. Let $r: I \longrightarrow M$ be a parameterized oriented curve C from p to q, so that, if $I = [a, b], r(a) = p$, and $r(b) = q$. We compute $\int_C \varphi$.

We know that pullback commutes with exterior derivative, so

$$r^*(\varphi) = r^*(dF) = d(r^*(F)).$$

For simplicity, set $\tilde{\varphi} = r^*(\varphi)$ and $\tilde{F} = r^*(F)$, so that $\tilde{\varphi} = d\tilde{F}$. On the one hand, since $\tilde{\varphi}$ is a 1-form on I, we can write it as

$$\tilde{\varphi} = f(t)dt$$

for some function $f(t)$. On the other hand, since $\tilde{F}$ is a function, its exterior derivative is given by

$$d\tilde{F} = \tilde{F}'(t)dt.$$

Comparing these two expressions we see that

$$f(t) = \tilde{F}'(t).$$

But then, applying the Fundamental Theorem of Calculus,

$$
\begin{aligned}
\int_C \varphi &= \int_s \int_I r^*(\varphi)(\mathbf{e}^1)d\ell_t \\
&= \int_s \int_I \big(f(t)dt(\mathbf{e}^1)\big)d\ell_t \\
&= \int_a^b f(t)dt \\
&= \tilde{F}(b) - \tilde{F}(a) \\
&= \big(r^*(F)\big)(b) - \big(r^*(F)\big)(a) \\
&= F(r_*(b)) - F(r_*(a)) \\
&= F(q) - F(p).
\end{aligned}
$$

Since this formula only depends on the points p and q, and not the curve C, this shows that $\int_C \varphi$ is path-independent.

Now suppose $\int_C \varphi$ is path-independent. Fix a point p in M and define a function F on M as follows: Let q be any point in R and set

$$F(q) = \int_C \varphi,$$

where C is any path from p to q. Note that this function is well defined (i.e., depends only on q) precisely because of path independence—no matter what path we choose from p to q, the integral has the same value.

We compute dF. Let q_0 and q be any two points of M. We first compute $F(q) - F(q_0)$. By definition,

$$F(q) - F(q_0) = \int_C \varphi - \int_{C_0} \varphi$$

for any paths C_0 from p to q_0 and C from p to q. We choose C_0 arbitrarily, and then $C = C_0 \cup C_1$, where C_1 is any path from q_0 to q.

Then we see

$$F(q) - F(q_0) = \int_C \varphi - \int_{C_0} \varphi$$

$$= \int_{C_0 \cup C_1} \varphi - \int_{C_0} \varphi$$

$$= \int_{C_0} \varphi + \int_{C_1} \varphi - \int_{C_0} \varphi$$

$$= \int_{C_1} \varphi.$$

We first carry the computation further in the special case that M is a region in $\mathbb{R}^n$. For any point q_0 in M,

$$dF(q_0) = F_{x_1}(q_0)dx_1 + \cdots + F_{x_n}(q_0)dx_n$$

and so

$$F_{x_i}(q_0) = dF(q_0)\left(\mathbf{e}_{q_0}^i\right)$$

for each i. Fix a value of i. Let q_0 be the point $q_0 = (x_0^1, \ldots, x_0^{i-1}, x_0^i, x_0^{i+1}, \ldots, x_0^n)$ and let q_1 be the point $q = (x_0^1, \ldots, x_0^{i-1}, x^i, x_0^{i+1}, \ldots, x_0^n)$.

We compute $F(q) - F(q_0)$ as follows: Choose the path C_1 to be the straight line segment joining q_0 to q_1, parameterized by

$$q = r(t) = q_0 + t(q - q_0) = (x_0^1, \ldots, x_0^{i-1}, t, x_0^{i+1}, \ldots, x_0^n)$$

for $t \in I = [x_0^i, x^i]$, so that

$$\mathbf{r}'(t) = \begin{bmatrix} 0 \\ \vdots \\ 0 \\ 1 \\ 0 \\ \vdots \\ 0 \end{bmatrix}_q = \mathbf{e}_q^i.$$

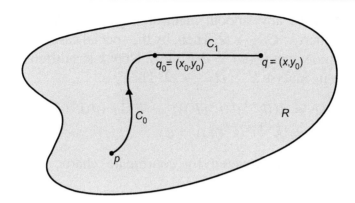

Write
$$\varphi = A^1(q)dx_1 + \cdots + A^n(q)dx_n.$$

Then
$$\varphi(\mathbf{r}'(t)) = A^i(r(t)).$$

Then
$$F(q) - F(q_0) = \int_{x_0^i}^{x^i} A^i\left(x_0^1, \ldots, x_0^{i-1}, t, x_0^{i+1}, \ldots, x_0^n\right)dt.$$

Denote this integral by $a(x^i)$.

Now by the definition of the partial derivative,
$$F_{x_i}(q_0) = \lim_{x^i \to x_0^i} \frac{1}{x^i - x_0^i}\left(F(q) - F(q_0)\right).$$

Substituting, and noting that $a(x_0^i) = 0$, we see that
$$F_{x_i}(q_0) = \lim_{x^i \to x_0^i} \frac{a(x^i) - a(x_0^i)}{x^i - x_0^i}$$
$$= a'(x_0)$$
$$= A^i(q_0)$$

by the Fundamental Theorem of Calculus. Thus we see that
$$dF = \varphi$$

in this case, and in particular that φ is exact.

Now let M be any smooth manifold, and consider an arbitrary coordinate chart $k\colon \mathcal{O} \longrightarrow \mathcal{U}$. Then, by the special case of M a region in $\mathbb{R}^n$, we know that $k^*(\varphi) = d(k^*(F))$. Now k is a diffeomorphism, so we have its inverse $k^{-1}\colon \mathcal{U} \longrightarrow \mathcal{O}$. Then

$$dF = d((k^{-1})^*(k^*(F))) = (k^{-1})^*(d(k^*(F)))$$
$$= (k^{-1})^*(k^*(\varphi)) = \varphi$$

on $\mathcal{U}$. Since M is covered by coordinate charts, this equality holds on M. $\qquad\square$

The proof of Theorem 6.4.12 not only shows that $\int_C \varphi$ is path-independent when φ is exact, but also gives the value of this integral.

COROLLARY 6.4.13. *Let $\varphi = dF$ be an exact 1-form on a smooth manifold M. Let p and q be any two points in M. Then for any path C in M from p to q,*

$$\int_C \varphi = F(q) - F(p).$$

Proof. This is taken verbatim from the proof of Theorem 6.4.12. $\qquad\square$

If we begin with an exact 1-form φ, the proof of Theorem 6.4.12 also shows us how to explicitly construct a function F with $dF = \varphi$.

COROLLARY 6.4.14. *Let φ be an exact 1-form on a smooth manifold M. Fix a point p of M and for any point q of M, set*

$$F(q) = \int_C \varphi,$$

where C is any smooth path from p to q. Then F is a primitve for φ, i.e., $dF = \varphi$.

Proof. This is also taken verbatim from the proof of Theorem 6.4.12. $\qquad\square$

REMARK 6.4.15. Note that Corollaries 6.4.13 and 6.4.14 are generalizations of the Fundamental Theorem of Calculus (FTC).

Let $M = I = [a, b] \subset \mathbb{R}^1$, and regard this as a path from a to b (which gives it an orientation). Let f be a function on I and let F be an antiderivative of f, so $f(t) = F'(t)$. If $\varphi = f(t)dt$, then $\varphi = dF$, so we get

$$\int_I f(t)dt = F(b) - F(a),$$

which is just the Fundamental Theorem of Calculus.

Set $p_0 = a$. Then, for any x in I, let C be the subinterval $[a, x]$, regarded as path from a to x. Write $\varphi = f(t)dt$, so

$$F(x) = \int_C \varphi = \int_a^x f(t)dt$$

and $f(t)dt = \varphi = d(F(t)) = F'(t)dt$, so $F'(x) = f(x)$, again as given to us by the Fundamental Theorem of Calculus.

(This remark should come as no surprise, since the key step in the proof of Theorem 6.4.12 was provided by the FTC.)   ◇

REMARK 6.4.16. Note that in the proof of Theorem 6.4.12, and in the statement of Corollary 6.4.14, we picked an arbitrary point p at which to start our paths. You may wonder what would change if we picked a different point p'. Suppose we do, and we define $G(q) = \int_{C'} \varphi$ where C' is a path from p' to q.

Since we may choose any path C', let us choose $C' = C'' \cup C$, where C'' is a path from p' to p.

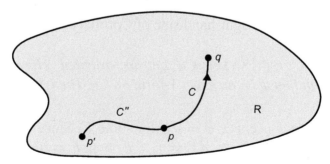

Then

$$G(q) = \int_{C'} \varphi = \int_{C''} \varphi + \int_C \varphi = \int_{C''} \varphi + F(q).$$

Notice, however, that $\int_{C''} \varphi$ is a *constant*–it does not depend on q. Thus we see $G(q) = F(q) + c$, for some constant c.

This is only to be expected. We are looking for a primitive F of φ, and we note that a primitive is only unique up to addition of a closed form, which, in the case of a 0-form, is a constant function. $\diamond$

Let us now reformulate Corollary 6.4.14 in a way that looks more mysterious, but actually points us toward future developments.

THEOREM 6.4.17. *Let C be an oriented smooth 1-manifold with boundary and let F be a 0-form on C. Then*

$$\int_C dF = \int_{\partial C} F.$$

Proof. It suffices to consider the case when C is connected (i.e., consists of a single curve) because if C has several components, we can simply add up the integrals on each of them.

The left-hand side of the equality is simply the left-hand side of Corollary 6.4.14, in the special case that $M = C$.

As for the right-hand side, $\partial C = \{q\} \cup -\{p\}$, where C is a path from p to q. But now recall from Section 6.3 what the right-hand side is:

$$\int_{\partial C} F = \int_{\{q\} \cup -\{p\}} F = \int_{\{q\}} F + \int_{-\{p\}} F = F(q) - F(p),$$

and this is simply the right-hand side of Corollary 6.4.14. $\square$

COROLLARY 6.4.18. *Let C_0 be an oriented closed smooth 1-manifold and let φ be an exact 1-form on C_0. Then $\int_{C_0} \varphi = 0$.*

Proof. Since φ is exact, $\varphi = dF$ for some F. Since C_0 is closed, $\partial C_0 = \emptyset$. Then $\int_{C_0} \varphi = \int_{C_0} dF = \int_{\partial C_0} F = \int_{\emptyset} F = 0$. $\square$

REMARK 6.4.19. As we have observed, Corollary 6.4.14 is a generalization of the Fundamental Theorem of Calculus (FTC) from intervals in $\mathbb{R}$ to curves, i.e., 1-manifolds.

If you compare Corollary 6.4.14 with the introduction, you will see that we have the first case of the Generalized Stokes's Theorem (GST)! This theorem is the leitmotiv of this book. From this point of view you can see that the GST is a vast, higher-dimensional analog of the FTC, and, as you will see, its proof reduces (after enough work) to the FTC on bricks (as defined in Definition 4.1.46), a brick being the higher-dimensional version of an interval.   ◇

We now return to Theorem 6.4.12 and its consequences.

COROLLARY 6.4.20. *Let S^1 be the unit circle in $\mathbb{R}^2$, and let f_0:* $S^1 \longrightarrow M$ *be a smooth map that is smoothly null-homotopic. Let* $C_0 = f_0(S^1) \subseteq M$ *(so that C_0 is a closed curve in M). Then for any closed 1-form φ on M, $\int_{C_0} \varphi = 0$.*

Proof. First, we know that $\int_{C_0} \varphi = \int_{S^1} f_0^*(\varphi)$.

Now let $c\colon S^1 \longrightarrow M$ be a constant map. Then, by Corollary 4.7.4, $\psi = f^*(\varphi) - c^*(\varphi)$ is exact. Thus,

$$
\int_{C_0} \varphi = \int_{S^1} f^*(\varphi)
$$
$$
= \int_{S^1} c^*(\varphi) + \int_{S^1} \psi
$$
$$
= 0 + 0 = 0,
$$

where the first integral on the right-hand side is 0 as $c^*(\varphi) = 0$, and the second integral on the right-hand side is 0 by Theorem 6.4.12, as ψ is exact.   □

Although, strictly speaking, we will not need the following property, it is a standard definition that the reader should be aware of.

DEFINITION 6.4.21. Let M be a connected smooth manifold with boundary. Then M is *smoothly simply connected* if it has the following property: Regard S^1 as the unit circle in $\mathbb{R}^2$ and let $p_0 = (1, 0) \in S^1$. Let $f_0\colon S^1 \longrightarrow M$ be any smooth map and set $q_0 = f_0(p_0)$. Then f_0 is smoothly null-homotopic rel q_0, i.e., there is a null-homotopy as in Definition 4.7.2 with $F(p_0, t) = q_0$ for all $0 \le t \le 1$.   ◇

Here are some examples of simply connected/non-simply connected manifolds. In the non-simply connected cases we have drawn loops that cannot be shrunk to a point.

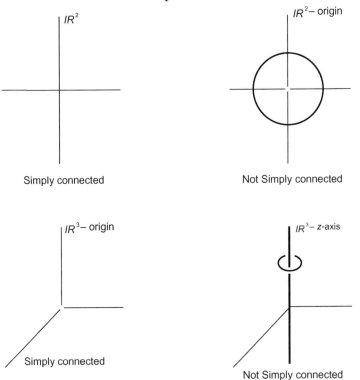

COROLLARY 6.4.22. *Let M be a smooth manifold with boundary such that every component of M is smoothly simply connected, or, more generally, let M be a smooth manifold with boundary with the property that every smooth map $f_0: S^1 \longrightarrow M$ is smoothly null-homotopic. Then every closed 1-form φ on M is exact.*

Proof. By Corollary 6.4.20, $\int_{C_0} \varphi = 0$ for every closed curve C_0, so $\int_C \varphi$ is path-independent, by Lemma 6.4.10. But then, by Theorem 6.4.12, φ is exact.   □

We can now apply these results in two different directions.

COROLLARY 6.4.23. *The closed form $\varphi^1 = (-y/(x^2 + y^2))dx + (x/(x^2 + y^2))dy$ is not exact on $\mathbb{R}^2 - \{(0,0)\}$.*

Proof. We introduced φ^1 in Example 1.2.17, and we computed $\int_{S^1} \varphi^1$ in Example 6.4.2, for S^1 the unit circle, oriented counterclockwise, and found

$$\int_{S^1} \varphi^1 = 2\pi \neq 0. \qquad \square$$

REMARK 6.4.24. In Proposition 1.2.19 we proved that φ^1 is not exact on $\mathbb{R}^2 - \{(0, 0)\}$ "by hand", and that involved a lot of work. But with the theory we have developed, the proof is a simple computation. $\diamond$

COROLLARY 6.4.25. *The inclusion $i \colon S^1 \longrightarrow \mathbb{R}^2 - \{(0, 0)\}$ is not smoothly null-homotopic. In particular, $\mathbb{R}^2 - \{(0, 0)\}$ is not smoothly simply connected.*

Proof. If i were smoothly null-homotopic, then we would have $\int_{S^1} \varphi = 0$ for any closed 1-form φ, by Corollary 6.4.22. But

$$\int_{S^1} \varphi^1 = 2\pi \neq 0. \qquad \square$$

REMARK 6.4.26. Of course, it is geometrically "obvious" that the unit circle, which winds around the origin, cannot be deformed to a single point without passing through the origin, i.e., while remaining in $\mathbb{R}^2 - \{(0, 0)\}$. But it is not always the case that things that are "obvious" are easy to prove (or even true). $\diamond$

REMARK 6.4.27. Suppose φ is a closed 1-form on a smooth manifold M, and we wish to determine whether φ is exact.

Of course, if M is smoothly contractible, this is automatic, by Corollary 4.7.6. More generally, if every smooth curve in M is smoothly null-homotopic, this is also automatic, by Corollary 6.4.22.

Otherwise, Lemma 6.4.10 looks like a good method for showing that φ is *not* exact—all we have to do is find a single smooth closed curve C_0 in M with $\int_{C_0} \varphi \neq 0$.

On the other hand, it looks hopeless as a method for showing that φ *is* exact, as we would have to investigate $\int_{C_0} \varphi$ for *every* smooth closed curve C_0 in R. However, that turns out not to be the case. Rather, we need only investigate $\int_{C_0} \varphi$ for smooth closed curves C_0 that form a "test family," which often consists of only finitely many curves. We shall defer consideration of what constitutes such a family in general, but give an example here.   ◇

THEOREM 6.4.28.   *Let φ be a closed 1-form on $\mathbb{R}^2 - \{(0,0)\}$. Let S^1 denote the unit circle in $\mathbb{R}^2$, oriented counterclockwise, and set*

$$s_0 = \frac{1}{2\pi} \int_{S^1} \varphi,$$

where φ^1 is the 1-form of Example 1.2.17. Then $\psi = \varphi - s_0 \varphi^1$ is exact on $\mathbb{R}^2 - \{(0,0)\}$.

Proof.   Although we won't need this fact until later in the proof, let us first make the statement of the theorem a little less arcane by seeing where the definition of s_0 arises. Namely, if $\varphi = \varphi^1$, by Example 6.4.2, we have

$$s_0 = \frac{1}{2\pi} \int_{S^1} \varphi^1 = \frac{1}{2\pi}(2\pi) = 1.$$

Indeed, if φ is a multiple of φ^1, $\varphi = s\varphi^1$, this computation shows $s_0 = s$.

Let's also make an observation, for future reference. Let R be a region in the plane obtained by deleting any ray (i.e., any closed half-line). Then R is star-shaped. For example, if $R = \mathbb{R}^2 - \{(x,0) \mid x \leq 0\}$, i.e., the plane with the nonpositive part of the real line removed, then any point of R can be joined to the point $(1,0)$ by a line lying entirely in R, so R is star-shaped with $(1,0)$ as center.

The consequence of this observation that we will use is Theorem 2.4.3: Any closed 1-form on such a region R is exact.

Now, let $C(a)$ denote the circle of radius a with center at the origin, oriented counterclockwise. Then the unit circle S^1 is $C(1)$. We claim that, for any a, and for any closed 1-form φ,

$$\int_{C(a)} \varphi = \int_{C(1)} \varphi.$$

To see this, consider the following picture (where we have chosen $a > 1$, the case $a < 1$ being similar):

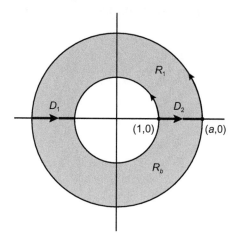

We have divided the region R between these two circles into two halves, R_t, the top half, and R_b, the bottom half, and we have labeled the two oriented straight line segments D_1 and D_2.

We give R the standard orientation. Then, keeping track of orientations,

$$\partial R = C(a) - C(1).$$

On the other hand, also keeping track of orientations,

$$\partial R_t = C_t(a) + D_1 - C_t(1) + D_2$$

and

$$\partial R_b = C_b(a) - D_2 - C_b(1) - D_1,$$

where C_t and C_b denote the top and bottom halves of the curves, so

$$\partial R = \partial R_t + \partial R_b$$

and hence

$$\int_{\partial R} \varphi = \int_{\partial R_t} \varphi + \int_{\partial R_b} \varphi.$$

But each of ∂R_t and ∂R_b is a closed curve contained in a region of the form $\mathbb{R}^2$ minus a ray (as described above), so the closed form φ is exact on each of these star-shaped regions, and hence, by Theorem 2.4.3,

$$\int_{\partial R} \varphi = \int_{\partial R_t} \varphi + \int_{\partial R_b} \varphi = 0 + 0 = 0.$$

But then

$$0 = \int_{\partial R} \varphi = \int_{C(a) - C(1)} \varphi = \int_{C(a)} \varphi - \int_{C(1)} \varphi,$$

so

$$\int_{C(a)} \varphi = \int_{C(1)} \varphi$$

as claimed.

Now let s_0 be as defined in the theorem, and set $\psi = \varphi - s_0 \varphi^1$. We wish to show that ψ is exact, and to this end we will find an explicit function F, defined on $\mathbb{R}^2 - \{(0, 0)\}$, with $dF = \psi$. Actually, we will begin by defining two functions F_1 and F_2, each on part of $\mathbb{R}^2 - \{(0, 0)\}$, and see that they match up to give a function F.

First let us consider the complement of the nonpositive x-axis in $\mathbb{R}^2$,

$$R_1 = \{(x, y) \in \mathbb{R}^2 \mid (x, y) \neq (a, 0) \text{ for some } a \leq 0\}.$$

For a point (x, y) in R_1, we let

$$F_1(x, y) = \int_{C_1} \psi,$$

where C_1 is the path obtained by starting at the point $(1, 0)$, going along the positive x-axis to $(a, 0)$, where $a = \sqrt{x^2 + y^2}$, and then counterclockwise along $C(a)$ from $(a, 0)$ to (x, y), if $y > 0$, or clockwise along $C(a)$ from $(a, 0)$ to (x, y) if $y < 0$, as shown. (Note this path always lies in R_1.)

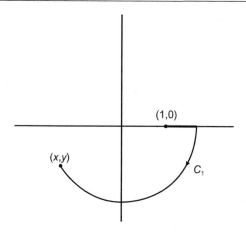

Next let us consider the complement of the nonpositive y-axis in $\mathbb{R}^2$,

$$R_2 = \{(x, y) \in \mathbb{R}^2 | (x, y) \neq (0, a) \text{ for some } a \geq 0\}.$$

For a point (x, y) in R_2, we let

$$F_2(x, y) = \int_{C_2} \psi,$$

where C_2 is the path obtained by starting at the point $(1, 0)$, going along the positive x-axis to $(a, 0)$, where $a = \sqrt{x^2 + y^2}$, and then clockwise along $C(a)$ from $(a, 0)$ to (x, y), if the point (x, y) is in the fourth quadrant, or counterclockwise from $(a, 0)$ to (x, y) otherwise, as shown. (Note this path always lies in R_2.)

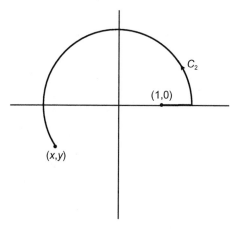

Now F_1 and F_2 are each continuous functions—if you vary the endpoints, the paths C_1 and C_2 don't change by much and hence neither do the integrals. (Of course, this needs proof, but it is easy to verify, and we omit the verification.) These are each defined on open subsets R_1 and R_2 of $\mathbb{R}^2 - \{(0,0)\}$, and so piece together to give a continuous function F on $\mathbb{R}^2 - \{(0,0)\} = R_1 \cup R_2$, providing they agree on their overlap $R_1 \cap R_2$.

In a moment we'll see that this is indeed the case, but first let's see that, assuming it, we are done. We need to show that ψ is exact, and we claim that in fact $\psi = dF$. But we already know that!

We have that $d\psi = d(\varphi - s_0\varphi^1) = 0$ so ψ is closed, and since R_1 and R_2 are each star-shaped, ψ is exact on R_1 and exact on R_2, as well.

Consider a point (x, y) of $\mathbb{R}^2 - (0, 0)$. Then (x, y) is in R_1 or R_2 (or perhaps both). Suppose $(x, y) \in R_1$. Then $F(x, y) = F_1(x, y)$. But $dF_1 = \psi$ on R_1—this is exactly the situation of Corollary 6.4.14. Similarly, if $(x, y) \in R_2$, then $F(x, y) = F_2(x, y)$ and $dF_2 = \psi$ on R_2 by Corollary 6.4.14.

Now to check agreement on the overlap: Note that $R_1 \cap R_2$ is everything not on one of the heavy lines shown here.

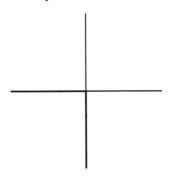

There are two possibilities for a point (x, y) of $R_1 \cap R_2$: Either it's in the third quadrant or it's not. If it's not, we certainly have $F_2(x, y) = F_1(x, y)$, as in this case C_2 and C_1 are the *same* path. Thus the crucial case to consider is when (x, y) is in the third quadrant, so C_2 and C_1 are *different* paths. (This is the case illustrated.) For such a point, $C_2 - C_1 = C(a)$, and, by the definition of s_0,

$$\int_{C(1)} \varphi = 2\pi s_0.$$

Then

$$F_2(x, y) - F_1(x, y) = \int_{C_2} \psi - \int_{C_1} \psi = \int_{C(a)} \psi$$

$$= \int_{C(a)} \varphi - s_0 \varphi_1$$

$$= \int_{C(a)} \varphi - \int_{C(a)} r \varphi_1$$

$$= \int_{C(a)} \varphi - s_0 \int_{C(a)} \varphi_1$$

$$= \int_{C(1)} \varphi - s_0 \int_{C(1)} \varphi_1$$

$$= 2\pi s_0 - s_0(2\pi) = 0,$$

and so $F_2(x, y) = F_1(x, y)$, completing the proof. $\qquad\qquad \square$

COROLLARY 6.4.29. *Let φ be a closed 1-form on $\mathbb{R}^2 - \{(0, 0)\}$ such that $\int_{S^1} \varphi = 0$. Then φ is exact on $\mathbb{R}^2 - \{(0, 0)\}$.*

Proof. This is just the special case of Theorem 6.4.28 when $s_0 = 0$. $\qquad\qquad \square$

We now relate integration of 1-forms in $\mathbb{R}^3$ to classical line integrals. As will be clear, this is a poor name. A much better name would be "curve integrals," but "line integrals" is the standard term, so we're stuck with it.

First we recall what line integrals are.

DEFINITION 6.4.30. Let C be an oriented curve in $\mathbb{R}^3$ parameterized by $r(t) = (f(t), g(t), h(t)), a \le t \le b$, and let $\mathbf{F}(x, y, z) = A(x, y, z)\mathbf{i} + B(x, y, z)\mathbf{j} + C(x, y, z)\mathbf{k}$ be a vector field defined on C. Then the *(line) integral of $\mathbf{F}$ along C* is

$$\int_C \mathbf{F} \cdot d\mathbf{r} = \int_a^b \mathbf{F}(t) \cdot \mathbf{r}'(t)dt$$

$$= \int_a^b \begin{bmatrix} A(r(t)) \\ B(r(t)) \\ C(r(t)) \end{bmatrix} \cdot \begin{bmatrix} f'(t) \\ g'(t) \\ h'(t) \end{bmatrix} dt$$

$$= \int_a^b \Big(A(f(t), g(t), h(t))f'(t) + B(f(t), g(t), h(t))g'(t) + C(f(t), g(t), h(t))h'(t) \Big) dt. \qquad \diamond$$

This has an alternative formulation. Note that $\mathbf{r}'(t)$ is the tangent vector to C at the point $r(t)$, and so $\mathbf{r}'(t)/\|\mathbf{r}'(t)\|$ is a unit vector pointing in the same direction as $\mathbf{r}'(t)$. This we call, for obvious reasons, the unit tangent vector to C at $r(t)$, and denote it by $\mathbf{T}(t)$. Then

$$\mathbf{T}(t) = \mathbf{r}'(t)/\|\mathbf{r}'(t)\| = \frac{1}{\sqrt{f'(t)^2 + g'(t)^2 + h'(t)^2}} \begin{bmatrix} f'(t) \\ g'(t) \\ h'(t) \end{bmatrix}.$$

PROPOSITION 6.4.31. *In the situation of Definition 6.4.30,*

$$\int_C \mathbf{F} \cdot d\mathbf{r} = \int_0^\ell \begin{bmatrix} A(r(t)) \\ B(r(t)) \\ C(r(t)) \end{bmatrix} \cdot \mathbf{T}(t)ds,$$

where the integral is taken with respect to arc length, and ℓ denotes the length of C.

Proof. Recall that $ds = \sqrt{f'(t)^2 + g'(t)^2 + h'(t)^2}dt = \|\mathbf{r}'(t)\|dt$. When this is substituted in the above integral, the $\|\mathbf{r}'(t)\|$ factors in the numerator and denominator cancel, leaving the integral in the original Definition 6.4.30.   □

In actual fact, though, line integrals are nothing new! This is one instance of our basic theme, that the constructions of vector calculus all come from differential forms. In this case the relationship is given by the following result:

THEOREM 6.4.32. *Let C be parameterized by $r(t) = (f(t), g(t),$ $h(t))$, $a \leq t \leq b$, and let $\mathbf{F}(x, y, z) = A(x, y, z)\mathbf{i} + B(x, y, z)\mathbf{j} +$ $C(x, y, z)\mathbf{k}$ be a vector field defined on C. Let φ be the 1-form that corresponds to the vector field $\mathbf{F}$ under the Fundamental Correspondence (Definition 1.3.6). Then:*

$$\int_C \mathbf{F} \cdot d\mathbf{r} = \int_C \varphi.$$

Proof. The 1-form φ that corresponds to $\mathbf{F}$ is given by $\varphi(x, y, z) =$ $A(x, y, z)dx + B(x, y, z)dy + C(x, y, z)dz$. Then the left-hand side of this equation is given by Definition 6.4.30, and the right-hand side by Theorem 6.4.4, and these are identical. □

COROLLARY 6.4.33.

(1) $\int_C \mathbf{F} \cdot d\mathbf{r}$ depends only on the oriented curve C, not on its parameterization.

(2) $\int_{-C} \mathbf{F} \cdot d\mathbf{r} = -\int_C \mathbf{F} \cdot d\mathbf{r}$.

Proof. Given Theorem 6.4.32, (1) and (2) are immediate from Lemma 6.2.7 and Corollary 6.2.12, respectively. □

Let us now restate some of our earlier results in the language of line integrals. There is no more work to do—we just have a direct translation from 1-forms to vector fields.

DEFINITION 6.4.34. Let $\mathbf{F}$ be a vector field on a region R of $\mathbb{R}^3$. Then $\int_C \mathbf{F} \cdot d\mathbf{r}$ is *path-independent* if the value of this line integral depends only on the endpoints (in order) of the path C in R. ◇

LEMMA 6.4.35. *Let $\mathbf{F}$ be a vector field on a region R of $\mathbb{R}^3$. Then $\int_C \mathbf{F} \cdot d\mathbf{r}$ is path-independent if and only if for every closed curve C_0 in R, $\int_{C_0} \mathbf{F} \cdot d\mathbf{r} = 0$.*

Recall that a vector field $\mathbf{F}$ is conservative if and only if the corresponding 1-form φ is exact.

THEOREM 6.4.36. *Let* **F** *be a vector field on a region R of* $\mathbb{R}^3$. *Then* **F** *is conservative on R if and only if* $\int_C \mathbf{F} \cdot d\mathbf{r}$ *is path-independent on R.*

COROLLARY 6.4.37. *The vector field*

$$\mathbf{F}(x, y) = \frac{-y}{x^2 + y^2}\mathbf{i} + \frac{x}{x^2 + y^2}\mathbf{j}$$

is not conservative on $\mathbb{R}^2 - \{(0, 0)\}$.

Recall that **curl**(**F**) $= 0$ if and only if φ is closed, where φ is the 1-form corresponding to **F**.

COROLLARY 6.4.38. *Let R be a region of* $\mathbb{R}^3$ *such that every component of R is smoothly simply connected, or, more generally, let R be a region of* $\mathbb{R}^3$ *with the property that every smooth map* $f: S^1 \longrightarrow R$ *is smoothly null-homotopic. Let* **F** *be a vector field on R with* **curl**(**F**) $= 0$. *Then* $\int_C \mathbf{F} \cdot d\mathbf{r}$ *is path-independent.*

Recall that f is a potential for **F** if **grad**(f) $=$ **F**, i.e., if $df = \varphi$, where φ is 1-form corresponding to **F**.

COROLLARY 6.4.39. *Let* **F** *be a conservative vector field on a region R in* $\mathbb{R}^3$ *and let f be a potential for* **F**. *Then for any path C in R from p to q,* $\int_C \mathbf{F} \cdot d\mathbf{r} = f(q) - f(p)$.

REMARK 6.4.40. We have stated these in $\mathbb{R}^3$ purely for convenience. It is clear that these results hold unchanged in $\mathbb{R}^n$, except for Corollary 6.4.38, as there is no notion of the curl of a vector field in $\mathbb{R}^n$ unless $n = 3$. ◇

Let us pause in our mathematical development to consider some physical questions.

We consider the following physical situation: *We have a moving particle of mass m in* $\mathbb{R}^n$, *whose position at time t is given by r(t).*

Then the velocity of this particle at time t is given by $\mathbf{v}(t) = \mathbf{r}'(t)$, and its acceleration at time t is given by $\mathbf{a}(t) = \mathbf{v}'(t)$.

DEFINITION 6.4.41. In this situation, the *kinetic energy* $E_k(t)$ of the particle at time t is

$$E_k(t) = \frac{1}{2}m\|\mathbf{v}(t)\|^2. \qquad \diamond$$

Notice that the kinetic energy of the particle depends on how fast the particle is moving, but not on where it is.

Now suppose that we have a force acting on our particle, with the force depending only on the position of the particle, so that it is given by $\mathbf{F}(r(t))$ for some vector field $\mathbf{F}$. Let the particle move along a piecewise smooth curve C, from point p (at time $t = a$) to point q (at time $t = b$); i.e., the position of the particle at time t is $r(t)$ with $r(a) = p$ and $r(b) = q$ for some parameterization of C.

DEFINITION 6.4.42. In this situation, the *work* W done on the particle by the force $\mathbf{F}$ is

$$W = \int_C \mathbf{F} \cdot d\mathbf{r}. \qquad \diamond$$

Of course, this is just the sort of line integral we have been considering. Already, Corollary 6.4.33 tells us something interesting—the work done depends only on the curve C, not on its parameterization.

Our first result is that the work done is equal to the change in the kinetic energy of the particle.

THEOREM 6.4.43. *Let a particle move along a curve C, subject to a force $\mathbf{F}$, as above, according to Newton's laws of motion. Then, if W is defined as above,*

$$W = E_k(b) - E_k(a).$$

Proof. Newton's laws state that $\mathbf{F} = m\mathbf{a}$, or, to be more precise in our situation here where $\mathbf{F}$ (but not m) is varying,

$$\mathbf{F}(t) = m\mathbf{a}(t).$$

By the definition of the line integral,

$$W = \int_C \mathbf{F} \cdot d\mathbf{r} = \int_a^b \mathbf{F}(t) \cdot \mathbf{r}'(t) dt.$$

As $\mathbf{a}(t) = \mathbf{v}'(t)$ and $\mathbf{r}'(t) = \mathbf{v}(t)$, we see

$$W = \int_a^b m\mathbf{v}'(t) \cdot \mathbf{v}(t) dt.$$

It is easy to check that $\frac{d}{dt}(\mathbf{v}(t) \cdot \mathbf{v}(t)) = 2\mathbf{v}'(t) \cdot \mathbf{v}(t)$, so

$$W = \frac{1}{2}m\mathbf{v}(t) \cdot \mathbf{v}(t)\big|_a^b = \frac{1}{2}m\mathbf{v}(b) \cdot \mathbf{v}(b) - \frac{1}{2}m\mathbf{v}(a) \cdot \mathbf{v}(a).$$

But for any vector $\mathbf{v}$, $\mathbf{v} \cdot \mathbf{v} = \|\mathbf{v}\|^2$, so we see

$$\begin{aligned} W &= \frac{1}{2}m\|\mathbf{v}(b)\|^2 - \frac{1}{2}m\|\mathbf{v}(a)\|^2 \\ &= E_k(b) - E_k(a). \end{aligned} \qquad \square$$

Now let us suppose in addition that our force field $\mathbf{F}$ is conservative. By Theorem 6.4.36, this tells us immediately that the work W is independent of path. But in this situation something even more interesting is true.

To say that $\mathbf{F}$ is conservative means that $\mathbf{F}$ has a potential; i.e., there is a function f with $\mathbf{F} = \mathbf{grad}(f)$. We fix such a function f.

DEFINITION 6.4.44. In this situation, where $\mathbf{F}$ is conservative with potential f, the *potential energy* of the particle at time t is

$$E_p(t) = -f(r(t)).$$

(Note the minus sign.)   ◇

Notice that the potential energy of the particle depends on where the particle is, but not on how fast it is moving. (In light of this, we could forget about the path $r(t)$ and directly define the potential energy in terms of position simply by defining potential energy at $(x, y, z) = -f(x, y, z)$.)

DEFINITION 6.4.45. In this situation, where **F** is a conservative force, the *energy* $E(t)$ of the particle at time t is the sum of its potential and kinetic energies, i.e.,

$$E(t) = E_p(t) + E_k(t). \qquad \diamond$$

THEOREM 6.4.46 (*Conservation of Energy*). *Let a particle move along a curve C, subject to a conservative force* **F**, *as above, according to Newton's laws of motion. Then*

$$E(b) = E(a).$$

Proof. By Theorem 6.4.43,

$$W = \int_C \mathbf{F} \cdot d\mathbf{r} = E_k(b) - E_k(a).$$

On the other hand, since **F** is conservative, Corollary 6.4.39 gives

$$\begin{aligned} W = \int_C \mathbf{F} \cdot d\mathbf{r} &= f(q) - f(p) \\ &= f(r(b)) - f(r(a)) \\ &= -E_p(b) + E_p(a) \end{aligned}$$

so

$$\begin{aligned} E_k(b) - E_k(a) &= -E_p(b) + E_p(a) \\ E_p(b) + E_k(b) &= E_p(a) + E_k(a) \\ E(b) &= E(a) \end{aligned}$$

as claimed.   □

6.5 The integral of a 2-form over a surface

In this section we investigate integrals of 2-forms over oriented 2-manifolds with boundary, i.e., over oriented surfaces. In addition to developing the theory of these in general, we will see that in the special case of surfaces in $\mathbb{R}^3$, they correspond to classical "flux integrals."

The initial setup here is entirely parallel to that in the last section.

Let $\mathbb{R}^2 = \{(u, v)\}$, and first consider an oriented surface S in $\mathbb{R}^2$. S is the archetype of an oriented surface. Consider a 2-form φ on S and write $\varphi = f(u, v)dudv$. Note that $dudv$ is the volume form on $\mathbb{R}^2$ in the standard orientation of $\mathbb{R}^2$. Then, directly from Lemma 6.2.2, we see that

$$\int_S \varphi = {}_s\!\!\int_S f(u, v)dA_{uv}.$$

We observe that if S has the standard orientation, then the sign in the signed definite integral is positive.

Now we deal with surfaces in $\mathbb{R}^n$ for $n > 2$. The theory works for any value of n, but for simplicity we will choose our examples in $\mathbb{R}^3$. Recall our theory from Section 6.2: Let S be an oriented surface in $\mathbb{R}^n$, parameterized by $k(u, v)$ for $(u, v) \in R$, where R is some oriented compact region in $\mathbb{R}^2$, and $k \colon R \longrightarrow S$ is orientation-preserving.

Let φ be a 2-form defined on S. Then

$$\int_S \varphi = \int_R k^*(\varphi)$$

and, furthermore:

(a) $\int_S \varphi$ depends only on the oriented surface S, but is independent of the parameterization k; and

(b) $\int_{-S} \varphi = -\int_S \varphi$ where $-S$ denotes S with the opposite orientation.

Also, we can concretely evaluate this integral as follows:

$$\int_S \varphi = {}_s\!\!\int_R k^*(\varphi)(\mathbf{e}^1, \mathbf{e}^2)dA_{uv} = {}_s\!\!\int_R (\varphi)(k_*(\mathbf{e}^1), k_*(\mathbf{e}^2))dA_{uv}$$

$$= {}_s\!\!\int_R (\varphi)(\mathbf{k}_u(u, v), \mathbf{k}_v(u, v))dA_{uv},$$

where we use the fact that $k_* = k'$. We also recall that we can evaluate $k^*(\varphi)$ by "substitution," and that if $k^*(\varphi) = A(u, v)dudv$, then $k^*(\varphi)(\mathbf{e}^1, \mathbf{e}^2) = A(u, v)$.

Before proceeding further we observe that in this situation we have tiled S with a single tile.

First we give a couple of examples of integrals on surfaces in $\mathbb{R}^2$.

EXAMPLE 6.5.1.

(a) If T is the rectangle $\{(x, y) | 1 \leq x \leq 3, 1 \leq y \leq 2\}$ with its standard orientation, and $\varphi = 6x^2 y dx dy$, then

$$\int_T \varphi = + \iint_T 6x^2 y dA_{xy}.$$

(Note the positive sign on the right.)

To evaluate this double integral we use Fubini's theorem, which tells us we can evaluate it as an iterated integral (in any order)

$$= \int_1^2 \int_1^3 6x^2 y dx dy = \int_1^3 \int_1^2 6x^2 y dy dx$$

and you can easily check that the common value of these integrals is 78.

(b) If T is the top half of the unit disk with its standard orientation, and $\varphi = dx dy$, then

$$\int_T \varphi = + \iint_T dA_{xy}.$$

(Again note the positive sign on the right).

The value of this integral is simply the area of T, which is $\pi/2$. Of course, we may also compute it by Fubini's theorem

$$= \int_{-1}^1 \int_0^{\sqrt{1-x^2}} dy dx = \int_0^1 \int_{-\sqrt{1-y^2}}^{\sqrt{1-y^2}} dx dy$$

and you may check that the common value of these integrals is $\pi/2$. ◇

DEFINITION 6.5.2. Let $z = h(x, y)$ be a smooth function on an oriented surface T in $\mathbb{R}^2$. Then the *graph* of this function is

$$S = \{(x, y, z) \subset \mathbb{R}^3 | z = h(x, y)\},$$

regarded as a smooth surface with parameterization $k: T \longrightarrow S$ defined by $(x, y, z) = k(x, y) = (x, y, h(x, y))$, with the orientation induced by k from the orientation on T. If T has the standard orientation of a surface in $\mathbb{R}^2$ then we will say that S has the standard orientation of a graph.   ◇

(Strictly speaking, we should use different variables and write $(x, y, z) = k(u, v) = (u, v, h(u, v))$ but in the situation of a graph it is simpler to "reuse" the same variables.)

EXAMPLE 6.5.3.

(a) Let T be the rectangle $\{(x, y) | 1 \le x \le 3, 1 \le y \le 2\}$. Let S be the graph of $z = h(x, y) = x^3 + xy^2$ on T with the standard orientation of a graph. Let $\varphi = 6(2x^3 - y^3)dy\,dz + z\,dz\,dx + 2xyz\,dx\,dy$.

We compute $\int_S \varphi$.

We will do this two ways—first, directly from the definition of the integral, and second, using "substitution."

The first method: By definition,

$$\int_S \varphi = + \iint_T \varphi(\mathbf{k}_x, \mathbf{k}_y)\,dA_{xy}.$$

(Note the sign is $+$ as S has the standard orientation of a graph.)
We compute

$$\mathbf{k}_x = \begin{bmatrix} 1 \\ 0 \\ 3x^2 + y^2 \end{bmatrix} \text{ and } \mathbf{k}_y = \begin{bmatrix} 0 \\ 1 \\ 2xy \end{bmatrix}.$$

Then

$$\varphi(\mathbf{k}_x, \mathbf{k}_y) = 6(2x^3 - y^3)dy\,dz \left(\begin{bmatrix} 1 \\ 0 \\ 3x^2 + y^2 \end{bmatrix}, \begin{bmatrix} 0 \\ 1 \\ 2xy \end{bmatrix} \right)$$

$$+ (x^3 + xy^2)dz\,dx \left(\begin{bmatrix} 1 \\ 0 \\ 3x^2 + y^2 \end{bmatrix}, \begin{bmatrix} 0 \\ 1 \\ 2xy \end{bmatrix} \right)$$

$$+ 2xy(x^3 + xy^2)dx\,dy \left(\begin{bmatrix} 1 \\ 0 \\ 3x^2 + y^2 \end{bmatrix}, \begin{bmatrix} 0 \\ 1 \\ 2xy \end{bmatrix} \right).$$

Now recall that, in general, $dydz(\mathbf{v}, \mathbf{w}) = dy(\mathbf{v})dz(\mathbf{w}) - dy(\mathbf{w})dz(\mathbf{v})$, and similarly for $dzdx(\mathbf{v}, \mathbf{w})$ and $dxdy(\mathbf{v}, \mathbf{w})$. Thus we see

$$
\begin{aligned}
\varphi(\mathbf{k}_x, \mathbf{k}_y) &= 6(2x^3 - y^3)(0 \cdot 2xy - 1 \cdot (3x^2 + y^2)) \\
&\quad + (x^3 + xy^2)((3x^2 + y^2) \cdot 0 - 2xy \cdot 1) \\
&\quad + (2xy)(x^3 + xy^2)(1 \cdot 0 - 0 \cdot 1) \\
&= -6(2x^3 - y^2)(3x^2 + y^2) - (x^3 + xy^2)(2xy) \\
&\quad + (2xy)(x^3 + xy^2) \\
&= -36x^5 - 12x^3y^2 + 18x^2y^3 + 6y^5
\end{aligned}
$$

and hence

$$
\int_S \varphi = \iint_T (-36x^5 - 12x^3y^2 + 18x^2y^3 + 6y^5) dA_{xy}.
$$

You will see shortly that we get exactly the same integral for the second method (as we must), and we evaluate it there.

The second method: Here $x = x$, $y = y$, and $z = x^3 + xy^2$. Note that $dx = dx$, $dy = dy$ (of course), and $dz = h_x dx + h_y dy = (3x^2 + y^2)dx + 2xydy$. Then

$$
\begin{aligned}
k^*(\varphi) &= 6(2x^3 - y^3)dy((3x^2 + y^2)dx + 2xydy) \\
&\quad + (x^3 + xy^2)((3x^2 + y^2)dx + 2xydy)dx \\
&\quad + 2xy(x^3 + xy^2)dxdy \\
&= -6(2x^3 - y^3)(3x^2 + y^2)dxdy - (x^3 + xy^2)(2xy)dxdy \\
&\quad + 2xy(x^3 + xy^2)dxdy \\
&= (-36x^5 - 12x^3y^2 + 18x^2y^3 + 6y^5)dxdy,
\end{aligned}
$$

so

$$
\begin{aligned}
\int_S \varphi &= \iint_T (-36x^5 - 12x^3y^2 + 18x^2y^3 + 6y^5) dA_{xy} \\
&= \int_1^2 \int_1^3 (-36x^5 - 12x^3y^2 + 18x^2y^3 + 6y^5) dxdy
\end{aligned}
$$

$$= \int_1^2 (-6x^6 - 3x^4y^2 + 6x^3y^3 + 6xy^5)|_1^3 dy$$

$$= \int_1^2 (-4368 - 240y^2 + 156y^3 + 12y^5)dy$$

$$= -4368y - 80y^3 + 39y^4 + 2y^6|_1^2 = -4217.$$

(b) Let T be the top half of the unit disk, and let S be the graph of $z = h(x, y) = x^2 + y^2$ on T, with the standard orientation of the graph. Let $\varphi = dzdx + 8ydxdy$.

We compute $\int_S \varphi$.

We simply use the second method. Again $dx = dx$ and $dy = dy$. This time $dz = 2xdx + 2ydy$.

Then $k^*(\varphi) = (2xdx + 2ydy)dx + 8ydxdy = 6ydxdy$, so

$$\int_S \varphi = \iint_T 6y\,dA_{xy}$$

$$= \int_{-1}^1 \int_0^{\sqrt{1-x^2}} 6y\,dy\,dx$$

$$= \int_{-1}^1 3y^2|_0^{\sqrt{1-x^2}}\,dx$$

$$= \int_{-1}^1 3(1 - x^2)dx = 3x - x^3|_{-1}^1 = 4.$$

(Note that in part (b) of this example we evaluated the iterated integral in the opposite order to that in part (a), for variety.)

(c) Let T be the triangular region bounded by the coordinate axes and the line $u + v = 1$, with its standard orientation. Let S be the surface given by

$$(x, y, z) = k(u, v) = (u - v, uv, u^2 + (v + 1)^2), \ (u, v) \in T$$

with orientation induced by k from the orientation on T. Let φ be the 2-form $\varphi = dydz - (1 + x^2)ddy$.

We compute $\int_S \varphi$. We use both methods.

The first method: We have the parameterized surface $k(u, v) = (u - v, uv, u^2 + (v + 1)^2)$, and so we compute

$$\mathbf{k}_u = \begin{bmatrix} 1 \\ v \\ 2u \end{bmatrix} \text{ and } \mathbf{k}_v = \begin{bmatrix} -1 \\ u \\ 2(v + 1) \end{bmatrix}.$$

We have $\varphi = dydz - (1 + x^2)dxdy$, so

$$\varphi(\mathbf{k}_u, \mathbf{k}_v) = dydz \left(\begin{bmatrix} 1 \\ v \\ 2u \end{bmatrix}, \begin{bmatrix} -1 \\ u \\ 2(v + 1) \end{bmatrix} \right)$$

$$- (1 + (u - v)^2)dxdy \left(\begin{bmatrix} 1 \\ v \\ 2u \end{bmatrix}, \begin{bmatrix} -1 \\ u \\ 2(v + 1) \end{bmatrix} \right)$$

$$= dy \left(\begin{bmatrix} 1 \\ v \\ 2u \end{bmatrix} \right) dz \left(\begin{bmatrix} -1 \\ u \\ 2(v + 1) \end{bmatrix} \right)$$

$$- dy \left(\begin{bmatrix} -1 \\ u \\ 2(v + 1) \end{bmatrix} \right) dz \left(\begin{bmatrix} 1 \\ v \\ 2u \end{bmatrix} \right)$$

$$- (1 + (u - v)^2) \left[dx \left(\begin{bmatrix} 1 \\ v \\ 2u \end{bmatrix} \right) dy \left(\begin{bmatrix} -1 \\ u \\ 2(v + 1) \end{bmatrix} \right) \right.$$

$$\left. - dx \left(\begin{bmatrix} -1 \\ u \\ 2(v + 1) \end{bmatrix} \right) dy \left(\begin{bmatrix} 1 \\ v \\ 2u \end{bmatrix} \right) \right]$$

$$= v(2(v + 1)) - u(2u)$$
$$- (1 + (u - v)^2)[1 \cdot u - (-1) \cdot v]$$
$$= 2v(v + 1) - 2u^2 - (1 + (u - v)^2)(u + v)$$
$$= -u^3 + uv^2 + u^2v - v^3 - 2u^2 + 2v^2 - u + v$$

and so, noting that the sign is $+$ here, as the orientation on S is induced from the standard orientation of T,

$$\int_S \varphi = + \int\int_T (-u^3 + uv^2 + u^2v - v^3 - 2u^2 + 2v^2 - u + v)dA_{uv}.$$

This is the same integral as in the second method (of course) and we find it there.

The second method: Note that $x = u - v$, so $dx = du - dv$; $y = uv$, so $dy = vdu + udv$; and $z = u^2 + (v + 1)^2$, so $dz = 2udu + 2(v+1)dv$. We have $\varphi = dydz - (1+x^2)dxdy$. Substituting, we obtain

$$
\begin{aligned}
k^*(\varphi) &= (vdu + udv)(2udu + 2(v + 1)dv) \\
&\quad - (1 + (u - v)^2)(du - dv)(vdu + udv) \\
&= (2v(v + 1) - 2u^2)dudv - (1 + (u - v)^2)(u + v)dudv \\
&= (-u^3 + uv^2 + u^2v - v^3 - 2u^2 + 2v^2 - u + v)dudv
\end{aligned}
$$

so

$$
\begin{aligned}
\int_S \varphi &= \iint_T (-u^3 + uv^2 + u^2v - v^3 - 2u^2 + 2v^2 - uv)dA_{uv} \\
&= \int_0^1 \int_0^{1-u} (-u^3 + uv^2 + u^2v - v^3 - 2u^2 \\
&\qquad\qquad\qquad + 2v^2 - u + v)dvdu \\
&= \int_0^1 -u^3v + \frac{1}{3}uv^3 + \frac{1}{2}u^2v^2 - \frac{1}{4}v^4 - 2u^2v \\
&\qquad\quad + \frac{2}{3}v^3 - uv + \frac{1}{2}v^2 \Big|_0^{1-u} du \\
&= \int_0^1 \frac{11u^4 + 16u^3 - 6u^2 - 32u + 11}{12}du \\
&= \frac{1}{12}\left(\frac{11}{5}u^5 + 4u^4 - 2u^3 - 16u^2 + 11u\right)\Big|_0^1 = -\frac{1}{15}. \qquad \diamond
\end{aligned}
$$

We again write down a pair of propositions which simply codifies the procedure we have been using. We have two propositions, one for each method.

PROPOSITION 6.5.4. *Let T be a surface in $\mathbb{R}^2$ and let S be the surface in $\mathbb{R}^3$ that is the image of T under a diffeomorphism $k\colon T \longrightarrow S$, with the orientation of S induced from the orientation of T by k. Let*

k be given by $(x, y, z) = k(u, v) = (f(u, v), g(u, v), h(u, v))$. *Let* φ
be the 2-form on S given by $\varphi = A(x, y, z)dydz + B(x, y, z)dzdx + C(x, y, z)dxdy$. *Then*

$$
\int_S \varphi = \pm \iint_T \Big[A(k(u, v))\big(g_u(u, v)h_v(u, v) - g_v(u, v)h_u(u, v)\big)
$$

$$
+ B(k(u, v))\big(h_u(u, v)f_v(u, v) - h_v(u, v)f_u(u, v)\big)
$$

$$
+ C(k(u, v))\big(f_u(u, v)g_v(u, v) - f_v(u, v)g_u(u, v)\big) \Big] dA_{uv},
$$

where the sign is positive (resp. negative) if T has the standard (resp. nonstandard) orientation of a surface in $\mathbb{R}^2$.

Proof. We know that

$$
\int_S \varphi = \int_{S\int T} (\varphi)(k_*(\mathbf{e}^1), k_*(\mathbf{e}^2)) dA_{uv}
$$

and $k_*(\mathbf{e}^1) = \mathbf{k}_u$, $k_*(\mathbf{e}^2) = \mathbf{k}_v$, so we see that the proposition simply claims that $\varphi(\mathbf{k}_u, \mathbf{k}_v)$ is the above integrand. This is a matter of direct computation.

The form φ has three terms, and we will simply do the computation for one of them, the others being similar. We choose to do the third term. Thus, let $\varphi_1(x, y, z) = C(x, y, z)dxdy$. Then

$$
\begin{aligned}
\varphi_1(x, y, z)(\mathbf{k}_u, \mathbf{k}_v) &= C(x, y, z)dxdy(\mathbf{k}_u, \mathbf{k}_v) \\
&= C(k(u, v))dxdy(\mathbf{k}_u, \mathbf{k}_v) \\
&= C(k(u, v))\big(dx(\mathbf{k}_u)dy(\mathbf{k}_v) \\
&\qquad - dx(\mathbf{k}_v)dy(\mathbf{k}_u)\big).
\end{aligned}
$$

But we've computed $\mathbf{k}_u$ and $\mathbf{k}_v$. At the point $(x, y, z) = k(u, v)$, they are

$$
\mathbf{k}_u = \begin{bmatrix} f_u(u, v) \\ g_u(u, v) \\ h_u(u, v) \end{bmatrix} \text{ and } \mathbf{k}_v = \begin{bmatrix} f_v(u, v) \\ g_v(u, v) \\ h_v(u, v) \end{bmatrix}
$$

Thus, by the definitions of dx and dy,

$$dx(\mathbf{k}_u) = f_u(u, v) \qquad dx(\mathbf{k}_v) = f_v(u, v)$$
$$dy(\mathbf{k}_u) = g_u(u, v) \qquad dy(\mathbf{k}_v) = g_v(u, v)$$

so

$$dx(\mathbf{k}_u)dy(\mathbf{k}_v) - dy(\mathbf{k}_u)dx(\mathbf{k}_v)$$
$$= f_u(u, v)g_v(u, v) - f_v(u, v)g_u(u, v)$$

and substituting into the equation above yields the proposition. □

PROPOSITION 6.5.5. *In the situation of Proposition 6.5.4, let ψ be the 2-form on T defined by*

$$\psi = A(k(u, v))dgdh + B(k(u, v))dhdf + C(k(u, v))dfdg.$$

Then

$$\int_S \varphi = \int_T \psi.$$

Proof. This is just the claim that $\psi = k^*(\varphi)$. But we have already seen that, as this is the method of calculating pullbacks by "substitution":

$$x = f(u, v) \text{ so } dx = df = f_u du + f_v dv,$$
$$y = g(u, v) \text{ so } dy = dg = g_u du + g_v dv,$$
$$z = h(u, v) \text{ so } dz = dh = h_u du + h_v dv,$$

where $f_u du + f_v dv$ means $f_u(u, v)du + f_v(u, v)dv$, etc.—we have merely omitted the arguments for the sake of brevity. □

We can carry this one step further.

COROLLARY 6.5.6. *In the situation of Proposition 6.5.5, suppose $\psi = D(u, v)dudv$. Then*

$$\int_S \varphi = \pm \iint_T D(u, v)dA_{uv},$$

where the sign is + (resp. −) if T has the standard (resp. nonstandard) orientation.

Proof. Once we have calculated that $\psi = D(u, v)dudv$, we immediately have that $\psi(\mathbf{e}^1, \mathbf{e}^2) = D(u, v)$.   □

Before proceeding further we want to more carefully consider the standard orientation of a graph.

REMARK 6.5.7. Consider the case where S is the surface in $\mathbb{R}^3$ which is the graph of the function $z = h(x, y)$ defined on a surface T in the plane. Then $(x, y, z) = k(x, y) = (x, y, h(x, y))$. Suppose T has the standard orientation (counterclockwise). We have called the orientation on S induced by k the standard orientation of a graph.

With a little bit of geometry, we can see that *the standard orientation of a graph is counterclockwise when viewed from above.*

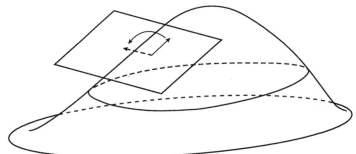

Even though the tangent plane at a point may not be horizontal, this still makes sense, as you can see from the figure. More precisely, in this situation

$$\mathbf{k}_x = \begin{bmatrix} 1 \\ 0 \\ h_x \end{bmatrix}, \quad \mathbf{k}_y = \begin{bmatrix} 0 \\ 1 \\ h_y \end{bmatrix}.$$

so, the direction of rotation from the projection $\begin{bmatrix} 1 \\ 0 \end{bmatrix}$ of the first of these vectors to the projection $\begin{bmatrix} 0 \\ 1 \end{bmatrix}$ of the second of these two vectors is counterclockwise, and that is what we see when we are viewing them from above.

Suppose furthermore that $\mathbb{R}^3$ has the standard orientation, given by the right-hand rule. We can then ask about the induced orientation in the normal direction; this is similar to the situation we have seen before, except that here we have no notion of "outward." However, the same logic applies: Given two tangent vectors $\mathbf{v}_1$ and $\mathbf{v}_2$ at a point, with the direction from $\mathbf{v}_1$ to $\mathbf{v}_2$ counterclockwise when viewed from above (i.e., given the standard orientation of a graph as we have just defined it), we want a third vector $\mathbf{w}$, normal to the surface, so that the triple $\{\mathbf{w}, \mathbf{v}_1, \mathbf{v}_2\}$ (in that order) defines a right-hand frame (i.e., gives the standard orientation on $\mathbb{R}^3$). A little geometry shows that the desired vector $\mathbf{w}$ is the *upward-pointing normal*. (You can check this in the above picture. If you put the base of your right hand on the tangent plane, with your thumb pointed up, then your fingers curl from the solid tangent vector to the dashed tangent vector.)   ◇

With this in hand we consider another pair of examples.

EXAMPLE 6.5.8.   Let S be the surface consisting of the graph of $z = h(x, y) = 36 - 4x^2 - 9y^2$ for $4x^2 + 9y^2 \leq 36$, together with the points $(x, y, 0)$ in the xy-plane with $4x^2 + 9y^2 \leq 36$, with the standard orientation on S.

Observe that S is *piecewise-smooth* but not smooth. This presents no problem—we shall merely integrate over the smooth pieces and add the values to get the integral over S. Note that S is *closed* surface, so the standard orientation is the one induced on it as the boundary of a solid body in $\mathbb{R}^3$ with the standard orientation. This body is a dome over an elliptical base in the xy-plane with part of an elliptical paraboloid as its top.

Write $S = S_1 \cup T$, where T is the portion of S that lies in the xy-plane and S_1 is the portion of S that is the graph of $z = h(x, y)$ over T. Now S_1 gets the orientation which is counterclockwise when viewed from the outside, that is from above (the point $(0, 0, 36)$ of S_1 being the highest point on S), and this is the standard orientation for the graph. Also, T gets the orientation that is counterclockwise when viewed from the outside, which in this case is from *below*, so this is the *opposite* of the standard orientation for the surface T in the plane.

Let φ be the 2-form $\varphi = 4y\,dy\,dz - 9x\,dz\,dx + 6x^2y^2\,dx\,dy$. Again we wish to compute, $\int_S \varphi$, and we see $\int_S \varphi = \int_{S_1} \varphi + \int_T \varphi$, so we compute each of these two. In each case we use "substitution."

First we compute $\int_{S_1} \varphi$. The surface S_1 is parameterized by $(x, y, z) = k_1(x, y) = (x, y, 36 - 4x^2 - 9y^2)$, so

$$dx = dx, dy = dy, dy = dy, \quad \text{and} \quad dz = -8x\,dx - 18y\,dy.$$

We further compute

$$dy\,dz = 8x\,dx\,dy, dz\,dx = 18y\,dx\,dy, \quad \text{and} \quad dx\,dy = dx\,dy.$$

Then, substituting, we find

$$
\begin{aligned}
k_1^*(\varphi) &= 4y(8x\,dx\,dy) - 9x(18y\,dx\,dy) + 6x^2y^2\,dx\,dy \\
&= (-130xy + 6x^2y^2)\,dx\,dy
\end{aligned}
$$

and so, bearing in mind the orientation of S_1, we see

$$\int_{S_1} \varphi = \iint_T (-130xy + 6x^2y^2)\,dA_{xy}.$$

Next we compute $\int_T \varphi$. The surface T is parameterized by $(x, y, z) = k_2(x, y) = (x, y, 0)$, so

$$dx = dx, dy = dy, dy = dy, \quad \text{and} \quad dz = 0,$$

and we see immediately that

$$k_2^*(\varphi) = 6x^2y^2\,dx\,dy$$

and so, bearing in mind the orientation of T, we see

$$\int_T \varphi = -\iint_T 6x^2y^2\,dA_{xy}.$$

Putting these together, we conclude

$$\int_S \varphi = \iint_T (-130xy)dA_{xy}$$

$$= \int_{-3}^3 \int_{-\frac{2}{3}\sqrt{9-x^2}}^{\frac{2}{3}\sqrt{9-x^2}} -130xy\,dy\,dx$$

$$= \int_{-3}^3 -65xy^2\Big|_{-\frac{2}{3}\sqrt{9-x^2}}^{\frac{2}{3}\sqrt{9-x^2}}dx = \int_{-3}^3 0\,dx = 0. \qquad \diamond$$

REMARK 6.5.9. Example 6.5.8 illustrates a situation we often encounter: A closed surface S is decomposed as a union of two tiles $S = S_{top} \cup S_{bot}$; here, each of S_{top} and S_{bot} is the graph of a function, where S_{top} is the top part of S and S_{bot} is the bottom part. In this situation, if S has the standard orientation of a closed surface, then S_{top} has the standard orientation of a graph and S_{bot} the nonstandard orientation. $\diamond$

EXAMPLE 6.5.10. Let S^2 be the unit sphere in 3-space with its standard orientation, i.e., the induced orientation it gets as the boundary of a solid ball in $\mathbb{R}^3$ with the standard orientation. Let φ^2 be the 2-form of Example 1.2.18,

$$\varphi^2 = \frac{x}{(x^2 + y^2 + z^2)^{3/2}}dy\,dz + \frac{y}{(x^2 + y^2 + z^2)^{3/2}}dz\,dx$$
$$+ \frac{z}{(x^2 + y^2 + z^2)^{3/2}}dx\,dy.$$

We compute $\int_{S^2} \varphi^2$.

In order to do so we use the tiling $\{k_+ : T \longrightarrow H_N, k_- : T \longrightarrow H_S\}$ of Example 6.2.16(b).

Now S^2 has its standard orientation. Let us first see what the two orientations on T, compatible with that on H_N via k_+ and on H_S via k_-, respectively, are.

Once again, the standard orientation for S^2 is the one induced on it as the boundary of the unit ball, where the unit ball inherits the standard (right-hand) orientation from $\mathbb{R}^3$. This orientation is counterclockwise when viewed from outside.

At the north pole ($p_+ = (0, 0, 1) = k_+(0, 0) \in H_N$), this means counterclockwise when viewed from above. It is routine to check, by direct computation of the partial derivatives, that at p_+, $(\mathbf{k}_+)_u = 2\mathbf{i}$ and $(\mathbf{k}_+)_v = 2\mathbf{j}$. When viewed from above, the direction of rotation from $(\mathbf{k}_+)_u$ to $(\mathbf{k}_+)_v$ is indeed counterclockwise, so the standard orientation of T is the one compatible with that on H_N. At the south pole ($p_- = (0, 0, -1) \in H_N$), this means counterclockwise when viewed from below. A similar computation shows that, at p_-, $(\mathbf{k}_-)_u = 2\mathbf{i}$ and $(\mathbf{k}_-)_v = -2\mathbf{j}$. When viewed from below, the direction of rotation from $(\mathbf{k}_-)_u$ to $(\mathbf{k}_-)_v$ is again counterclockwise (as when viewed from above, as usual, the direction is clockwise), and so the standard orientation of T is again the one compatible with that on H_S. Thus, to summarize, in both cases T gets its standard orientation, counterclockwise, as a subset of $\mathbb{R}^2$.

Now we come to computing $(k_+)^*(\varphi^2)$ and $(k_-)^*(\varphi^2)$. On the unit sphere, $x^2 + y^2 + z^2 = 1$, so (fortunately) we can ignore the denominators in the expression for φ^2.

We first find $(k_+)^*(\varphi^2)$. We have on H_N

$$x = \frac{2u}{u^2 + v^2 + 1}, \qquad y = \frac{2u}{u^2 + v^2 + 1}, \qquad z = \frac{1 - (u^2 + v^2)}{u^2 + v^2 + 1}$$

and routine but lengthy computations show

$$dx = \frac{2(-u^2 + v^2 + 1)du - 4uvdv}{(u^2 + v^2 + 1)^2},$$

$$dy = \frac{-4uvdu + 2(-u^2 + v^2 + 1)dv}{(u^2 + v^2 + 1)^2},$$

$$dz = \frac{-4udu - 4vdv}{(u^2 + v^2 + 1)^2}.$$

Substituting all these into our expression for φ^2 and doing a final routine (even lengthier) computation shows that on H_N

$$(k_+)^*(\varphi^2) = \frac{4}{(u^2 + v^2 + 1)^2} dudv$$

(a blessedly simple answer).

Now we also have to find $(k_-)^*(\varphi^2)$, but we can see, without doing much computation (fortunately) that it is the same. To see this, write

$$k_+(u, v) = (f_+(u, v), g_+(u, v), h_+(u, v)),$$
$$k_-(u, v) = (f_-(u, v), g_-(u, v), h_-(u, v)).$$

By comparing the formulas for k_+ and k_-, we see

$$f_-(u, v) = f_+(u, v), \ g_-(u, v) = -g_+(u, v),$$
$$h_-(u, v) = -h_+(u, v).$$

On H_N,

$$(k_+)^*(\varphi^2) = (f_+)(dg_+)(dh_+) + (g_+)(dh_+)(df_+)$$
$$+ (h_+)(df_+)(dg_+)$$

(omitting the arguments for brevity), while on H_S,

$$\begin{aligned}
(k_-)^*(\varphi^2) &= (f_-)(dg_-)(dh_-) + (g_-)(dh_-)(df_-) \\
&\quad + (h_-)(df_-)(dg_-) \\
&= (f_+)(d(-g_+))(d(-h_+)) + (-g_+)(d(-h_+)) \\
&\qquad (d(f_+)) + (-h_+)(d(f_+))(d(-g_+)) \\
&= (f_+)(-dg_+)(-dh_+) + (-g_+)(-dh_+)(df_+) \\
&\quad + (-h_+)(df_+)(-dg_+) \\
&= (f_+)(dg_+)(dh_+) + (g_+)(dh_+)(df_+) \\
&\quad + (h_+)(df_+)(dg_+)
\end{aligned}$$

which is the same expression we had before. Hence, also on H_S,

$$(k_-)^*(\varphi^2) = \frac{4}{(u^2 + v^2 + 1)^2} du\, dv.$$

Then, remembering that the unit disk T has the standard orientation in both cases,

$$\int_{S^2} \varphi^2 = \int_{H_N} \varphi^2 + \int_{H_S} \varphi^2$$

$$= \int_T (k_+)^*(\varphi^2) + \int_T (k_-)^*(\varphi^2)$$

$$= \iint_T \frac{4}{(u^2+v^2+1)^2} dA_{uv} + \iint_T \frac{4}{(u^2+v^2+1)^2} dA_{uv}$$

$$= \iint_T \frac{8}{(u^2+v^2+1)^2} dA_{uv}.$$

The easiest way to evaluate this integral is to change it into polar coordinates, which gives us

$$\int_{S^2} \varphi^2 = \int_0^{2\pi} \int_0^1 \frac{8}{(r^2+1)^2} r \, dr \, d\theta = \int_0^{2\pi} \frac{-4}{r^2+1} \bigg|_0^1 d\theta$$

$$= \int_0^{2\pi} 2 \, d\theta = 2\theta \big|_0^{2\pi} = 4\pi. \qquad \diamond$$

REMARK 6.5.11. We will see that the Generalized Stokes's Theorem will enable us to evaluate the integral in Example 6.5.8 with almost no computation at all, and the integral in Example 6.5.10 with very little computation. $\qquad \diamond$

Now let us relate our integrals to the usual surface integrals in $\mathbb{R}^3$, also known as *flux integrals*. First we need a bit of geometry.

LEMMA 6.5.12. *Let S be the graph of the function* $z = h(x, y)$, $(x, y) \in T$, *and let* $p_0 = (x_0, y_0, z_0)$ *be a point of S. Then the vector*

$$\mathbf{m}_{p_0} = \begin{bmatrix} -h_x(x_0, y_0) \\ -h_y(x_0, y_0) \\ 1 \end{bmatrix}_{p_0}$$

is an upward-pointing normal to S at p_0.

Proof. The vector $\mathbf{m}_{p_0}$ certainly points up, as the coefficient of $\mathbf{k}_{p_0}$ is $+1$. We need to see it that is normal to S.

Recall that if a surface S is given by $f(x, y, z) = 0$, a normal vector to it at the point p_0 is

$$\mathbf{grad}(f)(p_0) = \begin{bmatrix} f_x(p_0) \\ f_y(p_0) \\ f_z(p_0) \end{bmatrix}_{p_0}.$$

Since S is given by $z = h(x, y)$, we may also express the equation of S as

$$f(x, y, z) = z - h(x, y) = 0,$$

and then

$$\mathbf{grad}(f)(p_0) = \begin{bmatrix} -h_x(x_0, y_0) \\ -h_y(x_0, y_0) \\ 1 \end{bmatrix}_{p_0} = \mathbf{m}_{p_0}. \qquad \square$$

DEFINITION 6.5.13. Let S be the graph of the function $z = h(x, y)$, $(x, y) \in T$, with the standard orientation for a graph, and let $\mathbf{F}(x, y, z) = A(x, y, z)\mathbf{i} + B(x, y, z)\mathbf{j} + C(x, y, z)\mathbf{k}$ be a vector field defined on S. Then the *flux integral* of $\mathbf{F}$ over S is

$$\iint_T \mathbf{F}(x, y, z) \cdot \mathbf{m}(x, y, z) dA_{xy}. \qquad \diamond$$

This definition has an alternate, and more common, formulation.

Let us set $\mathbf{n}(x, y, z) = \mathbf{m}(x, y, z)/\|\mathbf{m}(x, y, z)\|$, the upward-pointing unit normal to the surface S.

PROPOSITION 6.5.14. *In the situation of Definition 6.5.13, the flux integral of $\mathbf{F}$ over S is*

$$\iint_S \mathbf{F}(x, y, z) \cdot \mathbf{n}(x, y, z) dA,$$

where the integral is taken with respect to the element of surface area on S.

Proof. Recall that $dA = \sqrt{h_x(x, y)^2 + h_y(x, y)^2 + 1}\, dA_{xy} = (\|\mathbf{m}(x, y, z)\|)dA_{xy}$; when this is substituted, the $\|\mathbf{m}(x, y, z)\|$ factors in the numerator and denominator cancel, leaving the integral in Definition 6.5.13.   □

One advantage of Proposition 6.5.14 is that it easily allows us to define flux integrals over closed surfaces. Recall from Definition 4.4.7 that the standard orientation of a closed surface in $\mathbb{R}^3$ is induced from the standard orientation of the solid body it bounds via the *outward-pointing* normal.

DEFINITION 6.5.15.   Let S be a closed surface in $\mathbb{R}^3$ with the standard orientation, and let $\mathbf{F}(x, y, z)$ be a vector field defined on S. Then the *flux integral* of $\mathbf{F}$ over S is

$$\iint_S \mathbf{F}(x, y, z) \cdot \mathbf{n}(x, y, z)\, dA,$$

where $\mathbf{n}(x, y, z)$ is the outward-pointing unit normal to S (and dA is the element of surface area on S).   ◇

We now show that flux integrals are given by integrals of 2-forms (just as line integrals were given by integrals of 1-forms in Theorem 6.4.32). For simplicity, we will prove this in the situation of Definition 6.5.13 only, although it is true in the situation of Definition 6.5.15 as well.

THEOREM 6.5.16.   *Let S be either a graph of a function $z = h(x, y)$ or a closed surface in $\mathbb{R}^3$, in each case with the standard orientation, and let $\mathbf{F}(x, y, z) = A(x, y, z)\mathbf{i} + B(x, y, z)\mathbf{j} + C(x, y, z)\mathbf{k}$ be a vector field on S. Let φ be the 2-form that corresponds to $\mathbf{F}$ under the Fundamental Correspondence (Definition 1.3.6). Then*

$$\iint_S \mathbf{F} \cdot \mathbf{n}\, dA = \int_S \varphi.$$

Proof.   We prove this in the case where S is a graph, so suppose S is the graph of $z = h(x, y)$, $(x, y) \in T$.

The left-hand side is, by Proposition 6.5.14, equal to

$$\iint_T \mathbf{F}(x, y, z) \cdot \mathbf{m}(x, y, z) \, dA_{xy},$$

and

$$\mathbf{F}(x, y, z) \cdot \mathbf{m}(x, y, z) = A(x, y, z)(-h_x(x, y))$$
$$+ B(x, y, z)(-h_y(x, y)) + C(x, y, z).$$

As for the right-hand side,

$$\int_S \varphi = \int_T k^*(\varphi),$$

where $k \colon T \longrightarrow S$ by $k(x, y) = (x, y, h(x, y))$.
The 2-form φ is given by

$$\varphi = A(x, y, z)dydz + B(x, y, z)dzdx + C(x, y, z)dxdy.$$

We evaluate $k^*(\varphi)$ by "substitution." Since $z = h(x, y)$,

$$dz = h_x(x, y)dx + h_y(x, y)dy,$$

and then

$$dydz = -h_x(x, y)dxdy,$$
$$dzdx = -h_y(x, y)dxdy,$$

so we see

$$k^*(\varphi) = \big(A(x, y, z)(-h_x(x, y)) + B(x, y, z)(-h_y(x, y))$$
$$+ C(x, y, z)\big)dxdy$$
$$= \mathbf{F}(x, y, z) \cdot \mathbf{m}(x, y, z)dxdy.$$

Since T has the standard orientation, we see immediately from
Corollary 6.5.6 that $\int_T k^*(\varphi)$ is equal to the left-hand side, as
claimed. $\square$

REMARK 6.5.17. The results in the first part of this section hold without change for surfaces in $\mathbb{R}^n$, not just a surfaces in $\mathbb{R}^3$.

As for the results in the second part of this section, beginning with Lemma 6.5.12, they generalize almost without change to $(n-1)$-manifolds in $\mathbb{R}^n$, for any n. The only difference is that we must be careful about orientation. In order to get the orientation correct no matter what the value of n is, we must redefine the notion of a graph slightly. We do that first, and then everything follows just as before.

(Recall the standard orientation of $\mathbb{R}^n$ is that given by the standard basis $\{\mathbf{e}^1, \ldots, \mathbf{e}^n\}$. The reason that everything works correctly in $\mathbb{R}^3$ with the definitions we have given is that the basis $\{\mathbf{k}, \mathbf{i}, \mathbf{j}\}$ of $\mathbb{R}^3$ gives the same orientation of $\mathbb{R}^3$ as the standard basis $\{\mathbf{i}, \mathbf{j}, \mathbf{k}\}$ does.)   ◇

DEFINITION 6.5.18. Let $x_1 = h(x_2, \ldots, x_n)$ be a smooth function on an oriented $(n-1)$-manifold T in $\mathbb{R}^{n-1} = \{(x_2, \ldots, x_n)\}$. Then the *graph* of this function is

$$S = \{(x_1, x_2, \ldots, x_n) \subset \mathbb{R}^n | x_1 = h(x_2, \ldots, x_n)\},$$

regarded as a smooth $(n-1)$-manifold with parameterization $k \colon T \longrightarrow S$ defined by $(x_1, x_2, \ldots, x_n) = k(x_2, \ldots, x_n) = (h(x_2, \ldots, x_n), x_2, \ldots, x_n)$, with the orientation induced by k from the orientation on T. If T has the standard orientation of an $(n-1)$-manifold in $\mathbb{R}^{n-1}$ then we will say that S has the standard orientation of a graph.   ◇

LEMMA 6.5.19. *Let S be the graph of the function $x_1 = h(x_2, \ldots, x_n)$, $(x_2, \ldots, x_n) \in T$, as in Definition 6.5.18, and let $p_0 = (x_0^1, x_0^2, \ldots, x_0^n)$ be a point of S. Then the vector*

$$\mathbf{m}_{p_0} = \begin{bmatrix} 1 \\ -h_{x_2}(x_2, \ldots, x_n) \\ \vdots \\ -h_{x_n}(x_2, \ldots, x_n) \end{bmatrix}_{p_0}$$

is a rightward-pointing normal to S at p_0.

DEFINITION 6.5.20. Let S be the graph of the function $x_1 = h(x_2, \ldots, x_n)$, $(x_2, \ldots, x_n) \in T$, as in Definition 6.5.18, with the standard orientation for a graph, and let $\mathbf{F}(x_1, \ldots, x_n)$ be a vector field defined on S. Then the *flux integral* of $\mathbf{F}$ over S is

$$\iint_T \mathbf{F}(x_1, \ldots, x_n) \cdot \mathbf{m}(x_1, \ldots, x_n) dA_{x_2 \ldots x_n}. \qquad \Diamond$$

Again let us set $\mathbf{n}(x_1, \ldots, x_n) = \mathbf{m}(x_1, \ldots, x_n) / \|\mathbf{m}(x_1, \ldots, x_n)\|$.

PROPOSITION 6.5.21. *In the situation of Definition 6.5.20, the flux integral of $\mathbf{F}$ over S is*

$$\iint_S \mathbf{F}(x_1, \ldots, x_n) \cdot \mathbf{n}(x_1, \ldots, x_n) dA,$$

where the integral is taken with respect to the element of volume on S.

Recall from Definition 4.5.32 that the standard orientation of a closed $(n-1)$-manifold in $\mathbb{R}^n$ is induced from the standard orientation of the n-manifold it bounds via the *outward-pointing* normal.

DEFINITION 6.5.22. Let S be a closed $(n-1)$-manifold in $\mathbb{R}^n$ with the standard orientation, and let $\mathbf{F}(x_1, \ldots, x_n)$ be a vector field defined on S. Then the *flux integral* of $\mathbf{F}$ over S is

$$\iint_S \mathbf{F}(x_1, \ldots, x_n) \cdot \mathbf{n}(x_1, \ldots, x_n) dA,$$

where $\mathbf{n}(x_1, \ldots, x_n)$ is the outward-pointing unit normal to S (and dA is the element of volume on S). $\qquad \Diamond$

THEOREM 6.5.23. *Let S be either a graph of a function or a closed surface in $\mathbb{R}^n$, in each case with the standard orientation, and let $\mathbf{F}$ be a vector field on S. Let φ be the $(n-1)$-form that corresponds to $\mathbf{F}$ under the Fundamental Correspondence (Definition 1.3.6). Then*

$$\iint_S \mathbf{F} \cdot \mathbf{n} \, dA = \int_S \varphi.$$

REMARK 6.5.24. To preclude confusion, we remind the reader that the correspondence between vector fields and $(n - 1)$-forms in $\mathbb{R}^n$ is given by

$$\varphi = \sum_{i=1}^{n} (-1)^{i-1} A^i(x_1, \ldots, x_n) dx_1 \widehat{dx_i} dx_n \longleftrightarrow$$

$$\mathbf{F} = \sum_{i=1}^{n} A^i(x_1, \ldots, x_n) \mathbf{e}^i,$$

where $\widehat{dx_i}$ denotes that dx_i is omitted. In particular, we want to remind the reader of the alternation of signs in this correspondence. ◇

REMARK 6.5.25. Note that in Chapter 4 we chose to identify R^{n-1} with the subset $\{(0, x_2, \ldots, x_n)\}$ of $\mathbb{R}^n$. We made that choice with foresight, to be consistent with what we are doing here, where this subset is really the graph of the function $x_1 = 0$ on $\mathbb{R}^{n-1}$. Then the vector field $\mathbf{m}$ is just the vector field $\mathbf{e}^1$. Also, with the same foresight, we chose to consider the subset $\mathbb{R}^n_- = \{(x_1, x_2, \ldots, x_n) | x_1 \leq 0\}$ there, as $\mathbf{m}$ is the outward-pointing normal to this n-manifold with boundary in $\mathbb{R}^n$. ◇

REMARK 6.5.26. We have now come to the end of this section. However, looking back, in the last section we had a long discussion of path independence and related questions. For example, in Corollary 6.4.18 we showed that if C is a closed oriented curve and φ is an exact 1-form on C, then $\int_C \varphi = 0$. You may reasonably ask whether the analog here is true. Indeed, it is true: If S is a closed oriented surface and φ is an exact 2-form on S, then $\int_S \varphi = 0$. (Furthermore, the analogous theorem is true in higher dimensions as well.) However, we do not want to get into this topic here, for the following reason: Corollary 6.4.18 followed from Theorem 6.4.17, which was a generalization of the Fundamental Theorem of Calculus. There is in fact an analogous theorem here. It is the Generalized Stokes's Theorem, to which we are headed. The entirety of Chapter 7 will be devoted to the Generalized Stokes's Theorem and its consequences. So we will simply delay our discussion of these questions until we get there. ◇

6.6 The integral of a 3-form over a solid body

There is little more to say about the integral of a 3-form over a solid body in $\mathbb{R}^3$. It exactly parallels the integral of a 1-form over a curve in $\mathbb{R}^1$ and the integral of a 2-form over a surface in $\mathbb{R}^2$. But since we live in 3-space, we will address this briefly.

Just as before, let $\mathbb{R}^3 = \{(x, y, z)\}$, and consider an oriented 3-manifold with boundary B in $\mathbb{R}^3$. Consider a 3-form φ on B and write $\varphi = f(x, y, z)dxdydz$. Note that $dxdydz$ is the volume form on $\mathbb{R}^3$ in the standard orientation of $\mathbb{R}^3$. Then, directly from Lemma 6.2.2, we see that

$$\int_B \varphi = {}_s\!\int_B f(x, y, z) dV_{xyz}.$$

We observe that if B has the standard orientation, then the sign in the signed definite integral is positive.

EXAMPLE 6.6.1.   Let B be the rectangular parallelopiped (i.e., the "brick") $\{(x, y, z)|0 \leq x \leq 1, 0 \leq y \leq 2, 0 \leq z \leq 3\}$ with its standard orientation:

Let φ be the 3-form $\varphi = (2z + xz + y)dxdydz$. Then

$$\int_B \varphi = \iiint_B (2z + xz + y)dV_{xyz}$$

which, by Fubini's theorem, is equal to the iterated integral

$$= \int_0^3 \int_0^2 \int_0^1 (2z + xz + y)dxdydz$$

$$= \int_0^3 \int_0^2 (2xz + x^2z/2 + xy \,|_0^1)dydz$$

$$= \int_0^3 \int_0^2 (5z/2 + y)dydz = \int_0^3 (5yz/2 + y^2/2)|_0^2 dz$$

$$= \int_0^3 (5z + 2)dz = 5z^2/2 + 2z \,|_0^2 = 57/2. \qquad \diamond$$

We now state how to translate between integrals of 3-forms over solid bodies in $\mathbb{R}^3$ and ordinary triple integrals over solid bodies in $\mathbb{R}^3$; i.e., we give the analog here of Theorem 6.4.32 (for 1-forms) and Theorem 6.5.16 (for 2-forms). Given our previous development, the result is obvious. We merely state it explicitly for completeness.

THEOREM 6.6.2. *Let B be a 3-manifold with boundary in $\mathbb{R}^3$ with the standard orientation and let φ be a 3-form defined on B. Let $F(x, y, z)$ be the function corresponding to φ under the Fundamental Correspondence (Definition 1.3.6). Then*

$$\int_B \varphi = \iiint_B F(x, y, z) dV_{xyz}.$$

REMARK 6.6.3. Clearly this result extends without change to n-manifolds in $\mathbb{R}^n$, for any n.   ◇

REMARK 6.6.4. Of course our theory has been developed for oriented n-manifolds with boundary in general, and in particular for oriented n-manifolds with boundary in $\mathbb{R}^N$ for any $N \geq n$, but this theory was based on the case of oriented n-manifolds with boundary in $\mathbb{R}^n$: We evaluated integrals in general by pulling them back to this case, with the values of the integrals being independent of the pullbacks chosen. (We remind the reader of Theorem 6.2.3, which guaranteed that.)   ◇

6.7   Chains and integration on chains

So far we have focused on integration of k-forms on k-manifolds. But in fact k-forms can be integrated over more general objects, smooth k-chains. We introduce these here.

DEFINITION 6.7.1. Let M be a manifold. A basic smooth k-chain c in M is a smooth map $k: B \longrightarrow M$ where B is an oriented k-brick. A smooth k-*chain* C in M is a finite formal sum

$$C = \sum_i r_i c_i,$$

where each c_i is a basic k-chain and each r_i is a real number.

The boundary ∂c of the basic chain c is the smooth map k: $\partial B \longrightarrow M$, where ∂B gets the induced orientation as the boundary of T. (Note that ∂B has $2k$ $(k-1)$-faces, so ∂B is a formal sum of $2k$ terms.) The *boundary* of the chain C is the formal sum

$$\partial C = \sum_i r_i \partial c_i.$$

A chain C is a *cycle* if $\partial C = 0$ (or $\emptyset$).   ◇

DEFINITION 6.7.2. Let φ be a k-form on M. Let $C = \sum_i r_i c_i$ be a smooth k-chain in M, where c_i is the basic chain $k_i \colon T_i \longrightarrow M$. Then the *integral* of φ over C is

$$\int_C \varphi = \sum_i r_i \int_{T_i} k_i^*(\varphi).$$

◇

REMARK 6.7.3. This integral is defined very similarly to the integrals we have been considering, but is more general in a number of ways:

1. There are no conditions, other than smoothness, being placed on the maps k_i.

2. The manifold M need not be orientable. Each basic chain in M is oriented, by definition.

3. The differential form φ need not have compact support, as each smooth k-chain is a finite sum, by definition.   ◇

REMARK 6.7.4. It is usual to start with a somewhat more restrictive notion than that of a basic smooth k-chain, that of a smooth singular k-cube, which is a smooth map $k \colon B \longrightarrow M$, where $B = \{(x_1, \ldots, x_k) \in \mathbb{R}^k \mid 0 \leq x_i \leq 1, i = 1, \ldots, k\}$ is the standard unit k-cube, with its standard orientation. But for our purposes here, developing the theory of integration in manifolds, it is more convenient to use this more general notion.   ◇

REMARK 6.7.5. Definition 6.7.1 has a slight ambiguity. Let c be a basic chain, given by a map $k \colon B \longrightarrow M$, and consider $-c$. This has two possible meanings:

(1) $-c = (-1)c$ is the formal sum given by (-1) times c.

(2) $-c$ is the chain given by $k \colon B \longrightarrow M$, where the orientation of B is reversed.

However, this ambiguity will never cause us any trouble, as in every situation we encounter these two meanings will give us the same answer. In particular:

(1) $\partial(-c) = -\partial c$ holds in either interpretation.

(2) $\int_{-c} \varphi = -\int_{c} \varphi$ in either interpretation. To see this, note that, regarding $-c$ as $(-1)c$, Definition 6.7.2 gives $\int_{(-1)c} \varphi = (-1) \int_{c} \varphi = -\int_{c} \varphi$, while regarding $-c$ obtained from c by reversing the orientation of B, $\int_{-c} \varphi = -\int_{c} \varphi$, as we have already seen that reversing orientation changes the sign of the integral. $\diamond$

REMARK 6.7.6. We have already mentioned that the theory of integration extends beyond smooth submanifolds, for example, to piecewise smooth submanifolds or to immersed submanifolds. This extension is via integration on chains. $\diamond$

We now define the "dual" objects to chains.

DEFINITION 6.7.7. Let M be a smooth manifold. A smooth k-cochain Γ on M is a linear function on {smooth k-chains on M}. $\diamond$

LEMMA 6.7.8. *Let M be a smooth manifold. For any k-form φ on M, Int_{φ} given by*

$$\mathrm{Int}_{\varphi}(C) = \int_{C} \varphi$$

is a smooth k-cochain on M.

Proof.   Immediate from Definition 6.7.2. $\square$

DEFINITION 6.7.9. Let Γ be a smooth $(k-1)$-cochain on M. Its *coboundary* $\delta\Gamma$ is the smooth k-cochain on M given by

$$\delta\Gamma(C) = \Gamma(\partial C)$$

for any smooth k-chain C on M.   ◇

REMARK 6.7.10. Let φ be a smooth $(k-1)$-form on M. We may obtain a smooth k-cochain on M from φ in two (apparently) different ways:

1. We have the exterior derivative $d\varphi$ of φ, a k-form on M, and hence we have the smooth k-cochain $\text{Int}_{d\varphi}$.

2. Since φ is a smooth $(k-1)$-form on M, it gives the smooth $(k-1)$-cochain Int_φ, and hence from Definition 6.7.9 we have its coboundary, the smooth k-cochain $\delta(\text{Int}_\varphi)$.

The Generalized Stokes's Theorem, which we present in the next chapter, tells us these are in fact equal!   ◇

6.8 Exercises

In each of the following exercises, unless otherwise stated, find $\int_M \varphi$.

1.

    (a) M is the oriented point $\{(1, 2, 3)\}$ and $\varphi = x^2 y^3 - z^2$.

    (b) M is the oriented point $-\{(2, 1, -1)\}$ and $\varphi = x(y - z)$.

2.

    (a) M is the parameterized curve $r(t) = (2t + 1, t^2)$, $1 \le t \le 2$, traversed in the direction of increasing t, and $\varphi = x\,dx + 4y\,dy$.

(b) M is the parameterized curve $r(t) = (\cos(t), \sin(t)), 0 \le t \le \pi/4$, traversed in the direction of increasing t, and $\varphi = 6xy^2dx$.

(c) M is the graph of $y = 12x^3 - 12x, 0 \le x \le 2$, traversed in the direction of increasing x, and $\varphi = 5xydx + xdy$.

(d) M is the graph of $y = e^{x^3}, -1 \le x \le 1$, traversed in the direction of increasing x, and $\varphi = x^2ydx$.

3. M is the closed curve that is the union of the top half of the unit circle in the (x, y)-plane together with the straight line segment from $(-1, 0)$ to $(1, 0)$, oriented counterclockwise.

(a) $\varphi = 3y^2dx$.

(b) $\varphi = 6xydy$.

(c) $\varphi = 3y^2dx + 6xydy$.

(d) Show that the form φ of part (c) is exact.

4.

(a) M is the parameterized curve $r(t) = (t, 2t^2, t^3 - 1), 1 \le t \le 3$, traversed in the direction of increasing t, and $\varphi = 4zdx + 2ydy + 3xydz$.

(b) M is the parameterized curve $r(t) = (t^2 + 1, 1 - t, (t + 1)^2), -1 \le t \le 2$, traversed in the direction of increasing t, and $\varphi = (z - x)dx + (x - y)dy + 2dz$.

(c) M is the parameterized curve $r(t) = (\cos(2\pi t), \sin(2\pi t), t)$, $0 \le t \le 1$, traversed in the direction of increasing t, and $\varphi = -yzdx + xzdy + xydz$.

5. Let φ be the 1-form $\varphi = 2xdx - zdy - ydz$.

(a) Show that $d\varphi = 0$, i.e., φ is closed.

(b) Since φ is defined on all of $\mathbb{R}^3$, this implies that φ is exact, i.e., $\varphi = df$ for some f. Find f.

(c) Let p be the point $(-1, 0, 1)$ and q the point $(0, 1, 4)$. Since φ is exact, we know that $\int_M \varphi$ is path-independent. Consider the following 3 paths from p to q:

$M_1 : r(t) = (t^2 - 1, t^3, (t+1)^2)$ where t goes from 0 to 1.

M_2: the union of the two paths

$M_2^1 : r(t) = (t, t+1, 1)$ where t goes from -1 to 0.

$M_2^2 : r(t) = (0, 1, t^2)$ where t goes from 1 to 2. (Note that M_2^1 ends at $(0, 1, 1)$, which is where M_2^2 begins.)

$M_3 : r(t) = (t, (t+1)^2, (t+2)^2)$ where t goes from -1 to 0.

Compute $\int_{M_1} \varphi$, $\int_{M_2} \varphi$, and $\int_{M_3} \varphi$.

(d) For the function you found in part (b), compute $f(q) - f(p)$.

(Of course, all three integrals in (c) should be equal, and their common value should be that computed in (d).)

6. Consider the following three curves:

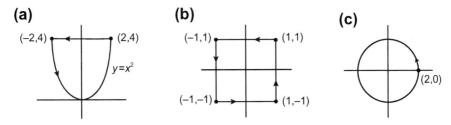

(a) $(-2,4)$ $(2,4)$ $y=x^2$

(b) $(-1,1)$ $(1,1)$ $(-1,-1)$ $(1,-1)$

(c) $(2,0)$

(a) Let φ be the exact 1-form $\varphi = 8x^3 dx + 3y^2 dy$. Verify that $\int_M \varphi = 0$ for each of these curves.

(b) Repeat part (a) for $\varphi = 6xy^3 dx + 9x^2 y^2 dy$.

7. In each part of this exercise, M has the standard orientation of a graph.

(a) M is the graph of $z = x^3 + 3xy^2$, $-1 \le x \le 1$, $-2 \le y \le 2$, and $\varphi = -dydz + dzdx$.

(b) M is the graph of $z = x^2 + y^3$, $0 \le x \le 1$, $0 \le y \le 2$, and $\varphi = 5xzdydz + 7yzdzdx + 31x^2y^3dxdy$.

(c) M is the graph of $z = x^2 - 2xy$ over the triangular region T bounded by the lines $y = 0$, $x = 3$, and $y = 2x$, and $\varphi = dydz + xdzdx + (y^2 + 4)dxdy$.

(d) M is the graph of $z = xy$ over the region T that is the top half of the unit disk (i.e., the region inside $x^2 + y^2 = 1$ which is on or above the x-axis), and $\varphi = x^2dydz + xydzdx + 7xzdxdy$.

(e) M is the portion of the plane $x + 2y + 3z = 6$ that lies inside the first octant (i.e., where $x \ge 0$, $y \ge 0$, and $z \ge 0$), and $\varphi = xdydz + ydzdx + 3zdxdy$.

(f) M is the graph of $z = 7 + (x^2 + y^2)$ over the disk $x^2 + y^2 \le 9$, and $\varphi = z^2dxdy$.

(g) S is the graph of $z = 25 - (x^2 + y^2)$ over the disk $x^2 + y^2 \le 9$, and $\varphi = z^2dxdy$.

8. In each part of this exercise, the parameterized surface M has orientation compatible with the standard orientation of the region T.

(a) M is the surface $(x, y, z) = k(u, v) = (u + v, u - v^2, uv)$ over T, where T is the region in the (u, v)-plane defined by $0 \le u \le 1$, $0 \le v \le 2$, and $\varphi = 6zdzdx$.

(b) M is the surface $(x, y, z) = k(u, v) = (u^3 \cos(v), u^3 \sin(v), u^2)$ over T, where T is the region in the (u, v)-plane defined by $1 \le u \le 2$, $0 \le v \le 2\pi$, and $\varphi = -xdydz - ydzdx + 2zdxdy$.

(c) M is the surface $(x, y, z) = k(u, v) = (v^2, u^3, 2u + v)$ over T, where T is the region in the first quadrant of the (u, v)-plane bounded by the coordinate axes and the curve $v = 4 - u^2$, and $\varphi = 2zdydz + 3ydzdx + dxdy$.

9. In each part of this exercise, M has the standard orientation of a closed surface in $\mathbb{R}^3$.

   (a) M is the sphere of radius a centered at the origin, given by $x^2 + y^2 + z^2 = a^2$, and $\varphi = xdydz + ydzdx + zdxdy$.

   (b) M is the closed surface whose "top" part is $z = x^2y^2 - x^2 - y^2$ and whose "bottom" part is $z = 1$, and $\varphi = y^3dzdx + 2y^2zdxdy$.

   (c) M is the surface of the cube with vertices $(0, 0, 0)$, $(0, 0, 1)$, $(0, 1, 0)$, $(0, 1, 1)$, $(1, 0, 0)$, $(1, 0, 1)$, $(1, 1, 0)$,   and $(1, 1, 1)$, and $\varphi = xdydz - y^2dzdx + 2xzdxdy$. (Note: You need to be careful about orientation.)

10. In each part of this exercise, M has its standard orientation as a solid body in $\mathbb{R}^3$.

    (a) M is the cube with vertices $(0, 0, 0)$, $(0, 0, 1)$, $(0, 1, 0)$, $(0, 1, 1)$, $(1, 0, 0)$, $(1, 0, 1)$, $(1, 1, 0)$,   and $(1, 1, 1)$,   and $\varphi = (1 + 2x - 2y)dxdydz$.

    (b) M is the body in the first octant bounded by the coordinate planes and the sphere $x^2 + y^2 + z^2 = 1$, and $\varphi = xydxdydz$.

    (c) M is the solid body bounded by the graphs of $z = 25 - (x^2 + y^2)$ and $z = 7 + (x^2 + y^2)$, and $\varphi = 2zdxdydz$.

    (d) M is the solid body bounded by the graphs of $z = x^2y^2 - x^2 - y^2$ and $z = -1$, and $\varphi = 5y^2dxdydz$.

11. Let S^n be the unit sphere in $\mathbb{R}^{n+1}$, with its standard orientation. Let φ^n be the n-form of Chapter 1, Exercise 16. Show that $\int_{S^n} \varphi^n > 0$. (As we will see in the next chapter, by the Generalized

Stokes's Theorem this implies that the closed form φ^n on S^n is not exact.)

12. Let T be the surface of Chapter 4, Exercise 7 and let α and β be the 1-forms of Chapter 3, Exercise 9. Let γ be the 2-form $\gamma = \alpha\beta$. Let C_1 be the closed curve on T given by $x^2 + y^2 = 3$, $z = 0$ and let C_2 be the closed curve on T given by $(x^2 - 4)^2 + z^4 = 1$, $y = 0$, $x > 0$. Show that

$$\int_{C_1} \alpha = 2\pi, \qquad \int_{C_2} \alpha = 0,$$

$$\int_{C_1} \beta = 0, \qquad \int_{C_2} \beta = 2\pi,$$

and

$$\int_T \gamma = 4\pi^2,$$

where the orientations are properly chosen. (As we will see in the next chapter, by the Generalized Stokes's Theorem this implies that the closed forms α, β, and γ on T are not exact.)

7 The Generalized Stokes's Theorem

In this chapter we arrive at our goal—the Generalized Stokes's Theorem (GST). We begin with the (beautifully simple) statement of the theorem. We then specialize it to obtain some classical results, which, translated into vector field terminology, yield the standard forms of Green's, Stokes's, and Gauss's theorems. We also give some applications here. In the final section we prove the Generalized Stokes's Theorem.

As we will be recapitulating some of our earlier work, we will be repeating ourselves in some places here. But this is deliberate, for two reasons: First, to reemphasize that these earlier results are part of the general framework, and second, to have all of these results in one place.

7.1 Statement of the theorem

We first give the Generalized Stokes's Theorem for smooth manifolds. Before stating this theorem we note that if M is compact, *every* form on M has compact support.

THEOREM 7.1.1 (*Generalized Stokes's Theorem (GST)*). *Let M be an oriented smooth k-manifold with boundary ∂M (possibly empty) and let ∂M be given the induced orientation. Let φ be a $(k-1)$-form on M with compact support. Then*

$$\int_M d\varphi = \int_{\partial M} \varphi.$$

Let us first observe that both sides of this equality make sense. If φ is a $(k-1)$-form, then $d\varphi$ is a k-form and so the left-hand side is the

Differential Forms, Second Edition. http://dx.doi.org/10.1016/B978-0-12-394403-0.00007-4

integral of a k-form over an oriented k-manifold, which makes sense. If M is a k-manifold, ∂M is a $(k-1)$-manifold, and φ, being defined on M, is also defined on ∂M; so the right-hand side is the integral of a $(k-1)$-form over an oriented $(k-1)$-manifold, which also makes sense. The GST says these two are equal.

Here are two immediate, but important, corollaries. Let us say that a form ψ on M with compact support is *compactly exact* if $\psi = d\varphi$ for some form φ with compact support.

COROLLARY 7.1.2. *Let ψ be a compactly exact k-form on the oriented k-manifold M. Then*

$$\int_M \psi = 0.$$

Proof. To say that ψ is exact means $\psi = d\varphi$ for some φ with compact support; to say that M is a manifold means that $\partial M = \emptyset$. But then

$$\int_M \psi = \int_M d\varphi = \int_{\partial M} \varphi = \int_\emptyset \varphi = 0. \qquad \square$$

COROLLARY 7.1.3. *Let φ be a closed $(k-1)$-form with compact support on the oriented k-manifold with boundary M. Then*

$$\int_{\partial M} \varphi = 0.$$

Proof. To say that φ is closed means $d\varphi = 0$. Then

$$\int_{\partial M} \varphi = \int_M d\varphi = \int_M 0 = 0. \qquad \square$$

We now give the version of the Generalized Stokes's Theorem, and its corollaries, for chains. Note that the underlying point set of a chain is always compact, so that it is unnecessary to restrict ourselves to forms with compact support.

THEOREM 7.1.4 (*Generalized Stokes's Theorem (GST)*). *Let M be a smooth manifold with boundary. Let φ be a $(k-1)$-form on M.*

Then for any smooth k-chain C on M,

$$\int_C d\varphi = \int_{\partial C} \varphi.$$

COROLLARY 7.1.5. *Let ψ be an exact k-form on the manifold M. Then for any smooth k-cycle C on M,*

$$\int_C \psi = 0.$$

COROLLARY 7.1.6. *Let φ be a closed $(k-1)$-form on the manifold M. Then for any smooth k-chain C on M,*

$$\int_{\partial C} \varphi = 0.$$

REMARK 7.1.7. These corollaries are best brought out if we use some nonstandard language. Call a chain C closed if $\partial C = 0$ and exact if $C = \partial D$ for some chain D. Then

$$\int_C \varphi = 0$$

if φ is closed and C is exact, or if φ is exact and C is closed.

(But since this language is nonstandard we will not use it further.) ◇

COROLLARY 7.1.8. *Let M be a smooth manifold with boundary. Let φ be a $(k-1)$-form on M with compact support, and let Int_φ be the $(k-1)$-cochain of Lemma 6.7.8. Then, if δ denotes the coboundary (as in Definition 6.7.9),*

$$\delta(\mathrm{Int}_\varphi) = \mathrm{Int}_{d\varphi}.$$

Proof. Immediate from the GST. □

REMARK 7.1.9. It is customary (though slightly imprecise) to simply identify the form φ with the cochain Int_φ. Then with this identification, Corollary 7.1.8 states that the coboundary δ is given by the exterior derivative d. ◇

7.2 The fundamental theorem of calculus and its analog for line integrals

In this section we specialize the GST to the case $k = 1$.

Let C be a 1-chain.

First consider the special case where C is a 1-manifold in $\mathbb{R}^1$. Then C is just an interval $I = [a, b]$, with some orientation. Let $F(x)$ be a 0-form, i.e., a function, on $[a, b]$. Then $dF = F'(x)dx$. Let us simply consider the case where C is oriented in the direction of increasing x. Then $\partial C = \{b\} \cup -\{a\}$. The GST then specializes to the following theorem:

THEOREM 7.2.1 (*Fundamental Theorem of Calculus (FTC)*).

$$\int_a^b F'(x)dx = F(b) - F(a).$$

To see this, note that in this case the GST says $\int_I dF = \int_{\partial I} F$. But $\int_I dF = \int_I F'(x)dx$ has the same meaning as the ordinary definite integral $\int_a^b F'(x)dx$ (see Lemma 6.2.2) while $\int_{\partial I} F = \int_{\{b\} \cup -\{a\}} F(x) = F(b) - F(a)$ by Definition 6.3.2.

Now let us suppose that C is an oriented curve in a smooth manifold M. The GST then specializes to:

THEOREM 7.2.2. *Let C be an oriented curve in a smooth manifold M with boundary from the point p of M to the point q of M, and let F be a 0-form, i.e., a smooth function, on C. Then*

$$\int_C dF = \int_{\partial C} F.$$

Now, since $\partial C = \{q\} \cup -\{p\}$,

$$\int_{\partial C} F = \int_{\{q\} \cup -\{p\}} F = F(q) - F(p)$$

by Definition 6.3.2, so we see that Theorem 7.2.2 is equivalent to:

THEOREM 7.2.3. *Let C be an oriented curve in a smooth manifold M with boundary from the point p of M to the point q of M, and let*

F be a 0-form, i.e., a smooth function, on C. Then

$$\int_C dF = F(q) - F(p).$$

More generally, let C be a 1-chain in M and let $\partial C = \sum_i r_i p_i$ where $\{r_i\}$ are real numbers and $\{p_i\}$ are points of M. Then

$$\int_C dF = \sum_i r_i F(p_i).$$

But we've already proved this case of the GST, as Theorem 7.2.2 is nothing other than Theorem 6.4.17, and Theorem 7.2.3 is nothing other than Corollary 6.4.13.

There are many results in Section 6.4 that we have derived as consequences of this case of the GST. We recapitulate some of those here (but for 1-chains, rather than just for curves).

COROLLARY 7.2.4. *Let φ be an exact 1-form defined on a closed oriented curve C, or, more generally, on a 1-cycle C in a smooth manifold M with boundary. Then*

$$\int_C \varphi = 0.$$

This is Corollary 7.1.2 of the GST, specialized to our case, and it is also Corollary 6.4.18.

COROLLARY 7.2.5. *Let φ be an exact 1-form defined on a smooth manifold M with boundary. Then*

$$\int_C \varphi = 0$$

for every oriented closed curve C, or more generally for every 1-cycle C, in M.

DEFINITION 7.2.6. Let φ be a 1-form on a smooth manifold M with boundary. Then $\int_C \varphi$ is *path-independent* if it depends only on the endpoints p and q of the oriented curve C in M from p to q, and not on the curve M itself. Equivalently, $\int_C \varphi$ is path-independent if it only depends on ∂C and not on the 1-chain C itself.   ◇

LEMMA 7.2.7. *Let M be a smooth manifold with boundary and let φ be a 1-form on M. Then $\int_C \varphi$ is path-independent if and only if $\int_C \varphi = 0$ whenever C is an oriented closed curve, or, equivalently, whenever C is a 1-cycle, on M.*

COROLLARY 7.2.8. *Let M be a smooth manifold and let φ be a 1-form on M. If φ is exact on M, $\int_C \varphi$ is path-independent. Moreover, if $\varphi = dF$ and C is an oriented curve in M from p to q, then*

$$\int_C \varphi = F(q) - F(p).$$

More generally, if $\varphi = dF$ and C is a 1-chain on M with $\partial C = \sum_i r_i p_i$, then

$$\int_C dF = \sum_i r_i F(p_i).$$

Recall also from Corollary 6.4.22 that in some cases, every closed 1-form on M is exact. We thus see the following result:

COROLLARY 7.2.9. *Let M be a smooth manifold with boundary such that every component of M is smoothly simply connected, or, more generally, let M be a smooth manifold with the property that every smooth closed curve in M is smoothly freely nullhomotopic. Let φ be any closed 1-form on M. Then $\int_C \varphi$ is path-independent. In particular, $\int_C \varphi = 0$ for every oriented closed curve, or, more generally, for every 1-cycle, in M.*

Next, in case M is a region in $\mathbb{R}^n$ for some n, we restate some of these results in the language of line integrals.

COROLLARY 7.2.10. *Let M be a region in $\mathbb{R}^n$. If $\mathbf{F}$ is a conservative vector field on M, then $\int_C \mathbf{F} \cdot d\mathbf{r}$ is path-independent. Moreover, if f is a potential for $\mathbf{F}$ and C is an oriented curve from p to q in M, then*

$$\int_C \mathbf{F} \cdot d\mathbf{r} = f(q) - f(p).$$

More generally, if C is a 1-chain on M with $\partial C = \sum_i r_i p_i$, then

$$\int_C \mathbf{F} \cdot d\mathbf{r} = \sum_i r_i f(p_i).$$

In particular, if C is an oriented closed curve in M, or more generally if C is a 1-cycle on M, then

$$\int_C \mathbf{F} \cdot d\mathbf{r} = 0.$$

Finally, we want to observe that the converse of Corollary 7.2.5 is also true. (Compare Theorem 6.4.12, which was an if and only if statement.) However, we should emphasize that this converse is *not* part of the GST–it needs a separate proof.

THEOREM 7.2.11. *Let M be a smooth manifold, and let φ be a 1-form on M. If $\int_C \varphi = 0$ for every oriented closed curve C (or every 1-cycle C) in M, then φ is exact on M.*   ◇

Proof.   Pick a point p_0 in M. For any point p in M, define

$$F(p) = \int_C \varphi$$

where C is any oriented path from p_0 to p. By Lemma 6.4.10, the hypothesis of this theorem is equivalent to the path-independence of the above integral, so this definition of F really makes sense (i.e., F depends only on p, not on C). But then $dF = \varphi$ by Corollary 6.4.14, so φ is exact.   □

We conclude this section with some examples.

EXAMPLE 7.2.12. The 1-form

$$\varphi^1 = \frac{-y}{x^2 + y^2} dx + \frac{x}{x^2 + y^2} dy$$

of Example 1.2.17, defined on $M = \mathbb{R}^2 - \{(0,0)\}$, is closed, as $d\varphi^1$ is readily computed to be zero. In Example 6.4.2 we computed that

$$\int_C \varphi^1 = 2\pi$$

for C the unit circle in the plane, oriented counterclockwise. Thus, from Corollary 7.2.5, we conclude that φ^1 is *not* exact on M.   ◇

EXAMPLE 7.2.13. Let $\varphi = ydx + xdy$ and C be the curve in the plane $\mathbb{R}^2$ given by $r(t) = (t, 1 + t^2)$, $-1 \leq t \leq 1$, oriented in the direction of increasing t. Thus C is a curve from the point $r(-1) = (-1, 2)$ to the point $r(1) = (1, 2)$. We computed in Example 6.4.1(c) that $\int_C \varphi = 4$. Observe that φ is exact, $\varphi = df$ where $f(x, y) = xy$, and that $f(1, 2) - f(-1, 2) = 2 - (-2) = 4$, which checks. $\diamond$

EXAMPLE 7.2.14. Let C be the oriented curve in the plane $\mathbb{R}^2$ parameterized by $r(t) = (5t^4 - t^2 + t, t^7 - 7t^6 + 3t^4 + 5t + 2)$, $0 \leq t \leq 1$, and let φ be the 1-form $\varphi = 2xydx + x^2dy$. We wish to compute $\int_C \varphi$.

First we note that $d\varphi = 0$ so φ is closed, and, as φ is defined on all of $\mathbb{R}^2$, which is smoothly contractible, φ must be exact. Thus $\int_C \varphi$ is path-independent, i.e., $\int_C \varphi = \int_{C'} \varphi$ for any oriented curve C' from $r(0) = (0, 2)$ to $r(1) = (1, 4)$. We choose C' to be $C_1' \cup C_2'$, where C_1' is the straight line segment from $(0, 2)$ to $(0, 4)$ and C_2' is the straight line segment from $(0, 4)$ to $(3, 4)$. Then $\int_{C'} \varphi = \int_{C_1'} \varphi + \int_{C_2'} \varphi$. We see immediately that $\int_{C_1'} \varphi = 0$ as $\varphi = 0$ on C_1'. Parameterize C_2' by $s(x) = (x, 4)$, $0 \leq x \leq 3$. On C_2', $y = 4$, so $dy = 0$ and $\varphi = 8xdx$. Then $\int_{C_2'} \varphi = \int_0^3 8xdx = 4x^2 \big|_0^3 = 36$, which is our final answer. (Note that it suffices to know that φ is exact to do this example—it is not necessary to find a function f with $\varphi = df$. But you can check that $\varphi = df$ for $f(x, y) = x^2y$, and you can check that indeed $f(3, 4) - f(0, 2) = 36$.) $\diamond$

7.3 Cap independence

In this section we generalize the notion of path independence to higher dimensions. There is no standard name for this property, so we have decided to call it cap independence (a name motivated by the case of dimension 2, which we consider in more detail in the next section).

DEFINITION 7.3.1. Let M be a smooth manifold and let φ be a k-form on M. Then $\int_C \varphi$ is *cap-independent* if the value of this integral only depends on the boundary ∂C of the k-chain C, and not on C itself. $\diamond$

LEMMA 7.3.2. *Let M be a smooth manifold and let φ be a k-form on M. Then $\int_C \varphi$ is cap-independent if and only if $\int_C \varphi = 0$ for every k-cycle C.*

Proof.   Although, logically speaking, it is unnecessary to do so, we will first consider the special case of submanifolds, rather than the general case of cycles, as the geometric intuition is clearer in this case. (Note that in the proof of this special case, C does not denote a k-cycle.)

First suppose $\int_S \varphi = 0$ for every closed oriented k-dimensional submanifold of M. Let C be an oriented $(k-1)$-dimensional submanifold of M and suppose that C is the boundary of two oriented k-dimensional submanifolds S_1 and S_2 of M, i.e., $C = \partial S_1 = \partial S_2$, where the orientations on S_1 and S_2 are chosen so that each induces the given orientation on C.

Consider $S = S_1 \cup -S_2$; i.e., S is $S_1 \cup S_2$ as a set, but the orientation on S *agrees* with the given orientation on S_1, but *disagrees* with the given orientation on S_2.

Observe that S is indeed oriented, as we see from the following argument, which we illustrate in the case of surfaces. Consider the induced orientation on C from S_1. It is indeed the orientation shown, for to get the induced orientation on C, we consider the *outward-pointing* normal to S_1 on C as the first vector and then choose the direction to traverse C by having this vector rotate to the tangent vector to C in the same direction as the direction of rotation on S_1. On the other hand, this orientation on C is also induced from the orientation on $-S_2$, as we see from the following argument. To get the induced orientation on C from S_2, we first consider the *outward-pointing* normal to S_2 on C; so to get the induced orientation on C from $-S_2$, we first consider the *inward-pointing* normal to S_2 on C, and then proceed as above. But of course, as we see from the picture, the *outward*-pointing normal to C from S_1 is the same as the *inward*-pointing normal to C from S_2, so the orientations match up.

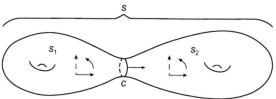

Then we see

$$0 = \int_S \varphi = \int_{S_1 \cup -S_2} \varphi = \int_{S_1} \varphi + \int_{-S_2} \varphi = \int_{S_1} \varphi - \int_{S_2} \varphi$$

so

$$\int_{S_1} \varphi = \int_{S_2} \varphi$$

as required for cap independence.

On the other hand, suppose that $\int_S \varphi$ is cap-independent and that S is a closed oriented k-dimensional submanifold of M. Let C be any oriented $(k-1)$-dimensional submanifold of S bounding part of S, as shown.

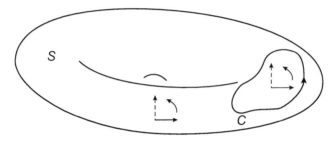

Then S is a union of two pieces, $S = S_1 \cup S_2$. Note that if we give S_1 and S_2 the orientations as shown, i.e., the orientations induced on them as submanifolds of S, $\partial S_1 = C$, and $\partial S_2 = -C$ (i.e., the orientation on S_1 induces the given orientation on C, and the orientation on S_2 induces the opposite orientation on C). Of course, $\partial S_2 = -C$ gives $\partial(-S_2) = C$. Then, by cap independence, we have

$$\int_{S_1} \varphi = \int_{-S_2} \varphi = -\int_{S_2} \varphi$$

and then

$$\int_S \varphi = \int_{S_1 \cup S_2} \varphi = \int_{S_1} \varphi + \int_{S_2} \varphi = -\int_{S_2} \varphi + \int_{S_2} \varphi = 0$$

as claimed.

Now for the general case: First suppose that $\int_C \varphi = 0$ for every k-cycle C. Let C_1 and C_2 be k-chains with $\partial C_1 = \partial C_2$. Call this common boundary B. Then $\partial(-C_2) = -B$, so $\partial(C_1 \cup -C_2) = B - B = 0$, so $C = C_1 \cup -C_2$ is a k-cycle. Thus

$$0 = \int_C \varphi = \int_{C_1 \cup -C_2} \varphi = \int_{C_1} \varphi + \int_{-C_2} \varphi = \int_{C_1} \varphi - \int_{C_2} \varphi$$

so

$$\int_{C_1} \varphi = \int_{C_2} \varphi$$

as claimed.

Next suppose that $\int_C \varphi$ is cap-independent. If $C = 0$ then certainly $\int_C \varphi = 0$. Otherwise, let C be a k-cycle, so that $C = \sum_i r_i C_i$ is a linear combination of basic k-chains with $\partial C = 0$. (If there is only one basic k-chain in the sum, rewrite C as $C = r_1' C_1' + r_2' C_2'$ where $r_1' = r_2' = r_1/2$ and $C_1' = C_2' = C_1$. Thus we may suppose that there is more than one basic k-chain in the sum.) Let $C' = r_1 C_1$ and $C'' = \sum_{i>1} r_i C_i$. Then $\partial C = \partial C' + \partial C''$, so, since C is a cycle, if $\partial C' = B$, then $\partial C'' = -B$. But then

$$\int_C \varphi = \int_{C' \cup C''} \varphi = \int_{C'} \varphi + \int_{C''} = 0$$

as these two integrals are negatives of each other, by cap independence.   □

THEOREM 7.3.3. *Let M be a smooth manifold. Let φ be an exact k-form on M. Then $\int_C \varphi$ is cap-independent.*

Proof.   This is an immediate consequence of the GST: Since φ is exact, $\int_C \varphi = 0$ for every k-cycle C, by Corollary 7.1.5.   □

We also have the following theorem:

THEOREM 7.3.4. *Let M be a smooth manifold. Let φ be a k-form on M such that $\int_C \varphi$ is cap-independent. Then φ is an exact k-form on M.*

REMARK 7.3.5.   In case $k = 1$ we proved this in Theorem 7.2.11. But, as we observed there, this follows from the converse of the GST, not the GST, so it needs a separate proof. The proof for general k is highly nontrivial and uses ideas from "homology theory" that are far beyond the scope of this book, so we do not give it here.   ◇

7.4 Green's and Stokes's theorems

In this section we specialize to the case of dimension $k = 2$, and we consider the special case of surfaces, i.e., 2-manifolds (as there is nothing special to say about 2-cycles in general).

First we restate the GST in the case of surfaces.

THEOREM 7.4.1. *Let S be a compact oriented surface with boundary ∂S (possibly empty) and let ∂S be given the induced orientation. Let φ be a 1-form on S. Then*

$$\int_S d\varphi = \int_{\partial S} \varphi.$$

COROLLARY 7.4.2. *Let φ be an exact 2-form defined on a closed oriented surface S. Then*

$$\int_S \varphi = 0.$$

COROLLARY 7.4.3. *Let φ be a closed 1-form defined on an oriented surface S. Then*

$$\int_{\partial S} \varphi = 0.$$

These are merely direct translations of Theorem 7.1.1, Corollaries 7.1.2 and 7.1.3, respectively.

Now let us look at some applications and examples. First, we will restrict our attention to surfaces (and forms) in the plane $\mathbb{R}^2$. Afterward, we will consider the situation in 3-space $\mathbb{R}^2$.

COROLLARY 7.4.4. *Let S be the surface that is the part of the plane bounded by a simple closed curve C, i.e., a closed curve C with no self-intersections. Suppose that S has the standard orientation and C is oriented counterclockwise. Then for any 1-form φ on S*

$$\int_S d\varphi = \int_C \varphi.$$

Proof. By comparison with Theorem 7.4.1 we see we need only check that the given orientation on C, counter-clockwise, is that induced on C from that on S. But we can see that from the following picture:

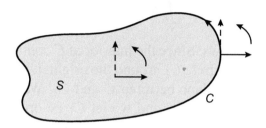

$\square$

COROLLARY 7.4.5. *Let S be the surface that is the part of the plane inside a simple closed curve C_0 and outside the simple closed curves $C_1, \ldots, C_k$. Suppose that S has the standard orientation and C_0 is oriented counterclockwise, while $C_1, \ldots, C_k$ are oriented clockwise. Then for any 1-form φ on S,*

$$\int_S d\varphi = \int_{C_0} \varphi + \int_{C_1} \varphi + \cdots + \int_{C_k} \varphi.$$

Proof. Again, by comparison with Theorem 7.4.1, we need only check that the given orientations on $C_0, C_1, \ldots, C_k$ are those induced from S, and we can see that from the following picture:

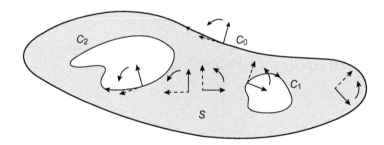

$\square$

EXAMPLE 7.4.6. Let C be any simple closed curve in the plane such that the origin is in its interior, and let C be oriented counterclockwise. Let

$$\varphi^1 = \frac{-y}{x^2 + y^2} dx + \frac{x}{x^2 + y^2} dy$$

be the 1-form of Example 1.2.17. Then

$$\int_C \varphi^1 = 2\pi.$$

We see this as follows: Since the interior of C contains the origin, it also contains a small circle C_1 around the origin. Let S be the surface that is the part of the plane between C and C_1. We are assuming C is oriented counterclockwise, and we let C_1 be oriented clockwise. Note that φ^1 is a closed form on $\mathbb{R}^2 - \{(0,0)\}$ (and hence on S, as S is a subset of $\mathbb{R}^2 - \{(0,0)\}$), i.e., that $d\varphi^1 = 0$. Thus we conclude, by the GST and Corollary 7.4.5, that

$$0 = \int_{S_0} 0 = \int_{\partial S} \varphi^1 = \int_C \varphi^1 + \int_{C_1} \varphi^1.$$

On the other hand, if we let C_0 be the unit circle, oriented counterclockwise, and S_0 be part of the plane between C_0 and C_1, exactly the same argument shows

$$0 = \int_{S_0} 0 = \int_{\partial S_0} \varphi^1 = \int_{C_0} \varphi^1 + \int_{C_1} \varphi^1.$$

Comparing these two equations we see that

$$\int_C \varphi^1 = \int_{C^0} \varphi^1 = 2\pi$$

where the last equality is Example 6.4.2.

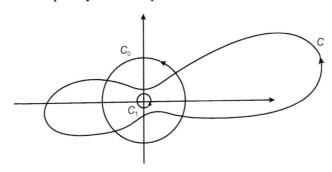

$$\diamondsuit$$

EXAMPLE 7.4.7.

(1) Let S be the part of the plane bounded by the curves $y = x^2$ and $y = 2x$, with its standard orientation.

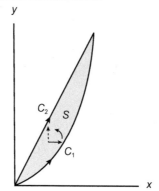

Note that, if C_1 and C_2 are oriented as indicated, $\partial S = C_1 \cup -C_2$. Let φ be the 1-form $\varphi = x^2 y \, dx + y \, dy$. Note that $d\varphi = -x^2 dx \, dy$. Then, on the one hand, you may easily compute that

$$\int_S d\varphi = \int_S -x^2 dx \, dy = \int\!\!\int_S -x^2 d A_{xy}$$

$$= \int_0^2 \int_{x^2}^{2x} -x^2 dy \, dx = -8/5,$$

and on the other hand, you may also easily compute that

$$\int_{C_1} \varphi = \int_{C_1} x^2 y \, dx + y \, dy = \int_0^2 (x^4 + 2x^3) dx = 72/5$$

and

$$\int_{-C_2} x^2 y \, dx + y \, dy = -\int_{C_2} x^2 y \, dx + y \, dy$$

$$= -\int_0^2 (2x^3 + 4x) dx = -16,$$

and $-8/5 = 72/5 - 16$, so this checks.

(2) Let C be the oriented curve parameterized by $r(\theta) = (\cos\theta, \sin\theta)$, $0 \le \theta \le \pi/2$. (You'll recognize that C is one quarter of the unit circle.)

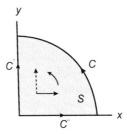

Let $\varphi = xy\,dx + x^2\,dy$. We wish to compute $\int_C \varphi$.
The GST tells us that

$$\int_S d\varphi = \int_{\partial S} \varphi = \int_C \varphi + \int_{C'} \varphi + \int_{C''} \varphi.$$

On C', $\varphi = 0$, so $\int_{C'} \varphi = 0$. Also, on C'', $\varphi = 0$, so $\int_{C''} \varphi = 0$. Thus
we see that

$$\int_C \varphi = \int_S d\varphi.$$

We compute $d\varphi = x\,dx\,dy$ and

$$\int_S \varphi = \int_S x\,dx\,dy = \int\int_S x\,dA_{xy} = \int_0^1 \int_0^{\sqrt{1-x^2}} x\,dy\,dx = 1/3,$$

which is our final answer.   ◇

Note that a 1-form φ in $\mathbb{R}^2$ can be written as $\varphi = A(x, y)dx + B(x, y)dy$, and then we compute that $d\varphi = (B_x(x, y) - A_y(x, y))\,dx\,dy$. Then the GST reads

THEOREM 7.4.8. *Let S be an oriented surface in the plane and let $A(x, y)dx + B(x, y)dy$ be a 1-form defined on S. Suppose that ∂S has the induced orientation from S. Then*

$$\int_S (B_x(x, y) - A_y(x, y))dx\,dy = \int_{\partial S} A(x, y)dx + B(x, y)dy.$$

This theorem holds whatever orientation S has. In particular, it holds when S has the standard orientation (counterclockwise). This leads us to the following useful application:

COROLLARY 7.4.9. *Let S be a surface in the plane with the standard orientation. Let φ be any 1-form on S with $d\varphi = dxdy$. (For example, $\varphi = xdy$ or $\varphi = -ydx$ or $\varphi = \frac{1}{2}(-ydx + xdy)$.) Then, if ∂S has the induced orientation from S,*

$$\int_{\partial S} \varphi = Area(S).$$

Proof.   We have

$$\int_{\partial S} \varphi = \int_S d\varphi = \int_S dxdy = \int_S \int dA_{xy} = Area(S)$$

where the third equality is Lemma 6.2.2.   □

EXAMPLE 7.4.10.   We compute the area of the region in the plane bounded by the curves $y = 1 + x^2$ and $y = 2$.

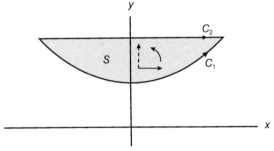

Note that $\partial S = C_1 \cup -C_2$ if C_1 and C_2 are oriented as indicated. First let us compute the area of S directly:

$$Area(S) = \int_S dxdy = \int_S \int dA_{xy} = \int_{-1}^1 \int_{1+x^2}^2 dydx = 4/3.$$

Second, let us compute $\int_{\partial S} xdy = \int_{C_1} xdy + \int_{-C_2} xdy = \int_{C_1} xdy - \int_{C_2} xdy$. Now, parameterizing C_1 by $r(x) = (x, 1+x^2)$, $-1 \le x \le 1$ we find

$$\int_{C_1} xdy = \int_{-1}^1 2x^2dx = 4/3,$$

and parameterizing C_2 by $s(x) = (x, 2)$, $-1 \le x \le 1$ we find

$$\int_{C_2} x\,dy = \int_{-1}^{1} 0 = 0,$$

and indeed $4/3 = 4/3 - 0$.

Third, let us compute $\int_{\partial S} -y\,dx = \int_{C_1} -y\,dx + \int_{-C_2} -y\,dx = \int_{C_1} -y\,dx - \int_{C_2} -y\,dx$. Using the same parameterizations, we find

$$\int_{C_1} -y\,dx = \int_{-1}^{1} -(1 + x^2)dx = -8/3$$

and

$$\int_{C_2} -y\,dx = \int_{-1}^{1} -2dx = -4,$$

and indeed $4/3 = -8/3 - (-4)$. $\diamond$

EXAMPLE 7.4.11. We compute the area of the plane region S bounded by the ellipse

$$\frac{x^2}{a^2} + \frac{y^2}{b^2} = 1.$$

To do so we observe that this curve, oriented counterclockwise, has the parameterization $r(\theta) = (a\cos\theta, b\sin\theta)$, $0 \le \theta \le 2\pi$. If $\varphi = \frac{1}{2}(-y\,dx + x\,dy)$ then computation shows that $r^*(\varphi) = \frac{1}{2}ab$ $(\sin^2\theta + \cos^2\theta)d\theta = \frac{1}{2}ab\,d\theta$, so

$$Area(S) = \int_{\partial S} \varphi = \int_{0}^{2\pi} \frac{1}{2}ab\,d\theta = \pi ab. \qquad \diamond$$

We conclude our discussion here with the classical statement of Green's theorem.

THEOREM 7.4.12 (*Green's Theorem*). *Let* $\mathbf{F}(x, y) = A(x, y)\mathbf{i} + B(x, y)\mathbf{j}$ *be a vector field on a plane region* S. *If* S *is the region bounded by a simple closed curve* C, *orient* C *counterclockwise. Then*

$$\iint_{S} (B_x(x, y) - A_y(x, y))dA_{xy} = \int_{C} \mathbf{F} \cdot d\mathbf{r}.$$

If S is the region inside C_0 and outside $C_1, \ldots, C_k$, orient C_0 counterclockwise and $C_1, \ldots, C_k$ clockwise. Then

$$\int_S\!\!\int (B_x(x, y) - A_y(x, y))dA_{xy}$$

$$= \int_{C_0} \mathbf{F} \cdot d\mathbf{r} + \int_{C_1} \mathbf{F} \cdot d\mathbf{r} + \cdots + \int_{C_k} \mathbf{F} \cdot d\mathbf{r}.$$

This follows immediately from Theorem 7.4.8, using Lemma 6.2.2 (resp. Theorem 6.4.3) to translate the left-hand (resp. right-hand) side of Theorem 7.4.8 into classical language and recalling, once again, that in this situation S has the standard orientation.

Now consider the vector field $\mathbf{F}(x, y) = A(x, y)\mathbf{i} + B(x, y)\mathbf{j}$. We may easily compute $\mathbf{curl}(\mathbf{F}) = (B_x(x, y) - A_y(x, y))\mathbf{k}$, so $\mathbf{curl}(\mathbf{F}) \cdot \mathbf{k} = (B_x(x, y) - A_y(x, y))$ is the integrand on the left-hand side of Theorem 7.4.12. Thus we see that Green's theorem may be reformulated as follows:

THEOREM 7.4.13. *Let $\mathbf{F}(x, y) = A(x, y)\mathbf{i} + B(x, y)\mathbf{j}$ be a vector field on a plane region S. If S is the region bounded by a simple closed curve C, orient C counterclockwise. Then*

$$\int_S\!\!\int (\mathbf{curl}(\mathbf{F}) \cdot \mathbf{k})dA_{xy} = \int_C \mathbf{F} \cdot d\mathbf{r}.$$

If S is the region inside C_0 and outside $C_1, \ldots, C_k$, orient C_0 counterclockwise and $C_1, \ldots, C_k$ clockwise. Then

$$\int_S\!\!\int (\mathbf{curl}(\mathbf{F}) \cdot \mathbf{k})dA_{xy} = \int_{C_0} \mathbf{F} \cdot d\mathbf{r} + \int_{C_1} \mathbf{F} \cdot d\mathbf{r} + \cdots + \int_{C_k} \mathbf{F} \cdot d\mathbf{r}.$$

REMARK 7.4.14. Philosophically speaking, this is a terrible way to formulate Green's theorem, for Green's theorem is about curves and surfaces in the plane, and this formulation introduces a totally extraneous third dimension. But we state it so you will be able to see that Stokes's theorem is a direct generalization of Green's theorem. (Also, you will occasionally see Green's theorem stated this way.) ◇

Now we consider examples of applications of the GST to surfaces in $\mathbb{R}^3$.

EXAMPLE 7.4.15. Let

$$\varphi^2 = \frac{x\,dy\,dz}{(x^2 + y^2 + z^2)^{3/2}} + \frac{y\,dz\,dx}{(x^2 + y^2 + z^2)^{3/2}} + \frac{z\,dx\,dy}{(x^2 + y^2 + z^2)^{3/2}}$$

be the 2-form of Example 1.2.18. Then φ^2 is a closed 2-form on $\mathbb{R}^3 - \{(0, 0, 0)\}$.

Let S^2 be the unit sphere in $\mathbb{R}^3$ with its standard orientation. We computed in Example 6.5.10 that

$$\int_{S^2} \varphi^2 = 4\pi,$$

so by Corollary 7.4.2 we see that φ^2 is not exact.   ◇

EXAMPLE 7.4.16.

(1) Let $\varphi = 4y\,dy\,dz - 9x\,dz\,dx + 6x^2y^2\,dx\,dy$. Note that $d\varphi = 0$, so φ is closed. Since φ is defined on *all* of $\mathbb{R}^3$, the converse of Poincaré's lemma tells us that φ is exact. (In fact, $\varphi = d(-9xz\,dx + 2x^3y^3\,dy + 2y^2\,dz)$, although we don't need to know this to work this example.) Then, by Corollary 7.2,

$$\int_S \varphi = 0$$

for any oriented closed surface S in $\mathbb{R}^3$. In particular, let S be the piecewise-smooth surface in $\mathbb{R}^3$ consisting of the portion of the graph of $z = 36 - 4x^2 - 9y^2$ lying above the xy-plane, together with the portion of the plane inside the curve $4x^2 + 9y^2 = 36$, and give S its standard orientation. We computed $\int_S \varphi$ in Example 6.5.8, and we found it was indeed equal to 0.

(2) Let S be the portion of the graph of $z = x^2 + y^2$ lying over the half-disk T as shown, with its standard orientation.

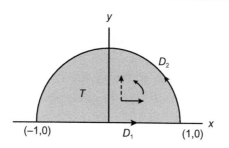

Recall that the standard orientation on S comes from the standard orientation on T (counterclockwise). Then $\partial S = C_1 \cup C_2$ is a union of two smooth curves, where C_1 is the graph of $z = x^2 + y^2$ over the curve D_1 in ∂T, and D_2 is the graph of $z = x^2 + y^2$ over the curve D_2 in ∂T, each with the orientations shown. Thus C_1 is the oriented curve parameterized by $r_1(x) = (x, 0, x^2)$, $-1 \le x \le 1$ (i.e. part of the parabola $z = x^2$ in the vertical plane $y = 0$) and C_2 is the oriented curve parameterized by $r_2(\theta) = (\cos\theta, \sin\theta, 1)$, $0 \le \theta \le \pi$ (i.e., a semicircle in the horizontal plane $z = 1$). Let ψ be the 1-form $\psi = z\,dx + 8xy\,dy$. By the GST, we have

$$\int_S d\psi = \int_{\partial S} \psi = \int_{C_1 \cup C_2} \psi = \int_{C_1} \psi + \int_{C_2} \psi.$$

Let us compute the two integrals on the right:

$$\int_{C_1} z\,dx + 8xy\,dy = \int_{C_1} x^2\,dx = \int_{-1}^{1} x^2\,dx = 2/3,$$

$$\int_{C_2} z\,dx + 8xy\,dy = \int_{0}^{\pi} (-\sin\theta + 8\sin\theta(\cos^2\theta))\,d\theta$$

$$= \cos\theta - (8/3)\cos^3\theta \,\Big|_0^{\pi} = 10/3.$$

We conclude

$$\int_S d\psi = 2/3 + 10/3 = 4.$$

Let $\varphi = d\psi = d\psi = dz\,dx + 8xy\,dx\,dy$. We computed $\int_S \varphi$ in Example 6.5.3(b), and found that this integral is indeed equal to 4.   ◇

We conclude our discussion here with the classical statement of Stokes's theorem.

THEOREM 7.4.17 (*Stokes's Theorem*). *Let S be a surface in* $\mathbb{R}^3$ *that is the graph of a function on a surface T in* $\mathbb{R}^2$. *Suppose* ∂S *is a simple closed curve C, and let C be oriented counterclockwise when viewed from above. Let* **n** *denote the vector field on S that is the upward-pointing unit normal to S at each point of S. Then*

$$\int_S\!\!\int (\mathbf{curl}(\mathbf{F}) \cdot \mathbf{n}) dA = \int_C \mathbf{F} \cdot d\mathbf{r}$$

(where the left-hand side is the integral with respect to the element of surface area dA on S).

If T is the region inside a curve D_0 and outside curves $D_1, \ldots, D_k$ and C_i is the image of D_i in S, $i = 0, \ldots, k$, with C_0 oriented counterclockwise when viewed from above and $C_1, \ldots, C_k$ oriented clockwise when viewed form above, then

$$\int_S\!\!\int (\mathbf{curl}(\mathbf{F}) \cdot \mathbf{n}) dA = \int_{C_0} \mathbf{F} \cdot d\mathbf{r} + \int_{C_1} \mathbf{F} \cdot d\mathbf{r} + \cdots + \int_{C_k} \mathbf{F} \cdot d\mathbf{r}.$$

This follows immediately from Theorem 7.4.1, using Theorem 6.5.15 (resp. Theorem 6.4.3) to translate the left-hand (resp. right-hand) side of the equality into classical language.

One point here needs care. We are using the correspondence between forms and vector fields. This is immediate for the right-hand side: **F** corresponds to a 1-form φ. But then for the left-hand side, **curl(F)** corresponds to the 2-form $d\varphi$. Thus, in applying Theorem 6.5.16, the relevant vector field is not **F**, but rather **curl(F)**.

REMARK 7.4.18. Suppose S is actually a surface in the plane. Then we may take $S = T, C = D$ (or $C_i = D_i, i = 0, \ldots, k$), regarding S as the graph of $z = 0$ on T. Then the unit upward normal to S is clearly the vector field **n** = **k**. Substituting this in Theorem 7.4.17 yields Theorem 7.4.13, which is an alternative formulation of Green's theorem, showing, as we said in Remark 7.4.14, that Green's theorem is a special case of Stokes's theorem; in other words, Stokes's theorem is a generalization of Green's theorem.   ◇

7.5 Gauss's theorem

In this section we specialize to the case of dimension $k = 3$.

First we restate the GST in this case.

THEOREM 7.5.1. *Let B be a compact oriented 3-manifold with boundary ∂B (possibly empty) and let ∂B be given the induced orientation. Let φ be a 2-form on B. Then*

$$\int_B d\varphi = \int_{\partial B} \varphi.$$

COROLLARY 7.5.2. *Let φ be an exact 3-form defined on a closed oriented 3-manifold B. Then*

$$\int_B \varphi = 0.$$

COROLLARY 7.5.3. *Let φ be a closed 2-form defined on an oriented 3-manifold B with boundary. Then*

$$\int_{\partial B} \varphi = 0.$$

These are merely direct translations of Theorem 7.1.1, Corollary 7.1.2, and Corollary 7.1.3, respectively.

Now let us look at some applications and examples in $\mathbb{R}^3$.

Recall that the standard orientation of a 3-manifold with boundary in $\mathbb{R}^3$ is that inherited from the standard orientation of $\mathbb{R}^3$, i.e., that given by the right-hand rule. Also recall that the standard orientation of a closed surface in $\mathbb{R}^3$ is that given by counterclockwise when viewed from outside.

COROLLARY 7.5.4. *Let B be the 3-manifold with boundary that is the part of $\mathbb{R}^3$ bounded by a closed surface S. Suppose that B and S both have the standard orientation. Then for any 2-form φ on B,*

$$\int_B d\varphi = \int_S \varphi.$$

If B is the 3-manifold with boundary that is the part of $\mathbb{R}^3$ inside the closed surface S_0 and outside the closed surfaces $S_1, \ldots, S_k$, and

B has its standard orientation, S_0 has its standard orientation, and $S_1, \ldots, S_k$ have their nonstandard orientations, then

$$\int_B d\varphi = \int_{S_0} \varphi + \int_{S_1} \varphi + \cdots + \int_{S_k} \varphi.$$

Proof.   This is the analog of Corollaries 7.4.4 and 7.4.5. To prove it we need only see that the standard orientation on B induces the standard orientation on S (or the orientations as stated above on $S_0, S_1, \ldots, S_k$). This is completely analogous to the situation of 2-manifolds with boundary in $\mathbb{R}^2$, which we considered in the last section.   □

Now let us see some applications and examples.

EXAMPLE 7.5.5.   Let S^2 be the unit sphere in $\mathbb{R}^3$ with its standard orientation. Let

$$\varphi^2 = \frac{x\,dy\,dz}{(x^2 + y^2 + z^2)^{3/2}} + \frac{y\,dz\,dx}{(x^2 + y^2 + z^2)^{3/2}} + \frac{z\,dx\,dy}{(x^2 + y^2 + z^2)^{3/2}}$$

be the 2-form of Example 1.2.18. We quite laboriously computed in Example 6.5.10 that $\int_{S^2} \varphi^2 = 4\pi$. This was a particularly interesting computation as we used it to conclude in Example 7.4.15 that this closed 2-form, defined on $\mathbb{R}^3 - \{(0, 0, 0)\}$, is not exact.

Let us use the GST to do this computation much more easily. Note that, on S^2, $(x^2 + y^2 + z^2)^{3/2} = (1)^{3/2} = 1$. Hence, on S^2,

$$\varphi^2 = \psi = x\,dy\,dz + y\,dz\,dx + z\,dx\,dy.$$

(Before proceeding further, we want to be clear about what this means. These two forms φ^2 and ψ take the same values on all pairs of tangent vectors to $\mathbb{R}^3$ based at the same point of S^2. But this does *not* mean that $d\varphi^2$ and $d\psi$ are equal, even on S^2, as to compute the exterior derivative, we must consider normal directions, i.e., what happens as we move *off S^2*.)

Thus, since φ^2 and ψ agree on S^2,

$$\int_{S^2} \varphi^2 = \int_{S^2} \psi.$$

Now it is easy to compute $d\psi$, and we see that

$$d\psi = 3dxdydz.$$

Also, $S^2 = \partial B^3$, where B^3 is the unit ball in $\mathbb{R}^3$, with its standard orientation. But then

$$\int_{S^2} \varphi^2 = \int_{S^2} \psi = \int_{B^3} d\psi = \int_{B^3} 3dxdydz = \iiint_{B^3} 3dV_{xyz}$$

$$= 3Vol(B^3) = 3(4\pi/3) = 4\pi,$$

where the second equality is the GST and the fourth equality is Lemma 6.2.2 (and the fifth integral is an ordinary triple integral). ◇

EXAMPLE 7.5.6.

(1) Let B be the 3-manifold with boundary of Example 6.6.1 and let ψ be the 2-form

$$\psi(x, y, z) = xydydz + xyzdzdx + z^2dxdy.$$

Then $d\psi = \varphi$, where φ is the 3-form of Example 6.6.1, so the GST says

$$\int_{\partial B} \psi = \int_B \varphi.$$

We computed in Example 6.6.1 that the right-hand side is equal to $57/2$. Let us compute the left-hand side. Clearly ∂B consists of the six faces of the "brick" B. You can easily check that each of the three terms in ψ is nonzero on only *one* face of the brick: $xydydz$ is nonzero only on the "front" S_1 (in the plane $x = 1$); $xyzdzdx$ is nonzero only on the "right side" S_2 (in the plane $y = 2$); z^2dxdy is nonzero only on the "top" S_3 (in the plane $z = 3$). The induced orientation of ∂B is counterclockwise when viewed from outside, which for S_1 (resp. S_2, S_3) is counterclockwise when viewed from in front (resp. on the right, above). Thus for S_1 the direction of rotation is from $\mathbf{j}$ to $\mathbf{k}$, for S_2 from $\mathbf{k}$ to $\mathbf{i}$, and for S_3 from $\mathbf{i}$ to $\mathbf{j}$. Hence, with these orientations we have

$$\int_{\partial B} \psi = \int_{S_1} xydydz + \int_{S_2} xyzdzdx + \int_{S_3} z^2dxdy.$$

We compute these three integrals in turn. We let T_1, T_2, and T_3 be the "obvious" surfaces with boundary in the (u, v)-plane: T_1 is the rectangle with vertices at $(0, 0)$, $(2, 0)$, $(2, 3)$, $(0, 3)$; T_2 is the rectangle with vertices at $(0, 0)$, $(3, 0)$, $(3, 1)$, $(0, 1)$; T_3 is the rectangle with vertices at $(0, 0)$, $(1, 0)$, $(1, 2)$, $(0, 2)$; all with their standard orientations.

We parameterize S_1 by $x = 1$, $y = u$, $z = v$, whence

$$\int_{S_1} xy\,dy\,dz = \int_{T_1} u\,du\,dv = \int_{T_1}\int u\,dA_{uv} = \int_0^3 \int_0^2 u\,du\,dv$$

$$= \int_0^3 2\,dv = 6.$$

We parameterize S_2 by $x = v$, $y = 2$, $z = u$, whence

$$\int_{S_2} xyz\,dz\,dx = \int_{T_2} 2uv\,du\,dv = \int_{T_2}\int 2uv\,dA_{uv}$$

$$= \int_0^1 \int_0^3 2uv\,du\,dv = \int_0^1 9/2\,dv = 9/2.$$

We parameterize S_3 by $x = u$, $y = v$, $z = 3$, whence

$$\int_{S_3} z^2\,dx\,dy = \int_{T_3} 9\,du\,dv = \int_{T_3}\int 9\,dA_{uv} = \int_0^2 \int_0^1 9\,du\,dv$$

$$= \int_0^2 9\,dv = 18.$$

Note that $6 + 9/2 + 18 = 57/2$, which checks.

(2) Let B be the solid cylinder $B = \{(x, y, z) \mid x^2 + y^2 \le 4,\ 0 \le z \le 6\}$.

Let φ be the 2-form $\varphi(x, y, z) = z^2\,dx\,dy + x\,dy\,dz$.

Now $\partial B = S = S_1 \cup S_2 \cup S_3$ where S_1 is the top, S_2 the bottom, and S_3 the side. Each of these is a single tile, parameterized by

$$S_1 = k_1(T_1) \quad k_1(u, v) = (u, v, 6) \quad T_1 = \{(u, v) \mid u^2 + v^2 \le 4\},$$
$$S_2 = k_2(T_2) \quad k_2(u, v) = (u, v, 0) \quad T_2 = \{(u, v) \mid u^2 + v^2 \le 4\},$$
$$S_3 = k_3(T_3) \quad k_3(u, v) = (2\cos u, 2\sin u, v)$$
$$T_3 = \{(u, v) \mid 0 \le u \le 2\pi, 0 \le v \le 6\}.$$

Here, T_1 and T_3 have the standard orientations, and T_2 has the nonstandard orientation, as regions in the plane. Then

$$\int_{S_1} \varphi = \int_{T_1} 36\,du\,dv = \int_{T_1}\int 36\,dA_{uv} = 36\,Area(T_1) = 144\pi,$$

while $\varphi = 0$ on S_2, so certainly

$$\int_{S_2} \varphi = 0,$$

and for S_3,

$$k_3^*(\varphi) = v^2(-2\sin u\,du)(2\cos u\,du) + (2\cos u)(2\cos u\,du)dv$$
$$= 4\cos^2 u\,du\,v,$$

so

$$\int_{S_3} \varphi = \int_{T_3} 4\cos^2 u\,du\,dv = \int_{T_3}\int 4\cos^2 u\,d\,A_{uv}$$
$$= \int_0^{2\pi} \int_0^6 4\cos^2 u\,dv\,du$$
$$= 6\int_0^{2\pi} 4\cos^2 u\,du = 6\int_0^{2\pi} (2 + 2\cos 2u)du$$
$$= 6(2u + \sin 2u)\,|_0^{2\pi} = 24\pi,$$

and so we obtain

$$\int_{\partial B} \varphi = 144\pi + 0 + 24\pi = 168\pi.$$

On the other hand, $d\varphi = (2z + 1)dx\,dy\,dz$, so

$$\int_B d\varphi = \int_B (2z + 1)dx\,dy\,dz = \int\int_B\int (2z + 1)dV_{xyz}$$
$$= \int\int_B\int 2z\,dV_{xyz} + \int\int_B\int dV_{xyz}.$$

Letting T denote the disk in the xy-plane of radius 2 around the origin, the first of these integrals is

$$\iint_B \int 2z \, dV_{xyz} = \int_T \int \int_0^6 2z \, dz \, dA_{xy}$$
$$= \int_T \int 36 \, dA_{xy} = 36 \, Area(T) = 144\pi,$$

and the second of these integrals is

$$\iint_B \int dV_{xyz} = Vol(B) = 24\pi,$$

and so we obtain

$$\iint_B \int d\varphi = 144\pi + 24\pi = 168\pi,$$

which checks.

We conclude this section with the classical statement of Gauss's theorem (also known as the *divergence theorem*).

THEOREM 7.5.7 (*Gauss's Theorem*). *Let B be the 3-manifold with boundary that is the part of* $\mathbb{R}^3$ *bounded by a closed surface S. Then for any vector field* **F** *on B,*

$$\iint_B \int \operatorname{div}(\mathbf{F}) dV_{xyz} = \int_S \int \mathbf{F} \cdot \mathbf{n} \, dA.$$

If B is the 3-manifold with boundary that is the part of $\mathbb{R}^3$ *inside the closed surface* S_0 *and outside the closed surfaces* $S_1, \ldots, S_k$, *then*

$$\iint_B \int \operatorname{div}(\mathbf{F}) dV_{xyz} = \int_{S_0} \int \mathbf{F} \cdot \mathbf{n} \, dA$$
$$- \int_{S_1} \int \mathbf{F} \cdot \mathbf{n} \, dA - \cdots - \int_{S_k} \int \mathbf{F} \cdot \mathbf{n} \, dA.$$

This follows immediately from Theorem 7.5.1. If B is given the standard orientation, the integral on the left-hand side is as stated by Theorem 6.6.2 because if **F** corresponds to the 2-form φ, then $\mathrm{div}(\mathbf{F})$ corresponds to $d\varphi$. Then, in the first case, S inherits its standard orientation, so the right-hand side is given by Theorem 6.5.15. In the second case, S_0 inherits its standard orientation and $S_1, \ldots, S_k$ inherit their nonstandard orientations, so, keeping track of signs, Theorem 6.5.15 again gives us the right-hand side.

7.6 Proof of the GST

In this section we prove the GST. It turns out that the key to proving it lies in proving it in a very simple special case, where the proof is merely an application of the FTC.

We have previously defined a brick (in Definition 4.1.46) and we begin by elaborating that definition slightly.

DEFINITION 7.6.1. A *standard brick B* in $\mathbb{R}^n$ is a subset of $\mathbb{R}^n$ of the form

$$B = \{(x_1, \ldots, x_n) | a_1 \leq x_1 \leq b_1, \ldots, a_n \leq x_n \leq b_n\}$$

for fixed real numbers $a_1 < b_1, \ldots, a_n < b_n$, with its standard orientation. ◇

Here is the key lemma:

LEMMA 7.6.2. *The GST is true whenever R is a standard k-brick.*

Proof. As we have observed, for $k = 1$ the GST for a standard brick R just *is* the FTC. Recall the situation: In this case, R is an interval $I = [a, b]$, traversed from a to b, and φ is a function $\varphi = A(x)$. Then $\partial R = \{b\} - \{a\}$, and $d\varphi = A'(x)dx$, so the GST reads

$$\int_I A'(x)dx = \int_{\{b\}-\{a\}} A(x) = A(b) - A(a)$$

which is *exactly* the FTC.

Before doing the general case, we will "warm up" by doing the case $k = 2$. Actually, this case already has all of the ideas of the general proof. The latter is simply more complicated as we have a lot more indices to keep track of. (Reading the proof for $k = 2$ first may help you to follow the general proof. However, if you're sufficiently bold, you may simply want to skip this and jump into the general case.)

We will change our notation to eliminate subscripts and superscripts here. Thus we let $\mathbb{R}^2$ be the xy-plane, R be a standard brick given by $a \leq x \leq b$ and $c \leq y \leq d$, and φ be a 1-form given by $\varphi = A(x, y)dy - B(x, y)dx$. (Note the order of the terms and the minus sign.) Then $d\varphi$ is the 2-form $d\varphi = (A_x(x, y) + B_y(x, y))dxdy$.

We want to show

$$(*) \qquad \int_R (A_x(x, y) + B_y(x, y))dxdy$$
$$= \int_{\partial R} A(x, y)dy - B(x, y)dx.$$

The left-hand side of $(*)$ is equal to

$$\int_R A_x(x, y)dxdy + \int_R B_y(x, y)dxdy$$

and the right-hand side of $(*)$ is equal to

$$\int_{\partial R} A(x, y)dy - \int_{\partial R} B(x, y)dx.$$

Thus, to show $(*)$ it suffices to show the two equalities

$$(**_1) \qquad \int_R A_x(x, y)dxdy = \int_{\partial R} A(x, y)dy$$

and

$$(**_2) \qquad \int_R B_y(x, y)dxdy = -\int_{\partial R} B(x, y)dx.$$

Here R has its standard orientation (counterclockwise) and ∂R has the orientation it inherits from R, as follows:

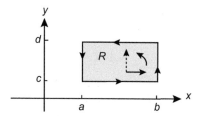

Note ∂R is the union of four curves (in fact, straight lines), given, in the obvious notation, as

$$\partial R = C_{bottom} \cup C_{right} \cup C_{top} \cup C_{left}$$

where these curves are parameterized by

$$C_{bottom} = \{r(t) = (t, c)|\, a \le t \le b\}$$
$$C_{right} = \{r(t) = (b, t)|\, c \le t \le d\}$$
$$-C_{top} = \{r(t) = (t, d)|\, a \le t \le b\}$$
$$-C_{left} = \{r(t) = (a, t)|\, c \le t \le d\}$$

Note in particular the minus signs for the last two, as they are traversed in the direction of *decreasing* t, the *opposite* of the standard orientation. The first two of these are traversed in the direction of *increasing* t, which is the standard orientation, so they count positive.

First we will show $(**_1)$. Since R is a 2-manifold in $\mathbb{R}^2$ with the standard orientation,

$$\int_R A_x(x, y)dxdy = \int_R\int A_x(x, y)dA_{xy}$$
$$= \int_c^d \left[\int_a^b A_x(x, y)dx \right] dy$$
$$= \int_c^d A(x, y)\big|_a^b \, dy$$
$$= \int_c^d (A(b, y) - A(a, y))dy.$$

On the other hand,

$$\int_{\partial R} A(x, y) dy = \int_{C_{bottom}} A(x, y) dy + \int_{C_{right}} A(x, y) dy$$
$$+ \int_{C_{top}} A(x, y) dy + \int_{C_{left}} A(x, y) dy.$$

Using our parameterizations of these four sides, and keeping track of orientations, we see that this is

$$= 0 + \int_{[c,d]} A(b, t) dt - 0 - \int_{[c,d]} A(a, t) dt$$
$$= \int_c^d A(b, t) dt - \int_c^d A(a, t) dt.$$

But these two values are equal, showing $(**_1)$.

The proof of $(**_2)$ is similar. Again,

$$\int_R B_y(x, y) dxy = \int_R \int B_y(x, y) dA_{xy}$$
$$= \int_a^b \left[\int_c^d B_y(x, y) dy \right] dx$$
$$= \int_a^b B(x, y)|_c^d dx$$
$$= \int_a^b (B(x, d) - B(x, c)) dx.$$

On the other hand,

$$-\int_{\partial R} B(x, y) dx = \int_{-\partial R} B(x, y) dx$$
$$= \int_{-C_{bottom}} B(x, y) dx + \int_{-C_{right}} B(x, y) dx$$
$$+ \int_{-C_{top}} B(x, y) dx + \int_{-C_{left}} B(x, y) dx.$$

Using our parameterizations and keeping track of orientations, this is

$$= - \int_{[a,b]} B(t,c)dt - 0 + \int_{[a,b]} B(t,d)dt + 0$$

$$= - \int_a^b B(t,c)dt + \int_a^b B(t,d)dt.$$

Again these two values are equal, showing $(**2)$, thus completing the proof for the case $k = 2$.

Now let us consider the general case, where k is arbitrary. Once again we choose to write an arbitrary $(k-1)$-form φ on R as

$$\varphi = \sum_{i=1}^k (-1)^{i-1} A^i(x_1, \dots, x_k) dx_1 \widehat{dx_i} dx_k.$$

In this expression, the signs *alternate*, with the sign of $A^1(x_1, \dots, x_k)$ $dx_2 \cdots dx_k$ (the *first* term) being *positive*, and dx_i is omitted from the ith term, $i = 1, \dots, k$. These signs are chosen, so that, as you may easily verify,

$$\psi = d\varphi = \left(\sum_{i=1}^k (A^i_{x_i}(x_1, \dots, x_k)) \right) dx_1 \cdots dx_k.$$

(Note here all signs are positive.)

Let φ_i be the ith term in φ and $\psi_i = d\varphi_i$, so that

$$\varphi^i = (-1)^{i-1} A^i(x_1, \dots, x_k) dx_1 \widehat{dx_i} dx_k$$

and

$$\psi_i = d\varphi_i = A^i_{x_i}(x_1, \dots, x_k) dx_1 \cdots dx_k.$$

With this notation in hand, we get down to work. Since $\psi = d\varphi$, what we are claiming is

$$\int_R \psi = \int_{\partial R} \varphi.$$

Of course, since $\varphi = \varphi^1 + \cdots + \varphi^k$ and $\psi = \psi^1 + \cdots + \psi^k$,

$$\int_{\partial R} \varphi = \int_{\partial R} \varphi^1 + \cdots + \int_{\partial R} \varphi^k$$

and

$$\int_R \psi = \int_R \psi^1 + \cdots + \int_R \psi^k.$$

We will prove the lemma by showing that, for each i,

$$\int_{\partial R} \varphi^i = \int_R \psi^i.$$

Let us first prove

$$\int_{\partial R} \varphi^1 = \int_R \psi^1.$$

Consider the right-hand side. Since R has the standard orientation,

$$\int_R \psi^1 = \int_R A^1_{x_1}(x_1, \ldots, x_k) dx_1 \cdots dx_k$$

$$= \overbrace{\int_R \cdots \int}^{k} A^1_{x_1}(x_1, \ldots, x_k) dV_{x_1 \ldots x_k}$$

$$= \overbrace{\int_{R^1} \cdots \int}^{k-1} \left(\int_{a_1}^{b_1} A^1_{x_1}(x_1, \ldots, x_k) dx_1 \right) dV_{x_2 \ldots x_k}$$

where R^1 is $\{(x_2, \ldots, x_k) | a_2 \le x_2 \le b_2, \ldots, a_k \le x_k \le b_k\}$, a subset of $\mathbb{R}^{k-1}$ (not the usual $\mathbb{R}^{k-1}$, with coordinates $(x_1, \ldots, x_{k-1})$, but another $\mathbb{R}^{k-1}$, with coordinates $(x_2, \ldots, x_k)$).

But now, by the FTC, we find

$$\int_{a_1}^{b_1} A^1_{x_1}(x_1, \ldots, x_k) dx_1 = A^1(x_1, \ldots, x_k) \Big|_{x_1=a_1}^{x_1=b_1}$$

$$= A^1(b_1, x_2, \ldots, x_k) - A^1(a_1, x_2, \ldots, x_k),$$

so, substituting, we find

$$\int_R \psi^1 = \overbrace{\int_{R^1} \cdots \int}^{k-1} \left(A^1(b_1, x_2, \ldots, x_k) \right.$$

$$\left. - A^1(a_1, x_2, \ldots, x_k) \right) dV_{x_2 \ldots x_k}.$$

Now consider the left-hand side. R has the face R_+^1 given by $x_1 = b_1$ and the face R_-^1 given by $x_1 = a_1$, where R_+^1 has the standard orientation and R_-^1 has the nonstandard orientation. To see this, note that the standard orientation on R_+^1 is given by the basic unit vectors $(\mathbf{e}_2, \ldots, \mathbf{e}_k)$, tangent to R_+^1, in that order. The outward normal to R_+^1 is $\mathbf{e}_1$; and, recalling that we get the induced orientation by putting the outward normal *first*, we get $(\mathbf{e}_1, \mathbf{e}_2, \ldots, \mathbf{e}_k)$, which is the standard orientation of $\mathbb{R}^k$. On the other hand, since the outward normal to R_-^1 is $-\mathbf{e}_1$, it gets the nonstandard orientation.

Consider the form φ^1. Note that every face other than R_+^1 and R_-^1 is given by $x_i = b_i$ or $x_i = a_i$ for some $i \geq 2$, so $dx_i = 0$ on each of these faces; and, since φ^1 contains a factor of dx_i for each $i \geq 2$, φ^1 is 0 there. Thus we see

$$\int_{\partial R} \varphi^1 = \int_{R_+^1 \cup R_-^1} \varphi^1 = \int_{R_+^1} \varphi^1 + \int_{R_-^1} \varphi^1.$$

Now on R_+^1, $x_1 = b_1$, so $\varphi^1 = A^1(x_1, \ldots, x_n) = A^1(b_1, x_2, \ldots, x_n)$ there. Similarly, on R_-^1, $x_1 = a_1$, so $\varphi^1 = A^1(x_1, \ldots, x_n) = A^1(a_1, x_2, \ldots, x_n)$ there. Hence we see that

$$\int_{R_+^1} = \overbrace{\int_{R^1} \cdots \int}^{k-1} A^1(b_1, x_2, \ldots, x_k) dV_{x_2 \ldots x_k}$$

and

$$\int_{R_-^1} = -\overbrace{\int_{R^1} \cdots \int}^{k-1} A^1(a_1, x_2, \ldots, x_k) dV_{x_2 \ldots x_k}$$

(since we may regard R_+^1 as the graph of $x_1 = b_1$ on R^1, and R_-^1 as the graph of $x_1 = a_1$ on R^1). Hence we find

$$\int_{\partial R} \varphi^1 = \overbrace{\int_{R^1} \cdots \int}^{k-1} \left(A^1(b_1, x_2, \ldots, x_k) \right.$$
$$\left. - A^1(a_1, x_2, \ldots, x_k) \right) dV_{x_2 \ldots x_k},$$

yielding the desired equality.

The proof that

$$\int_{\partial R} \varphi^i = \int_R \psi^i$$

for $i = 2, \ldots, k$ is essentially identical—we just use the index i everywhere we used the index 1. We will thus skip this proof almost entirely. The only part we will give explicitly is the determination of the orientations of R^i_+ and R^i_-. We claim that for i odd (resp. i even), R^i_+ has the standard orientation and R^i_- has the nonstandard orientation (resp. vice versa). But $\varphi_i = (-1)^i A^i(x_1, \ldots, x_n)$, which is positive for i odd (resp. negative for i even) so, after taking the orientations into account, we obtain the analogous equalities.

We obtain the induced orientation of R^i_+ by choosing the outward-pointing normal, which is $\mathbf{n} = \mathbf{e}_i$, and then choosing the order on the basis $\{\mathbf{e}_1, \ldots, \widehat{\mathbf{e}_i}, \ldots, \mathbf{e}_k\}$ of $\mathbb{R}^{k-1}$ so that $\mathbf{n}$ followed by the basis vectors in that order gives the standard orientation of $\mathbb{R}^k$. Transposing the order of any pair of vectors reverses the orientation given by an ordered basis. It takes $i - 1$ transpositions to move $\mathbf{n}$ from the first position to the ith position (first transpose $\mathbf{n}$ and $\mathbf{e}_1$, next transpose $\mathbf{n}$ and $\mathbf{e}_2, \ldots,$ and finally transpose $\mathbf{n}$ and $\mathbf{e}_{i-1}$). Thus if i is odd and we choose the standard order $(\mathbf{e}_1, \ldots, \widehat{\mathbf{e}_i}, \ldots, \mathbf{e}_k)$ on the basis of $\mathbb{R}^i_+$, we get the standard orientation of $\mathbb{R}^k$, so for i odd the induced orientation on $\mathbb{R}^i_+$ is its standard orientation. If i is even and we make this choice we get the nonstandard orientation of $\mathbb{R}^k$, so to get the standard orientation of $\mathbb{R}^k$ we must begin with the nonstandard orientation of $\mathbb{R}^i_+$; in other words for i even the induced orientation on $\mathbb{R}^i_+$ is its nonstandard orientation.

Since the outward-pointing normal to R^i_- is $-\mathbf{e}_i$, the situation for R^i_- is reversed.   □

Having proved this (very) special case, the proof of the GST for arbitrary chains is now almost immediate.

THEOREM 7.6.3. *The GST is true for an arbitrary chain.*

Proof.   Let M be a smooth manifold and let φ be a $(k-1)$-form on M. Let $C = \sum r_i c_i$ be a singular k-chain on manifold M, with boundary $\partial C = \sum r_i \partial c_i$. Then, by definition, $\int_C d\varphi = \sum r_i \int_{c_i} d\varphi$

and $\int_{\partial C} \varphi = \sum r_i \int_{\partial c_i} \varphi$. Thus it suffices to prove the theorem in the case that C is a single basic k-chain.

Let C be the basic k-chain given by $k : B \longrightarrow M$. We want to show that

$$\int_C d\varphi = \int_{\partial C} \varphi.$$

Now by definition,

$$\int_C d\varphi = \int_B k^*(d\varphi)$$

and

$$\int_{\partial C} \varphi = \int_{\partial B} k^*(\varphi),$$

so we need to show

$$\int_B k^*(d\varphi) = \int_{\partial B} k^*(\varphi).$$

Now by Lemma 7.6.2, the GST for B, applied to the form $k^*(\varphi)$, we have

$(*)$ $$\int_B d(k^*(\varphi)) = \int_{\partial B} k^*(\varphi).$$

This is almost, but not quite, what we need to show. However, recall Theorem 4.3.12, which states that pull-back and exterior derivative commute, i.e., that

$$d(k^*(\varphi)) = k^*(d\varphi).$$

Substituting this into $(*)$, we obtain the desired equality. $\qquad\qquad\square$

Now we come to the GST for smooth manifolds.

THEOREM 7.6.4. *The GST is true for an arbitrary compact smooth manifold M with boundary.*

FAKE PROOF. Let φ be a $(k-1)$-form on M. Tile M by submanifolds $M_1, \ldots, M_m$, each of which is the image of a standard brick, $M_i = k_i(B_i)$. Certainly

$$\int_M d\varphi = \int_{M_1} d\varphi + \cdots + \int_{M_m} d\varphi.$$

By the GST for chains, applied to $M_1, \ldots, M_m$ (each of which is a basic k-chain), we have that this is

$$= \int_{\partial M_1} \varphi + \cdots + \int_{\partial M_m} \varphi$$

$$= \int_{\partial M} \varphi$$

since the integrals over the parts of ∂M_i in the interior of M cancel, as in the following picture:

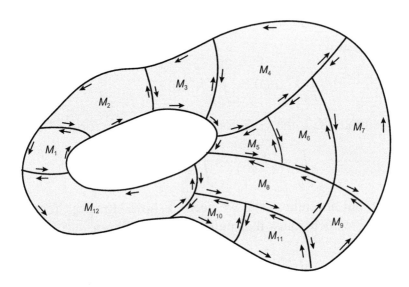

$\square$

REMARK 7.6.5. This is a fake proof because we have to show that any M *can* be tiled by submanifolds that are the images of bricks and that we *do* have the cancellation we claimed. Technically, this is difficult to carry out, and so we adopt a different strategy for our real proof. But we present this fake proof because it neatly illustrates the general idea. (Of course, if we happen to have a manifold which we have tiled in this fashion, this fake proof is indeed a proof.)   ◇

REAL PROOF. We shall see how to write φ as a sum $\varphi = \varphi_1 + \cdots + \varphi_m$ in such a way that the GST holds for each φ_j. This suffices, because in this case $d\varphi = d\varphi_1 + \cdots + d\varphi_m$ and then

$$\int_M d\varphi = \int_M d\varphi_1 + \cdots + \int_M d\varphi_m$$
$$= \int_{\partial M} \varphi_1 + \cdots + \int_{\partial M} \varphi_m = \int_{\partial M} \varphi.$$

To this end, let $\{\varepsilon_j : M \longrightarrow \mathbb{R}\}$ be the partition of unity on M constructed in Section 4.6. Since φ has compact support, there is a finite subset $\{\varepsilon_j : M \longrightarrow \mathbb{R}\}_{j=1,\ldots,m}$ with $\text{supp}(\varphi) \subset \cup_{j=1}^m \text{supp}(\varepsilon_j)$. Then let $\varphi_j = \varepsilon_j \varphi$, $j = 1, \ldots, m$, and let $M_j = \overline{V_j}$ (in the notation of the proof Theorem 4.6.3). By construction, $\text{supp}(\varphi_j) \subseteq M_j$ (and hence $\text{supp}(d\varphi_j) \subseteq M_j$ as well). In particular, this implies that $\int_M d\varphi_j = \int_{M_j} d\varphi_j$.

Now we verify that the GST holds for each φ_j. Referring back to the proof of Theorem 4.6.3, we see that there are two cases:

(a) $M_j \subseteq \text{int}(M)$: In this case, $\varphi_j = 0$ on the complement of V_j, and in particular, $\varphi_j = 0$ on ∂M_j, and $\varphi = 0$ on ∂M, so $\int_{\partial M_j} \varphi_j = 0$ and $\int_{\partial M_j} \varphi_j = 0$. Then

$$\int_M d\varphi_j = \int_{M_j} d\varphi_j = \int_{\partial M_j} \varphi_j = 0 = \int_{\partial M} \varphi_j,$$

where the second equality is the GST for M_j, which we have proved in Theorem 7.6.3. (Note M_j is a particularly nice basic n-chain.)

(b) $F_j = M_j \cap \partial M \neq \emptyset$. In this case, F_j is one "face" of ∂M_j, and $\varphi_j = 0$ on every other face of ∂M_j. (Compare Remark 4.6.4.) Hence $\int_{\partial M_j} \varphi = \int_{F_j} \varphi$. Also, $\varphi_j = 0$ on $\partial M - F_j$, so $\int_{\partial M} \varphi_j = \int_{F_j} \varphi_j$. Then

$$\int_M d\varphi_j = \int_{M_j} d\varphi_j = \int_{\partial M_j} \varphi_j = \int_{F_j} \varphi_j = \int_{\partial M} \varphi_j,$$

where the second equality is again the GST for M_j, which we have proved in Theorem 7.6.3. (Once again M_j is a particularly nice basic n-chain.)

With these two verifications, the proof is complete. □

7.7 The converse of the GST

You may recall that in the case $k = 1$ the converse of the GST is true as well—see Corollary 7.2.5 for the GST and Theorem 7.2.11 for its converse in this case. You may well ask whether this is true in general. The answer is yes, provided we allow ourselves to integrate over cycles, as defined in Definition 7.6.1.

The converse of the GST then reads as follows:

THEOREM 7.7.1 (*de Rham's Theorem*). *Let M be a smooth n-dimensional manifold with boundary and let φ be a closed k-form on M. If*

$$\int_C \varphi = 0$$

for all k-cycles C on M, then φ is exact.

We shall not prove this theorem here.

This theorem is apparently not very useful because it requires us to integrate φ over *all* k-cycles in M. But it has an equivalent and more useful form.

THEOREM 7.7.2. *Let M be a smooth n-manifold and let $\{C_i\}$ be a set of k-cycles with the following property: For every k-cycle C, there are real numbers $\{r_i\}$ such that*

$$D = C - \sum r_i C_i$$

is a boundary, $D = \partial E$ for some $(k + 1)$-chain E. If φ is a closed k-form on M such that

$$\int_{C_i} \varphi = 0 \ for \ each \ i,$$

then φ is exact.

PROOF OF EQUIVALENCE. We simply compute

$$\int_C \varphi = \left(\sum r_i \int_{C_i} \varphi \right) + \int_D \varphi.$$

But $\int_{C_i} \varphi = 0$ for each i, by assumption, and $\int_D \varphi = \int_{\partial E} \varphi = \int_E d\varphi = 0$, where the second equality is the GST and the last equality is true because φ is closed by hypothesis, so $d\varphi = 0$. $\quad\square$

This equivalence is useful because we can often find such a collection of cycles $\{C_i\}$. But this is also beyond the scope of this book.

However, we will prove the special case of top-dimensional forms. As in the case of the GST, the key to the theorem is to analyze the situation of a brick. We do this in two steps.

LEMMA 7.7.3. *Let B be a standard n-brick as in Definition 7.6.1, let F be the face of ∂B given by $x_1 = b_1$, and let $G = \overline{\partial B - F}$. Let φ be an n-form on B such that $\varphi = 0$ on (resp. in a neighborhood of) G. Then $\varphi = d\psi$ for some $(n-1)$-form ψ such that $\psi = 0$ on (resp. in a neighborhood of) G.*

Proof.   Write $\varphi = A(x_1, x_2, \ldots, x_n)dx_1 dx_2 \cdots dx_n$ and define ψ by $\psi = E(x_1, x_2, \ldots, x_n)dx_2 \cdots dx_n$ where

$$E(x_1, x_2, \ldots, x_n) = \int_{a_1}^{x_1} A(t, x_2, \ldots, x_n)dt.$$

Then $\frac{\partial E}{\partial x_1} = A$ by the FTC, so $d\psi = \varphi$. $\quad\square$

LEMMA 7.7.4. *Let B be a standard n-brick as in Definition 7.6.1. Let φ be an n-form on B such that $\varphi = 0$ on (resp. in a neighborhood of) ∂B. If*

$$\int_B \varphi = 0$$

then $\varphi = d\psi$ for some $(n-1)$-form ψ such that $\psi = 0$ on (resp. in a neighborhood of) ∂B.

Proof.   We use the notation of the previous lemma and its proof.

To shorten the proof slightly, we will just do the case $\varphi = 0$ on ∂B, noting that the case $\varphi = 0$ in a neighborhood of ∂B has exactly the same proof, simply replacing "on ∂B" by "in a neighborhood of ∂B" every place it appears.

We proceed by induction on n. If $n = 1$, define ψ as above. Then B is the interval $[a_1, b_1]$ and F is the point $x = b_1$. But under our hypotheses

$$E(b_1) = \int_{a_1}^{b_1} A(t)dt = \int_B \varphi = 0$$

so $\psi = 0$ on F, and if $\varphi = 0$ on a neighborhood of ∂B, so in particular $\varphi = 0$ for $d < x \leq b_1$ for some d with $a_1 < d < b_1$, then $\psi = 0$ for $x > d$, a neighborhood of F, as well.

The case $n = 2$ is crucial and we handle it separately, for clarity. We use coordinates (x, y).

Define the function $f(y)$ by

$$f(y) = \int_{a_1}^{b_1} A(x, y)dx.$$

Choose c and d with $a_1 < c < d < b_1$, and let $h(x)$ be a smooth function with $h(x) = 0$ for $a_1 \leq x \leq c$, $h(x) = 0$ for $d \leq x \leq b_1$, and $\int_{a_1}^{b_1} h(t)dt = 1$. Such a function exists by Lemma 4.6.1(2). Set $g(x, y) = h(x)f(y)$ and $A_1(x, y) = A(x, y) - g(x, y)$. Define the form φ_1 by

$$\varphi_1 = \varphi - g(x, y)dxdy = A_1(x, y)dxdy$$
$$= \big(A(x, y) - h(x)f(y)\big)dxdy.$$

Now define ψ_1 by $\psi_1 = E_1(x, y)dy$ where

$$E_1(x, y) = \int_{a_1}^{x} A_1(t, y)dt.$$

Then $\frac{\partial E_1}{\partial x_1} = A_1$ by the FTC, so $d\psi_1 = \varphi_1$. Clearly $E_1(x, y) = 0$ if $x = a_1$, and also $E_1(x, y) = 0$ if $y = a_2$ or $y = b_2$, as $A(x, y) = 0$ if $y = a_2$ or $y = b_2$ since by hypothesis $\varphi = A(x, y)dxdy = 0$ on ∂B. Furthermore,

$$E_1(b_1, y) = \int_{a_1}^{b_1} A_1(t, y)dt$$
$$= \int_{a_1}^{b_1} \big(A(t, y) - h(t)f(y)\big)dt$$

$$= \int_{a_1}^{b_1} A(t, y)dt - \int_{a_1}^{b_1} h(t)f(y)dt$$

$$= \int_{a_1}^{b_1} A(t, y)dt - f(y)\int_{a_1}^{b_1} h(t)dt$$

$$= f(y) - f(y) = 0,$$

so $E_1 = 0$ on ∂B.

Note that, by Fubini's theorem,

$$\int_{a_2}^{b_2} f(y)dy = \int_{a_2}^{b_2}\left[\int_{a_1}^{b_1} A(x, y)dx\right]dy$$

$$= \int_B\int A(x, y)dA_{xy} = \int_B \varphi = 0.$$

Let ρ be the 0-form (i.e., function) defined by

$$\rho = C(y) = \int_{a_2}^{y} f(t)dt$$

and let

$$\psi_2 = -h(x)\rho dx.$$

We claim that $\psi_2 = 0$ on ∂B. To see this, note that $\psi_2 = -h(x)C(y)$ dx. Certainly $C(a_2) = 0$, and we have just calculated that $C(b_2) = 0$. Also, $h(a_1) = h(b_1) = 0$ by the definition of h.

Furthermore, by the definition of the exterior derivative and by the FTC,

$$d\psi_2 = (-h(x)C_y dy)dx = h(x)C_y dxdy = h(x)f(y)dxdy$$
$$= g(x, y)dxdy = \varphi_2,$$

as required.

Now we prove the general case, by induction. We have proved the lemma in case $n = 1$. Assume the lemma is true in case $n - 1$, and let φ be an n-form as in the statement of the lemma. Write $\varphi = A(x_1, x_2, \ldots, x_n)dx_1 dx_2 \cdots dx_n$. Define the function $f(x_2, \ldots, x_n)$ by

$$f(x_2, \ldots, x_n) = \int_{a_1}^{b_1} A(x, x_2, \ldots, x_n)dx.$$

Again choose c and d with $a_1 < c < d < b_1$, and let $h(x)$ be a smooth function with $h(x) = 0$ for $a_1 \le x \le c$, $h(x) = 0$ for $d \le x \le b_1$, and $\int_{a_1}^{b_1} h(t)dt = 1$. Set $g(x_1, x_2, \ldots, x_n) = h(x_1)f(x_2, \ldots, x_n)$ and $A_1(x_1, x_2, \ldots, x_n) = A(x_1, x_2, \ldots, x_n) - g(x_1, x_2, \ldots, x_n)$. Define the form φ_1 by

$$
\begin{aligned}
\varphi_1 &= \varphi - g(x_1, x_2, \ldots, x_n)dx_1 dx_2 \cdots dx_n \\
&= A_1(x_1, x_2, \ldots, x_n)dx_1 dx_2 \cdots dx_n \\
&= \left(A(x_1, x_2, \ldots, x_n) - h(x_1)f(x_2, \ldots, x_n) \right) dx_1 dx_2 \cdots dx_n
\end{aligned}
$$

and define ψ_1 by $\psi_1 = E_1(x_1, x_2, \ldots, x_n)dx_2 \cdots dx_n$ where

$$
E_1(x_1, x_2, \ldots, x_n) = \int_{a_1}^{x} A_1(t, x_2, \ldots, x_n)dt.
$$

Again, $\frac{\partial E_1}{\partial x_1} = A_1$ by the FTC, so $d\psi_1 = \varphi_1$, and again

$$
\begin{aligned}
&E_1(b_1, x_2, \ldots, x_n) \\
&= \int_{a_1}^{b_1} A_1(t, x_2, \ldots, x_n)dt \\
&= \int_{a_1}^{b_1} \left(A(t, x_2, \ldots, x_n) - h(t)f(x_2, \ldots, x_n) \right) dt \\
&= \int_{a_1}^{b_1} A(t, x_2, \ldots, x_n)dt - \int_{a_1}^{b_1} h(t)f(x_2, \ldots, x_n)dt \\
&= \int_{a_1}^{b_1} A(t, x_2, \ldots, x_n)dt - f(x_2, \ldots, x_n) \int_{a_1}^{b_1} h(t)dt \\
&= f(x_2, \ldots, x_n) - f(x_2, \ldots, x_n) = 0
\end{aligned}
$$

so $E_1 = 0$ on ∂B.

To complete the proof we must show that, letting $\varphi_2 = g(x_1, x_2, \ldots, x_n)dx_1 dx_2 \cdots dx_n = h(x_1)f(x_2, \ldots, x_n)dx_1 dx_2 \cdots dx_n$, there is a form ψ_2 with $d\psi_2 = \varphi_2$, and with $\psi_2 = 0$ on ∂B.

Let B' be the brick in $\mathbb{R}^{n-1} = \{(x_2, \ldots, x_n)\}$ given by $a_i \le x_i \le b_i$, $i = 2, \ldots, n$. Let τ be the $(n-1)$-form $\tau = f(x_2, \ldots, x_n) dx_2 \cdots dx_n$. Observe that $\tau = 0$ on $\partial B'$ and observe that (once again by Fubini's theorem)

$$\int_{B'} \tau = \int_{B'} f(x_2, \ldots, x_n) dV_{x_2 \cdots x_n}$$

$$= \int_{B'} \left[\int_{a_1}^{b_1} A(x_1, x_2, \ldots, x_n) d\ell_{x_1} \right] dV_{x_2 \cdots x_n}$$

$$= \int_{B} \int A(x_1, x_2, \ldots, x_n) dV_{x_1 x_2 \cdots x_n} = \int_{B} \varphi = 0.$$

Then by the inductive hypothesis there is an $(n-2)$-form ρ, which is 0 on $\partial B'$, with $d\rho = \tau$. Let

$$\psi_2 = (-1)^{n-1} h(x_1) \rho dx_1.$$

Then $\psi_2 = 0$ on ∂B, and

$$d\psi_2 = (-1)^{n-1} h(x_1) d\rho dx_1$$
$$= (-1)^{n-1} h(x_1) f(x_2, \ldots, x_n) dx_2 \cdots dx_n dx_1$$
$$= h(x_1) f(x_2, \ldots, x_n) dx_1 dx_2 \cdots dx_n = \varphi_2,$$

as required.

(Strictly speaking, we should have let $\pi : B \longrightarrow B'$ be the projection onto the last $n-1$ coordinates and defined ψ_2 by $\psi_2 = (-1)^{n-1} h(x_1) \pi^*(\rho) dx_1$ but we decided to slightly simplify the proof at the cost of this slight inaccuracy.) $\qquad\square$

Now we transfer our results to a manifold. Before proving our main theorem, we prove a technical lemma, but one which is interesting in its own right.

LEMMA 7.7.5. *Let M be a connected smooth oriented n-manifold with boundary. Let p be any point of M and let U be any open neighborhood of p. Then there is a smooth n-form φ_0 on M with support contained in U and with $\int_M \varphi_0 = 1$. Furthermore, if φ is any n-form on M with compact support, then $\varphi = \left(\int_M \varphi \right) \varphi_0 + d\psi$ for some $(n-1)$-form ψ on M with compact support.*

Proof. We use the notation of Section 4.6 here.

For any $\varepsilon > 0$, let

$$f(x_1, \ldots, x_n) = f_4(2x_1/\varepsilon) \cdots f_4(2x_n/\varepsilon)$$

where $f_4(x)$ is the function of that name in Lemma 4.6.1. Let B be the brick in $\mathbb{R}^n$ defined by $B = \{(x_1, \ldots, x_n) \mid |x_i| \leq \varepsilon, \; i = 1, \ldots, n\}$ and let $I_0 = \int_B f(x_1, \ldots, x_n) dV_{x_1 \cdots x_n}$. Set $\tau_0 = \frac{1}{I_0} f(x_1, \ldots, x_n)$ $dx_1 \cdots dx_n$. Then τ_0 is a smooth n-form on $\mathbb{R}^n$ with $\mathrm{supp}(\tau_0) = B$ and with $\int_{\mathbb{R}^n} \tau_0 = \int_B \tau_0 = 1$. Let $k : \mathcal{O}_1 \longrightarrow M$ be a coordinate patch with $p \in \mathcal{U}_1 = k(\mathcal{O}_1)$. Clearly we may assume that $k(0, \ldots, 0) = p$. Then for ε sufficiently small, $k(B) \subset U \cap \mathcal{U}_1$. Define φ_0 by $\varphi_0 = (k^{-1})^*(\tau_0)$ on $\mathcal{U}_1$ and $\varphi_0 = 0$ on $M - k(B)$. Then φ_0 is the desired n-form.

Now let φ be an n-form as in the statement of the theorem. Let $I = \int_M \varphi$.

First consider the very special case that $\mathrm{supp}(\varphi) \subset \mathcal{V}_1$. Then

$$\varphi = I\varphi_0 + (\varphi - I\varphi_0)$$

Setting ρ equal to the second term, we have that $\mathrm{supp}(\rho) \subset \mathcal{V}_1$ and $\int_{\mathcal{V}_1} \rho = 0$. Let $\tau = k^{-1}(\rho)$. Then by Lemma 7.7.4, $\tau = d\sigma$ for some $(n-1)$ form σ with $\mathrm{supp}(\sigma) \subset \mathcal{V}_1$. Then the desired form ψ is defined by $\psi = (k^{-1})^*(\sigma)$ on $\mathcal{U}_1$ and $\psi = 0$ on $M - k(B)$.

Next consider the special case that $\mathrm{supp}(\varphi) \subset \mathcal{V}_2$ where $\mathcal{V}_1 \cap \mathcal{V}_2 \neq \emptyset$. Choose $p \in \mathcal{V}_1 \cap \mathcal{V}_2$ and a neighborhood U of p with $U \subseteq \mathcal{V}_1 \cap \mathcal{V}_2$. Write φ as above. Note that the second term ρ has $\mathrm{supp}(\rho) \subset \mathcal{V}_2$ and $\int_{\mathcal{V}_2} \rho = 0$, so we may find a form ψ as above. Also note that the first term has support contained in $\mathcal{V}_1$.

Next consider the case that $\mathrm{supp}(\varphi) \subset \mathcal{V}_j$ for some j. Since M is connected we can form a "chain" of neighborhoods $\mathcal{V}_1, \mathcal{V}_2, \mathcal{V}_3, \ldots, \mathcal{V}_j$ with $\mathcal{V}_1 \cap \mathcal{V}_2 \neq \emptyset, \mathcal{V}_2 \cap \mathcal{V}_3 \neq \emptyset, \ldots, \mathcal{V}_{j-1} \cap \mathcal{V}_j \neq \emptyset$ and successively perform the above construction to obtain ψ.

Finally consider the case of a general n-form φ with compact support. Then, taking a partition of unity, we may write φ as a finite sum $\varphi = \varepsilon_1 \varphi + \cdots + \varepsilon_i \varphi = \varphi_1 + \cdots + \varphi_i$ with, for each i, $\mathrm{supp}(\varphi_i) \subset \mathcal{V}_j$ for some j, and perform the above construction to obtain ψ_i for each i, and then $\psi = \psi_1 + \cdots + \psi_i$ is as required. $\qquad \square$

Here is the statement of our main theorem. The statement is simplified slightly by noting that any n-form on an n-manifold is automatically closed. Note that we need M to be oriented in order for integration over M to make sense. We also need the form φ to have compact support in order to be sure that $\int_M \varphi$ makes sense. Of course, this is no restriction if M itself is compact.

THEOREM 7.7.6. *Let M be a connected smooth oriented n-manifold with boundary, and let φ be an n-form on M with compact support.*

(1)

(a) If $\partial M \neq \emptyset$, then $\varphi = d\psi$ for some $(n-1)$-form ψ with compact support.

(b) If $\partial M \neq \emptyset$, then $\varphi = d\psi$ for some $(n-1)$-form ψ with compact support, and with $\psi = 0$ on ∂M, if and only if $\int_M \varphi = 0$.

(2) If $\partial M = \emptyset$, then $\varphi = d\psi$ for some $(n-1)$-form ψ with compact support if and only if $\int_M \varphi = 0$.

Proof. Choose an atlas $\{k_i : \mathcal{O}_i \longrightarrow \mathcal{U}_i\}$ as in our construction of a partition of unity in Theorem 4.6.3. In case $\partial M \neq \emptyset$, (re)number the sets in this atlas so that $\mathcal{V}_1 \cap \partial M \neq \emptyset$. In case $\partial M = \emptyset$, $\mathcal{V}_1$ is arbitrary. Apply Lemma 7.7.5 to φ to obtain $\varphi' = (\int_M \varphi)\varphi_0$ whose support is contained in $\mathcal{V}_1$. Note that $\int_{\mathcal{V}_1} \varphi' = \int_M \varphi' = \int_M \varphi$.

In case $\partial M \neq \emptyset$, apply Lemma 7.7.3 to obtain an $(n-1)$-form ψ defined by $\psi = (k_1^{-1})^*(\rho)$ where ρ is a $(n-1)$ form on $\overline{B}$ with $d\rho = k_1^*(\varphi')$. Then ψ extends to an $(n-1)$-form on M by setting $\psi = 0$ outside of $\mathcal{V}_1$, and $d\psi = \varphi$.

In case $\int_M \varphi = 0$, apply Lemma 7.7.4 to obtain an $(n-1)$-form ψ defined by $\psi = (k_1^{-1})^*(\rho)$ where ρ is a $(n-1)$ form on B with $d\rho = k_1^*(\varphi')$. (Note that $\int_B k_1^*(\varphi') = 0$ as, by hypothesis, $\int_M \varphi = 0$.) Then ψ extends to an $(n-1)$-form on M by setting $\psi = 0$ outside of $\mathcal{V}_1$, and $d\psi = \varphi$.

If $\int_M \varphi \neq 0$, then no such $(n-1)$-form ψ (with $\psi = 0$ on ∂M in case $\partial M \neq \emptyset$) can exist, as that would contradict the GST. □

7.8 Exercises

1. In each case, show that $\int_C \varphi$ is path-independent, and use the GST to find its value:

   (a) $\varphi = (2xy^3 + 4x)dx + (3x^2y^2 - 9y^2)dy$ and C is a path from $(2, 1)$ to $(4, 3)$.

   (b) $\varphi = (6x^2 - 3y)dx + (8y - 3x)dy$ and C is a path from $(1, 1)$ to $(2, 2)$.

   (c) $\varphi = 2xzdx + 4yz^2dy + (x^2 + 4y^2z - 9z^2)dz$ and C is a path from $(-1, 1, 2)$ to $(4, 0, 1)$.

   (d) $\varphi = -\sin(\pi x)\sin(\pi(y - z))dx + \cos(\pi x)\cos(\pi(y - z))dy - \cos(\pi x)\cos(\pi(y - z))dz$ and C is a path from $(1/4, 1, 1/4)$ to $(0, 1, 1/2)$.

2. In each case, show that $\int_C \varphi$ is path-independent, and find its value by using a "simpler" path C'.

   (a) $\varphi = 2xydx + x^2dy$ and C is the parameterized curve $r(t) = (t^3 - t^2 + t + 1, t^3 - 2t^2 + 2t + 1)$, $0 \leq t \leq 1$, traversed in the direction of increasing t.

   (b) $\varphi = (4x^3 + 6x^2y^2z)dx + (4x^3yz + 3z^2 - 4y^3)dy + (2x^3y^2 + 6yz - 9z^2)dz$ and C is the parameterized curve $r(t) = (t^4 + t^2 + 3, t^6 - 2t^2 + 7, t^8 - 4t^4 + 9)$, $-2 \leq t \leq 2$, traversed in the direction of increasing t.

3. In each case, use the GST to find $\int_{\partial M} \varphi$, where M has its standard orientation as a surface in $\mathbb{R}^2$.

   (a) M is the square $0 \leq x \leq \pi/3$, $0 \leq y \leq \pi/3$, and $\varphi = (\cos(x)\cos(y) + 3^{x^2})dx + (\sin(x)\sin(y) + \sqrt{y^4 + 1})dy$.

   (b) M is the region on and inside the ellipse $x^2 + 4y^2 = 9$ and on and above the line $x + 2y = 3$, and $\varphi = -2x^2y^2dx + 4x^3ydy$.

4. In each case, verify the GST, where M has its standard orientation as a surface in $\mathbb{R}^2$.

(a) M is the top half of the unit disk (i.e., the region on and inside $x^2 + y^2 = 1$ and on and above the x-axis), and $\varphi = xy\,dy$.

(b) M is the region bounded by the curves $y = x^2$ and $y = 2x+3$, and $\varphi = 6x^2y\,dy$.

(c) M is the triangular region bounded by the curves $y = 1$, $x = 2$, and $y = x$, and $\varphi = -ye^x\,dx + e^x\,dy$.

5. In each case, verify the GST, where the orientation on M is compatible with the standard orientation of T.

(a) M is the surface $(x, y, z) = k(u, v) = (u + v, u - v^2, uv)$ over T, where T is the region in the (u, v)-plane defined by $0 \le u \le 1, 0 \le v \le 2$, and $\varphi = 3z^2\,dx$. (See Chapter 6, Exercise 8(a).)

(b) M is the surface $(x, y, z) = k(u, v) = (u^3 \cos(v), u^3 \sin(v), u^2)$, $1 \le u \le 2, 0 \le v \le 2\pi$, and $\varphi = -yz\,dx + xz\,dy$. (See Chapter 6, Exercise 8(b).)

6. In each case, verify the GST, where M has the standard orientation of a graph.

(a) M is the surface $z = x^2 + y^2$ for $x^2 + y^2 \le 1$, and $\varphi = 4xy^2\,dy + z^2\,dz$.

(b) M is the surface $z = x^2 + y$ for $0 \le x \le 2, 0 \le y \le 3$, and $\varphi = z^2\,dx + xy^2\,dy + z\,dz$.

(c) M is the surface that is the part of the plane $4x + 2y + z = 4$ lying in the first octant (i.e., with $x \ge 0, y \ge 0$, and $z \ge 0$), and $\varphi = x^2\,dx + 3xz\,dy + (y^2 + 6x)\,dz$.

7. In each case show that $\int_S \varphi$ is cap-independent and use this fact to compute it.

(a) S is a surface whose boundary (with the induced orientation) is the curve $r(t) = (\cos(t), \sin(t), \sin(2t))$, $0 \le t \le 2\pi$, traversed in the direction of increasing t, and $\varphi = z\,dz\,dx - 6y^2\,dx\,dy$.

(b) S is a surface whose boundary (with the induced orientation) is the curve $r(t) = (\cos(t), \sin(t), 2)$, $0 \le t \le 2\pi$, traversed in the direction of increasing t, and $\varphi = -x^3\,dy\,dz - y^3\,dz\,dx + 3z(x^2 + y^2)\,dx\,dy$.

8. Let L_1 be the line segment from the point $(2, 2)$ to the point $(1, 4)$, and let C_1 be the oriented curve in $\mathbb{R}^3$ that is the graph of $z = x^2 y$ on L_1. Use the GST to find $\int_{C_1} \varphi$, where $\varphi = -2x^2 y\,dy + (6x - y^2)\,dz$, as follows: Consider the surface S that is the graph of $z = x^2 y$ on the triangle T bounded by L_1, L_2, and L_3, where L_2 is the line segment from $(1, 4)$ to $(1, 2)$ and L_3 is the line segment from $(1, 2)$ to $(2, 2)$. Let C_2 and C_3 be the oriented curves that are the graphs of $z = x^2 y$ on L_2 and L_3, respectively. It is easy to compute $\int_{C_2} \varphi$ and $\int_{C_3} \varphi$ directly.

9. In each case, use the GST to solve the problem.

(a) Chapter 6, Exercise 9(a).

(b) Chapter 6, Exercise 10(b). (Let $\varphi = xyz\,dx\,dy$. Note that ∂B has four parts.)

10. In each case, verify the GST, where M has its standard orientation as a solid body in $\mathbb{R}^3$.

(a) M is the solid body bounded by the surfaces $z = 25 - (x^2 + y^2)$ and $z = 7 + (x^2 + y^2)$, and $\varphi = z^2\,dx\,dy$. (See Chapter 6, Exercises 7(f), 7(g), and 10(c).)

(b) M is the cube $0 \le x \le 1$, $0 \le y \le 1$, $0 \le z \le 1$, and $\varphi = x\,dy\,dz - y^2\,dz\,dx + 2xz\,dx\,dy$. (See Chapter 6, Exercises 9(c) and 10(a).)

(c) M is the solid body bounded by $z = x^2y^2 - x^2 - y^2$ and $z = -1$, and $\varphi = y^3 dz dx + 2y^2 z dx dy$. (See Chapter 6, Exercises 9(b) and 10(d).)

11. Let S_1 be the surface that is the graph of $z = 5 - (x^2 + y^2)$ for $0 \le x^2 + y^2 \le 4$, with the standard orientation of a graph, and let $\varphi = (3x^2z + 2y^3z)dz dx + (6x^2z - 3y^2z^2)dx dy$. Use the GST to find $\int_{S_1} \varphi$ as follows: If B is the solid body bounded by $z = 5 - (x^2 + y^2)$ and $z = 1$, then ∂B has two parts, one of which is S_1 and the other is, say, S_2. $\int_{S_2} \varphi$ is easy to compute directly. (Be careful of the orientation of S_2.)

8 de Rham Cohomology

In this chapter we introduce de Rham cohomology, which in a precise way measures the difference between closed and exact forms. This is a topic that we have dealt with throughout this book, and we will see that many of our earlier results can be rephrased as results about de Rham cohomology.

We begin this chapter with some general algebraic considerations, and then we turn to de Rham cohomology. We will see that it appears in two related but distinct forms–de Rham cohomology per se, and de Rham cohomology with compact supports.

8.1 Linear and homological algebra constructions

We begin this section by discussing constructions in linear algebra. Then we move onto basic constructions in homological algebra.

DEFINITION 8.1.1. Let V be a vector space and let U and W be subspaces. Then V is the *direct sum* of U and W, $V = U \oplus W$, if $V = U + W$, i.e., if every $v \in V$ can be written as $v = u + w$ for some $u \in U$ and some $w \in W$, and $U \cap W = \{0\}$.

In this situation each of U and W is a *complement* of the other. ◇

LEMMA 8.1.2. *Let $V = U \oplus W$. If B_1 is a basis of U and B_2 is a basis of W, then $B = B_1 \cup B_2$ is a basis of V.*

Proof. Let $B_1 = \{u_1, \dots, u_j\}$ and $B_2 = \{w_1, \dots, w_k\}$. Let $v \in V$ be arbitrary. Then $v = u + w$ for some $u \in U$ and some $w \in W$. Then $u = c_1 u_1 + \cdots + c_j u_j$ and $w = d_1 w_1 + \cdots + d_k w_k$ so $v = c_1 u_1 + \cdots + c_j u_j + d_1 w_1 + \cdots + d_k w_k$, and hence B spans V. If $c_1 u_1 + \dots c_j u_j + d_1 w_1 + \cdots + d_k w_k = 0$, then $v = (c_1 u_1 + \dots c_j u_j) = -(d_1 w_1 + \cdots + d_k w_k) \in U \cap W$, so

Differential Forms, Second Edition. http://dx.doi.org/10.1016/B978-0-12-394403-0.00008-6

$v = 0$. But B_1 is linearly independent so $c_1 = \cdots = c_j = 0$ and B_2 is linearly independent so $d_1 = \cdots = d_k = 0$, and so B is linearly independent as well.   □

COROLLARY 8.1.3. *Let $V = U \oplus W$. Suppose that each of U and W is finite-dimensional. Then* $\dim(V) = \dim(U) + \dim(W)$.

Proof. In the notation of the proof of Lemma 8.1.2, $\dim(U) = j$, $\dim(W) = k$, and $\dim(V) = j + k$.   □

LEMMA 8.1.4. *Let V be a vector space and let U be a subspace. Then U has a complement W.*

Proof. Let U have basis $B_1 = \{u_1, \ldots, u_j\}$ and extend B_1 to a basis B of V, $B = \{u_1, \ldots, u_j, w_1, \ldots, w_k\}$. Then W, the subspace of V with basis $B_2 = \{w_1, \ldots, w_k\}$, is a complement of u.   □

DEFINITION 8.1.5. Let V be a vector space and let U be a subspace of V. Let $\sim$ be the equivalence relation on V defined by $v \sim v'$ if $v - v' \in U$, and let $[v]$ denote the equivalence class of v. Then the *quotient* V/U is the vector space $\{[v] | v \in V\}$ with vector space operations given by $[v_1] + [v_2] = [v_1 + v_2]$ and $c[v] = [cv]$.

The element v of V is a *representative* of the element $[v]$ of V/U.   ◇

(As usual, we write 0 for the zero element of any vector space; in particular we write 0 for $[0] \in V/U$.)

LEMMA 8.1.6. *Let V be a vector space and let U be a subspace of V. Let $\pi : V \longrightarrow V/U$ be the canonical projection, given by $\pi(v) = [v]$. If W is any complement of U, then $\pi : W \longrightarrow V/U$ is an isomorphism.*

Proof. Again we use the notation of the proof of Lemma 8.1.2. Consider an arbitrary element $[v] \in V/U$. Then $v = c_1 u_1 + \cdots + c_j u_j + d_1 w_1 + \cdots + d_k w_k$. But $v \sim w = d_1 w_1 + \cdots + d_k w_k$ as $u = v - w = c_1 u_1 + \cdots + c_j u_j \in U$. Hence $[v] = [w] = \pi(w)$. On the other hand, suppose $w \in W$ with $[w] = \pi(w) = 0$. Then $w \sim 0$, i.e., $w - 0 = w \in U$. Thus $w \in U \cap W = \{0\}$, i.e., $w = 0$.   □

COROLLARY 8.1.7. *Let V be a finite-dimensional vector space and let U be a subspace of V. Then* $\dim(V/U) = \dim(V) - \dim(U)$.

Proof. Let W be a complement of U. Then by Lemma 8.1.6 and Corollary 8.1.3, $\dim(V/U) = \dim(W) = \dim(V) - \dim(U)$. □

REMARK 8.1.8. For any subspace U of V, the quotient V/U is a well-defined vector space. However, except in the cases $U = \{0\}$ or $U = V$, a subspace U never has a well-defined complement, as there is always a choice to be made. Nevertheless, all complements are mutually isomorphic, as they are all isomorphic to V/U. ◇

As is common (indeed, universal), we abuse notation by writing 0 for the vector space $\{0\}$ consisting of the 0 element alone. Note there is a unique linear transformation from 0 to any vector space (the inclusion) and from any vector space to 0 (the zero map), so we do not bother to write these maps below.

DEFINITION 8.1.9. Let $T : V \longrightarrow W$ be a linear transformation. The *kernel* of T is

$$\mathrm{Ker}(T) = \{v \in V | T(v) = 0\}$$

and the *image* of T is

$$\mathrm{Im}(T) = \{w \in W | w = T(v) \text{ for some } v \in V\}.$$

If $f : U \longrightarrow V$ and $g : V \longrightarrow W$ are linear transformations, the sequence

$$U \xrightarrow{f} V \xrightarrow{g} W$$

is *exact* at V if $\mathrm{Ker}(g) = \mathrm{Im}(f)$. A sequence of vector spaces and linear transformations is an *exact sequence* if it is exact at every vector space in the sequence. An exact sequence

$$0 \longrightarrow U \xrightarrow{f} V \xrightarrow{g} W \longrightarrow 0$$

is a *short exact* sequence. ◇

LEMMA 8.1.10.

(1) The sequence $0 \longrightarrow U \xrightarrow{f} V$ *is exact if and only if f is an injection.*

(2) The sequence $V \xrightarrow{g} W \longrightarrow 0$ *is exact if and only if g is a surjection.*

(3) The sequence $0 \longrightarrow U \xrightarrow{f} V \longrightarrow 0$ *is exact if and only if f is an isomorphism.*

(4) The sequence $0 \longrightarrow U \xrightarrow{f} V \xrightarrow{g} W \longrightarrow 0$ *is exact if and only if f is an injection, g is a surjection, and* $\mathrm{Ker}(g) = \mathrm{Im}(f)$.

   Proof.   Immediate from the definitions.   □

   LEMMA 8.1.11.

(1) Let $T : V \longrightarrow W$ *be a linear transformation. Then the sequence*

$$0 \longrightarrow \mathrm{Ker}(T) \xrightarrow{i} V \xrightarrow{T} \mathrm{Im}(T) \longrightarrow 0$$

is exact, where i is the inclusion.

(2) Let U be a subspace of V. Let $i : U \longrightarrow V$ *be the inclusion, and let* $\pi : V \longrightarrow V/U$ *be the canonical projection. Then the sequence*

$$0 \longrightarrow U \xrightarrow{i} V \xrightarrow{\pi} V/U \longrightarrow 0$$

is exact.

   Proof.   Immediate from the definitions.   □

   LEMMA 8.1.12.

(1) (The five lemma) Given a commutative diagram of exact sequences

$$
\begin{array}{ccccccccc}
A_1 & \longrightarrow & B_1 & \longrightarrow & C_1 & \longrightarrow & D_1 & \longrightarrow & E_1 \\
\alpha\downarrow & & \beta\downarrow & & \gamma\downarrow & & \delta\downarrow & & \varepsilon\downarrow \\
A_2 & \longrightarrow & B_2 & \longrightarrow & C_2 & \longrightarrow & D_2 & \longrightarrow & E_2
\end{array}
$$

with α, β, δ, *and* ε *isomorphisms, then* γ *is an isomorphism.*

(2) (The short five lemma) Given a commutative diagram of exact sequences

$$0 \longrightarrow B_1 \longrightarrow C_1 \longrightarrow D_1 \longrightarrow 0$$
$$\beta\downarrow \qquad \gamma\downarrow \qquad \delta\downarrow$$
$$0 \longrightarrow B_2 \longrightarrow C_2 \longrightarrow D_2 \longrightarrow 0$$

with β and δ isomorphisms, then γ is an isomorphism.

Proof. This is a standard "diagram chase," and we omit it. □

We now embark on a sequence of definitions and constructions. These may appear unmotivated, but they are in fact exactly what we need for the next section, where we define de Rham cohomology and derive its basic properties.

DEFINITION 8.1.13. A *cochain complex* $\mathbf{V}^*$ is a sequence of vector spaces and maps

$$\mathbf{V}^* : \quad \cdots V^{i-1} \xrightarrow{\delta^{i-1}} V^i \xrightarrow{\delta^i} V^{i+1} \cdots$$

with the property that

$$\delta^i \delta^{i-1} = 0 \text{ (i.e., } \delta^i \delta^{i-1} : V^{i-1} \longrightarrow V^{i+1} \text{ is the 0 map)}$$

for each i. In this situation we write $\mathbf{V}^* = \{(V^i, \delta^i)\}$.

The elements of V^i are called i-*cochains*, and the map δ^i is called the ith *coboundary map*.

In this situation we let

$$Z^i = Z^i(\mathbf{V}^*) = \mathrm{Ker}\delta^i \text{ and } B^i = B^i(\mathbf{V}^*) = \mathrm{Im}\delta^{i-1}.$$

The elements of Z^i are called i-*cocycles* and the elements of B^i are called i-*coboundaries*.

We then let H^i be the quotient vector space

$$H^i = H^i(\mathbf{V}^*) = Z^i/B^i = \mathrm{Ker}\delta^i/\mathrm{Im}\delta^{i-1}.$$

H^i is called the ith *cohomology group* of $\mathbf{V}^*$. Elements of H^i are called *cohomology classes*. Two elements of Z^i that represent the same cohomology class are called *cohomologous*. ◇

REMARK 8.1.14.

(1) Note that the property $\delta^i \delta^{i-1} = 0$ is equivalent to $B^i \subseteq Z^i$, so it is precisely this property that enables us to define cohomology.

(2) Note that two i-cocycles z_1 and z_2 are cohomologous, i.e., are such that $[z_1] = [z_2]$, if and only if $z_2 = z_1 + \delta^{i-1}(v)$ for some $(i-1)$-chain v.

(3) Note that $H^i = 0$ if and only if $Z^i = B^i$, i.e., $H^i = 0$ if and only if every i-cocycle is an i-coboundary.   ◇

We will often simply write δ for δ^i and refer to cochains, cocycles, and coboundaries when i is understood.

DEFINITION 8.1.15.   Let $\mathbf{V}^* = \{(V^i, \delta^i_V)\}$ and $\mathbf{W}^* = \{(W^i, \delta^i_W)\}$ be two cochain complexes. A *map of cochain complexes* f : $\mathbf{V}^* \longrightarrow \mathbf{W}^*$ is a collection of linear transformations $\{f^i : V^i \longrightarrow W^i\}$ with the property that, for each i, the following diagram commutes:

$$
\begin{array}{ccc}
V^i & \xrightarrow{f^i} & W^i \\
\delta^i_V \downarrow & & \downarrow \delta^i_W \\
V^{i+1} & \xrightarrow{f^{i+1}} & W^{i+1}
\end{array}
$$
                                                                                        ◇

LEMMA 8.1.16.

(1) In the situation of Definition 8.1.15, there is a well-defined map

$$(f^*)^i : H^i(\mathbf{V}^*) \longrightarrow H^i(\mathbf{W}^*)$$

given by

$$(f^*)^i([z]) = [f^i(z)],$$

for each i.

(2)

(a) If $f : \mathbf{V}^* \longrightarrow \mathbf{V}^*$ is the identity map, then $(f^*)^i : H^i(\mathbf{V}^*) \longrightarrow H^i(\mathbf{V}^*)$ is the identity map, for each i.

(b) If $f : \mathbf{U}^* \longrightarrow \mathbf{V}^*$ and $g : \mathbf{V}^* \longrightarrow \mathbf{W}^*$ are maps of chain complexes, then $(gf^*)^i = (g^*)^i (f^*)^i : H^i(\mathbf{U}^*) \longrightarrow H^i(\mathbf{W}^*)$, for each i.

Proof.

(1) We need to check that $(f^*)^i$ is independent of the choice of representative, i.e., that if $[z'] = [z]$, then $[(f^*)^i(z')] = [f^i(z)]$. But $[z'] = [z]$ means $z' = z + \delta_V^{i-1}(v)$ for some v, and then $f^i(z') = f^i(z) + f^i(\delta_V^{i-1}(v)) = f^i(z) + \delta_W^{i-1}(f^{i-1}(v))$, so $[(f^*)^i(z')] = [f^i(z)]$ as claimed.

(2) (a) is obvious and (b) is easily verified. □

DEFINITION 8.1.17. The collection of maps $\{(f^*)^i : H^i(\mathbf{V}^*) \longrightarrow H^i(\mathbf{W}^*)\}$ as in Lemma 8.1.16 are the maps *induced on cohomology* by the map of cochain complexes $f : \mathbf{V}^* \longrightarrow \mathbf{W}^*$. ◇

DEFINITION 8.1.18. Let $f : \mathbf{V}^* \longrightarrow \mathbf{W}^*$ and $g : \mathbf{V}^* \longrightarrow \mathbf{W}^*$ be maps of cochain complexes. A *chain homotopy* between f and g is a collection of maps $\{H_i : V_i \longrightarrow W_{i-1}\}$ such that

$$f^i - g^i = \delta_W^{i-1} H_i + H_i \delta_V^i$$

for each i. If there is a chain homotopy between f and g, then f and g are *chain homotopic*. ◇

LEMMA 8.1.19. *Suppose that $f : \mathbf{V}^* \longrightarrow \mathbf{W}^*$ and $g : \mathbf{V}^* \longrightarrow \mathbf{W}^*$ are chain homotopic. Then*

$$(f^*)^i = (g^*)^i : H^i(\mathbf{V}^*) \longrightarrow H^i(\mathbf{W}^*)$$

for each i.

Proof. Consider an arbitrary element $[z]$ of $H^i(\mathbf{V})$. Note $\delta_V^i(z) = 0$ as z is a cocycle. But then $f^i(z) - g^i(z) = \left(\delta_W^{i-1} H_i + H_i \delta_V^i\right)(z) = \delta_W^{i-1}(H_i(z))$ so $[f^i(z)] = [g^i(z)]$, i.e., $(f^*)^i([z]) = (g^*)^i([z])$ as claimed. □

THEOREM 8.1.20.

(1) Let $\mathbf{U}^* = \{(U^i, \delta_U^i)\}$, $\mathbf{V}^* = \{(V^i, \delta_V^i)\}$, *and* $\mathbf{W}^* = \{(W^i, \delta_W^i)\}$
be cochain complexes. Let $f : \mathbf{U}^* \longrightarrow \mathbf{V}^*$ *and* $g : \mathbf{V}^* \longrightarrow \mathbf{W}^*$ *be*
maps of cochain complexes such that

$$0 \longrightarrow U^i \xrightarrow{f^i} V^i \xrightarrow{g^i} W^i \longrightarrow 0$$

is a short exact sequence, for each i ,in which case we say that

$$0 \longrightarrow \mathbf{U}^* \xrightarrow{f} \mathbf{V}^* \xrightarrow{g} \mathbf{W}^* \longrightarrow 0$$

is a short exact sequence of cochain complexes. Then there is a well-defined map
$$\delta^i : H^i(W) \longrightarrow H^{i+1}(U)$$

for each i, defined as follows: Let $c \in H^i(W)$. *Choose a representative*
z of c, so that $z \in Z^i(W)$ *with* $[z] = c$. *Then* $z = g^i(y)$ *for some*
$y \in C^i(V)$. *Then* $0 = \delta_W^i(z) = \delta_W^i(g^i(y)) = g^{i+1}(\delta_V^i(y))$ *so there is*
a unique $x \in C^{i+1}(U)$ *with* $f^{i+1}(x) = \delta_V^i(y)$. *Then* $x \in Z^{i+1}(U)$,
and we set $\delta^i(c) = a$ *where* $a = [x] \in H^{i+1}(U)$.

(2) In this situation, there is a (long) exact cohomology sequence

$$\cdots \longrightarrow H^i(U) \xrightarrow{(f^*)^i} H^i(V) \xrightarrow{(g^*)^i} H^i(W) \xrightarrow{\delta^i} H^{i+1}(U) \longrightarrow \cdots .$$

Proof.

(1) There are three things to check:

(a) $\delta_U^{i+1}(x) = 0$ so indeed $x \in Z^{i+1}(U)$.

(b) $a = \delta^i(c)$ is independent of the choice of z.

(c) $a = \delta^i(c)$ is independent of the choice of y.

These are proved by diagram-chasing, as follows:

Proof of (a):

$$f^{i+2}(\delta_U^{i+1}(x)) = \delta_V^{i+1}(f^{i+1}(x)) = \delta_V^{i+1}\delta_V^i(y) = 0$$

as $\delta_V^{i+1}\delta_V^i = 0$. But f^{i+2} is an injection, so $\delta_U^{i+1}(x) = 0$.

Proof of (b): Suppose $[z'] = [z]$. Then $z' = z + \delta_W^{i-1}(t)$ for some $t \in W^{i-1}$. Choose $s \in V^{i-1}$ with $g^{i-1}(s) = t$. Then

$$g^i(y + \delta_V^{i-1}(s)) = g^i(y) + g^i(\delta_V^{i-1}(s)) = g^i(y) + \delta_W^{i-1}(g^i(s))$$
$$= g^i(y) + \delta_W^{i-1}(t) = z'$$

and then

$$\delta_V^i(y + \delta_V^{i-1}(s)) = \delta_V^i(y) + \delta_V^i\delta_V^{i-1}(s) = \delta_V^i(y)$$

as $\delta_V^i\delta_V^{i-1} = 0$.

Proof of (c): Suppose that $z = g^i(y') = g^i(y)$. Then $0 = g^i(y') - g^i(y) = g^i(y' - y)$, i.e., $y' - y \in \text{Ker}(g^i)$. Hence $y' - y = f^i(r)$ for some $r \in U^i$, i.e., $y' = y + f^i(r)$. Define x' by $f^{i+1}(x') = \delta_V^i(y')$. Then

$$f^{i+1}(x' - x) = f^{i+1}(x') - f^{i+1}(x) = \delta_V^i(y') - \delta_V^i(y)$$
$$= \delta_V^i(y' - y) = \delta_V^i(f^i(r)) = f^{i+1}(\delta_U^i(r))$$

so $x' - x = \delta_U^i(r)$ and hence $[x'] = [x]$.

(2) This is a standard diagram chase (but far more complicated than those in the proof of (1)), which we omit. $\qquad\qquad\square$

COROLLARY 8.1.21. *Let*

$$0 \longrightarrow U_1^* \xrightarrow{f_1} V_1^* \xrightarrow{g_1} W_1^* \longrightarrow 0$$

and

$$0 \longrightarrow U_2^* \xrightarrow{f_2} V_2^* \xrightarrow{g_2} W_2^* \longrightarrow 0$$

be short exact sequences of cochain complexes. Suppose that there are maps of cochain complexes $\alpha : U_1^* \longrightarrow U_2^*, \beta : V_1^* \longrightarrow V_2^*,$ and $\gamma :

$\mathbf{W}_1^* \longrightarrow \mathbf{W}_2^*$ *such that, for each i, the following diagram commutes:*

$$
\begin{array}{ccccccccc}
0 & \longrightarrow & U_1^i & \xrightarrow{f_1^i} & V_1^i & \xrightarrow{g_1^i} & W_1^i & \longrightarrow & 0 \\
 & & \alpha^i \downarrow & & \beta^i \downarrow & & \gamma^i \downarrow & & \\
0 & \longrightarrow & U_2^i & \xrightarrow{f_2^i} & V_2^i & \xrightarrow{g_2^i} & W_2^i & \longrightarrow & 0
\end{array}
$$

Then the diagram

$$
\begin{array}{ccccccccc}
\cdots \longrightarrow & H^i(\mathbf{U}_1^*) & \xrightarrow{(f_1^*)^i} & H^i(\mathbf{V}_1^*) & \xrightarrow{(g_1^*)^i} & H^i(\mathbf{W}_1^*) & \xrightarrow{\delta_1^i} & H^{i+1}(\mathbf{U}_1^*) \longrightarrow & \cdots \\
 & (\alpha^*)^i \downarrow & & (\beta^*)^i \downarrow & & (\gamma^*)^i \downarrow & & (\alpha^*)^{i+1} \downarrow & \\
\cdots \longrightarrow & H^i(\mathbf{U}_2^*) & \xrightarrow{(f_2^*)^i} & H^i(\mathbf{V}_2^*) & \xrightarrow{(g_2^*)^i} & H^i(\mathbf{W}_2^*) & \xrightarrow{\delta_2^i} & H^{i+1}(\mathbf{U}_2^*) \longrightarrow & \cdots
\end{array}
$$

commutes.

Proof.   This is a routine verification. $\square$

THEOREM 8.1.22.   *Given a commutative diagram of long exact sequences:*

$$
\begin{array}{ccccccccc}
\cdots \longrightarrow & A^i & \xrightarrow{\alpha^i} & B^i & \xrightarrow{\beta^i} & C^{i+1} & \xrightarrow{\gamma^i} & A^{i+1} \longrightarrow & \cdots \\
 & f^i \downarrow & & g^i \downarrow & & h^{i+1} \downarrow & & f^{i+1} \downarrow & \\
\cdots \longrightarrow & D^i & \xrightarrow{\delta^i} & E^i & \xrightarrow{\varepsilon^i} & F^{i+1} & \xrightarrow{\zeta^i} & D^{i+1} \longrightarrow & \cdots
\end{array}
$$

with each map h^i an isomorphism, there is a long exact sequence

$$
\cdots \longrightarrow A^i \xrightarrow{p^i} B^i \oplus D^i \xrightarrow{q^i} E^i \xrightarrow{r^i} A^{i+1} \longrightarrow \cdots
$$

where the maps p^i, q^i, and r^i are defined as follows:

$$
\begin{aligned}
p^i(a) &= (\alpha^i(a), f^i(a)) \\
q^i(b, d) &= g^i(b) - \delta^i(d) \\
r^i(e) &= \gamma^i (h^{i+1})^{-1} \varepsilon^i(e)
\end{aligned}
$$

Proof.   This is a standard (but quite elaborate) diagram chase. $\square$

8.2 Definition and basic properties

In this section we define de Rham cohomology and derive some of its fundamental properties.

Recall that for a smooth manifold with boundary M, $\Omega^i(M)$ denotes the space of smooth i-forms on M, for any $i \geq 0$. We extend this definition to all integers i by setting $\Omega^i(M) = 0$ for $i < 0$. We have the exterior differentiation operator $d : \Omega^i(M) \longrightarrow \Omega^{i+1}(M)$ for every i (with d being the 0 operator for $i < 0$).

LEMMA 8.2.1. *Let M be a smooth manifold with boundary. Then $\Omega^*(M) = \{\Omega^i(M), d\}$ is a cochain complex.*

Proof. We have $d^2 = 0$ by Poincaré's Lemma, Theorem 1.2.7. □

With this lemma in hand, we may now define de Rham cohomology.

DEFINITION 8.2.2. Let M be a smooth manifold with boundary. The ith *de Rham cohomology group* $H^i_{dR}(M)$ is the group

$$H^i_{dR}(M) = H^i(\Omega^*(M)).$$ ◇

REMARK 8.2.3.

(1) We observe that the i-cocycles are

$$Z^i(M) = \{\text{closed } i-\text{forms on } M\}$$

and that the i-coboundaries are

$$B^i(M) = \{\text{exact } i-\text{forms on } M\}.$$

(2) We observe that

$H^i_{dR}(M) = 0$ *if and only if every closed i-form on M is exact.* ◇

LEMMA 8.2.4. *Let M and N be smooth manifolds with boundary and let $f : M \longrightarrow N$ be a smooth map. Then the pull-back $f^* : \Omega^*(N) \longrightarrow \Omega^*(M)$ is a map of cochain complexes.*

Proof. This is just the statement that pull-back commutes with exterior differentiation, which is Theorem 4.3.12.   □

With this lemma in hand, we may define the induced map on de Rham cohomology.

DEFINITION 8.2.5. Let M and N be smooth manifolds with boundary and let $f : M \longrightarrow N$ be a smooth map. Then the *map induced by f on de Rham cohomology* is the map $f^i : H^i_{dR}(N) \longrightarrow H^i_{dR}(M)$ for each i that is the map induced on cohomology by the map of cochain complexes $f^* : \Omega^*(N) \longrightarrow \Omega^*(M)$.   ◇

LEMMA 8.2.6.

(1) Let M be a smooth manifold with boundary and let $f : M \longrightarrow M$ be the identity map. Then $f^i : H^i_{dR}(M) \longrightarrow H^i_{dR}(M)$ is the identity map for each i.

(2) Let M, N, and P be smooth manifolds with boundary and let $f : M \longrightarrow N$ and $g : N \longrightarrow P$ be smooth maps. Then

$$(gf)^i = f^i g^i : H^i_{dR}(P) \longrightarrow H^i_{dR}(M)$$

for each i.

Proof. This is true on the cochain level by Theorem 4.3.5, and hence true for cohomology by Lemma 8.1.16.   □

LEMMA 8.2.7. *Let M and N smooth manifolds with boundary and let $f : M \longrightarrow N$ and $g : M \longrightarrow N$ be smoothly homotopic smooth maps. Then*

$$f^i = g^i : H^i_{dR}(N) \longrightarrow H^i_{dR}(M)$$

for each i.

Proof. In this situation we constructed a chain homotopy between f^* and g^* in Theorem 4.7.3, so the induced maps on cohomology agree by Lemma 8.1.19.   □

We now generalize the notion of a smooth manifold with boundary to that of a smooth pair.

DEFINITION 8.2.8. A *smooth pair* (M, A) consists of a smooth manifold M with boundary and a smooth submanifold A with boundary of M, where furthermore A is closed as a subset of M. A map of smooth pairs $f : (M, A) \longrightarrow (N, B)$ is a smooth map $f : M \longrightarrow N$ with $f(A) \subseteq B$. A smooth homotopy between maps of smooth pairs $f_i : (M, A) \longrightarrow (N, B), i = 0, 1$, is a smooth homotopy $F : (M \times I, A \times I) \longrightarrow (N, B)$ between f_0 and f_1. $\diamond$

REMARK 8.2.9. Note that this definition strictly generalizes the definition of a smooth manifold with boundary. Namely, a smooth manifold with boundary M is a smooth pair $(M, \emptyset)$, etc. $\diamond$

REMARK 8.2.10. Let (M, A) be a smooth pair. Then the inclusion $i : A \longrightarrow M$ is a smooth map, and so induces $i^* : \Omega^*(M) \longrightarrow \Omega^*(A)$. This map is nothing other than restriction, and we will often write it as $r(\varphi) = \varphi|_A$. Note in particular that $\varphi|_A$ is a differential form on A, i.e., a function on tuples of tangent vectors to A at points of A (rather than a function on tangent vectors to M at points of A). $\diamond$

DEFINITION 8.2.11. Let (M, A) be a smooth pair. Then $\Omega^*(M, A)$ is the kernel of the restriction map,

$$\Omega^*(M, A) = \text{Ker}(r) = \{\varphi \in \Omega^*(M) |\ \varphi|_A = 0\}.$$

$\Omega^*(M, A)$ is called the complex of differential forms on M relative to A. $\diamond$

Observe that if $\varphi \in \Omega^*(M, A)$, then $d\varphi \in \Omega^*(M, A)$. Thus $\{\Omega^i(M, A), d\}$ is indeed a cochain complex, and we may use it to define cohomology groups as well, exactly as above.

DEFINITION 8.2.12. Let (M, A) be a smooth pair. Then

$$Z^i(M, A) = \text{Ker}(d : \Omega^i(M, A) \longrightarrow \Omega^{i+1}(M, A))$$

is the group of *relative i-cocycles*,

$$B^i(M, A) = \text{Im}(d : \Omega^{i-1}(M, A) \longrightarrow \Omega^i(M, A)),$$

is the group of *relative i-coboundaries*, and the quotient

$$H^i_{dR}(M, A) = Z^i(M, A)/B^i(M, A)$$

is the ith *relative* de Rham cohomology group of (M, A). $\diamond$

REMARK 8.2.13. All of the results above stated for smooth manifolds with boundary go through without change for smooth pairs. We merely chose to state them for smooth manifolds with boundary rather than in this more general situation for the sake of directness and simplicity. In fact all these results go through without the condition that A be closed as a subset of M. But that condition is crucial for the next result.   ◇

LEMMA 8.2.14. *Let (M, A) be a smooth pair. Then there is a short exact sequence of chain complexes*

$$0 \longrightarrow \Omega^*(M, A) \xrightarrow{\text{incl}} \Omega^*(M) \xrightarrow{\text{rest}} \Omega^*(A) \longrightarrow 0$$

where incl *is the inclusion map and* rest *is the restriction map.*

Proof. The inclusion of $\Omega^*(M, A)$ into $\Omega^*(M)$ is certainly an injection, so the sequence is exact at $\Omega^*(M, A)$, and by definition, $\Omega^*(M, A)$ is the kernel of the restriction map from $\Omega^*(M)$ to $\Omega^*(A)$, so the sequence is exact at $\Omega^*(M)$. Thus we need to show that it is exact at $\Omega^*(A)$, i.e., that the restriction map from $\Omega^*(M)$ to $\Omega^*(A)$ is a surjection.

We begin by showing this in very special cases. First let us suppose we are dealing with 0-forms, i.e., functions. Consider the following cases:

Case (a): $M = \{(x_1, \ldots, x_n) \in \mathbb{R}^n \mid |x_i| < 3, \ i = 1, \ldots, n\}$, $A = \{(x_1, \ldots, x_n) \in \mathbb{R}^n \mid |x_i| < 3, \ i = 1, \ldots, k, \ x_i = 0, \ i = k+1, \ldots, n\}$, and $f(x_1, \ldots, x_k, 0, \ldots, 0) = g(x_1, \ldots, x_k)$ is a smooth function on A with $g(x_1, \ldots, x_k) = 0$ if $|x_i| \geq 2$ for some i, $i = 1, \ldots, k$. Then we may extend f to a function on M, which we still call f, by

$$f(x_1, \ldots, x_k, x_{k+1}, \ldots, x_n)$$
$$= g(x_1, \ldots, x_k) f_3(x_{k+1}) \cdots f_3(x_n)$$

where $f_3(x)$ is the function of that name constructed in Lemma 4.6.1. Note this extension satisfies $f(x_1, \ldots, x_k, x_{k+1}, \ldots, x_n) = 0$ if $|x_i| \geq 2$ for some $i, i = 1, \ldots, n$.

Case (b): $M = \{(x_1, \ldots, x_n) \in \mathbb{R}^n \mid |x_i| < 3, \ i = 1, \ldots, n\}$, $A = \{(x_1, \ldots, x_n) \in \mathbb{R}^n \mid -3 < x_1 \leq 0, |x_i| < 3, \ i = 2, \ldots, n\}$, and

$f(x_1, \ldots, x_n)$ is a smooth function on A with $f(x_1, \ldots, x_n) = 0$ if $|x_i| \geq 2$ for some $i, i = 1, \ldots, n$. Then by the very definition of a smooth function f extends to a smooth function, which we still call f, defined on a neighborhood of A in $\mathbb{R}^n$ (and hence in M). Indeed, using the compactness of the set $\{(0, x_2, \ldots, x_n) \mid |x_i| \leq 2, i = 2, \ldots, n\}$ we may assume that that neighborhood is of the form $\{(x_1, \ldots, x_n) \in \mathbb{R}^n \mid -3 < x_1 < \varepsilon, |x_i| < 3, i = 2, \ldots, n\}$ for some $\varepsilon > 0$. Then we may extend f to a function, which we again still call f, on M by

$$f(x_1, \ldots, x_n) = f(x_1, \ldots, x_n) f_2(-\varepsilon x_1/2) \text{ for } -3 < x_1 < 2,$$
$$= 0 \text{ for } x_1 > 1,$$

where $f_2(x)$ is the function of that name constructed in Lemma 4.6.1. Again this extension satisfies $f(x_1, \ldots, x_k, x_{k+1}, \ldots, x_n) = 0$ if $|x_i| \geq 2$ for some $i, i = 1, \ldots, n$.

Case (c): $M = \{(x_1, \ldots, x_n) \in \mathbb{R}^n \mid |x_i| < 3, i = 1, \ldots, n\}$, $A = \{(x_1, \ldots, x_n) \in \mathbb{R}^n \mid -3 < x_1 \leq 0, |x_i| < 3, i = 2, \ldots, k, x_i = 0, i = k+1, \ldots, n\}$, and $f(x_1, \ldots, x_n)$ is a smooth function on A with $f(x_1, \ldots, x_n) = 0$ if $|x_i| \geq 2$ for some $i, i = 1, \ldots, n$. First we extend f to $\{(x_1, \ldots, x_n) \in \mathbb{R}^n \mid |x_i| < 3, i = 1, \ldots, k, x_i = 0, i = k+1, \ldots, n\}$ as in case (b), and then to M as in case (a).

Thus in cases (a), (b), and (c), we have proved that restriction is surjective on 0-forms.

Next, let φ be an i-form on A, with A as above. Then φ is a sum of terms, of which a typical one is $f \, dx_1 \ldots dx_i$. An extension of the function f gives an extension of $f \, dx_1 \ldots dx_i$, so we have that in these cases restriction is surjective on i-forms.

Now let A and M be general, and let φ be an i-form on A. We apply a partition of unity argument, as usual, to write φ as a sum of i-forms, each of which has support contained in the interior of some coordinate patch, pull back to $\mathbb{R}^n$, perform the above construction there, and pull back to M to obtain that φ is the restriction of an i-form on M. Note the above cases are the "model" cases: Case (a) is the model for A a manifold in the interior of M and cases (a) and (c) for A a manifold with boundary in the interior of M. There are other model cases to consider in case A intersects the boundary of M, but since these merely involve complications without any essentially new ideas, we shall omit them. $\square$

COROLLARY 8.2.15. *Let (M, A) be a smooth pair. Then there is a long exact sequence in de Rham cohomology*

$$\cdots \longrightarrow H^i_{dR}(M, A) \longrightarrow H^i_{dR}(M) \longrightarrow H^i_{dR}(A)$$
$$\longrightarrow H^{i+1}_{dR}(M, A) \longrightarrow \cdots .$$

Proof.   Immediate from Lemma 8.2.14 and Theorem 8.1.20.   □

The exact sequence in the above corollary is known as the *exact sequence of the pair* (M, A).

COROLLARY 8.2.16. *Let $f : (M, A) \longrightarrow (N, B)$ be a map of smooth pairs. Then the following diagram commutes, where the horizontal sequences are the long exact sequences of the pairs (N, B) and (M, A), and the vertical maps are induced by f:*

$$
\begin{array}{ccccccccc}
\cdots \longrightarrow & H^i_{dR}(N, B) & \longrightarrow & H^i_{dR}(N) & \longrightarrow & H^i_{dR}(B) & \longrightarrow & H^{i+1}_{dR}(N, B) & \longrightarrow \cdots \\
& \downarrow & & \downarrow & & \downarrow & & \downarrow & \\
\cdots \longrightarrow & H^i_{dR}(M, A) & \longrightarrow & H^i_{dR}(M) & \longrightarrow & H^i_{dR}(A) & \longrightarrow & H^{i+1}_{dR}(M, A) & \longrightarrow \cdots
\end{array}
$$

Proof.   This follows directly from Corollary 8.2.16.   □

The property in the next lemma is known as the excision property of de Rham cohomology. Note that unless $U = \emptyset$, in which case the result is trivial, the hypothesis implies that the interior of A must be an open submanifold of the interior of M (and in particular that A and M must have the same dimension).

LEMMA 8.2.17 (*Excision*). *Let (M, A) be a smooth pair. Let U be a subset of A with the closure $\overline{U}$ of U contained in $\text{Int}(A)$, the interior of A, and such that $(M - U, A - U)$ is also a smooth pair. Then for each i, the restriction map*

$$H^i_{dR}(M, A) \longrightarrow H^i_{dR}(M - U, A - U)$$

is an isomorphism.

Proof.   Let r denote the restriction map. We show that r is both surjective and injective, hence an isomorphism.

Let $a \in H^i_{dR}(M - U, A - U)$ be a cohomology class. Choose a representative φ of a, so that $a = [\varphi]$. Then φ is an i-form on $M - U$ that is 0 on $A - U$, or, equivalently, φ is an i-form on $M - \overline{U}$ that is 0 on $A - \overline{U}$. Note that $\{\text{Int}(A), M - \overline{U}\}$ is an open cover of M, by our hypothesis on U, and $\varphi = 0$ on the intersection of these two sets. Hence we may define an i-form $\tilde{\varphi}$ on M by $\tilde{\varphi} = \varphi$ on $M - \overline{U}$ and $\tilde{\varphi} = 0$ on $\text{Int}(A)$. Then $\tilde{\varphi} = 0$ on A as well, so $\tilde{\varphi}$ represents a cohomology class $\tilde{a} \in H^i_{dR}(M, A)$ (i.e., $\tilde{a} = [\tilde{\varphi}]$) and $r(\tilde{a}) = a$. Hence r is surjective.

Now let $\tilde{a} \in H^i_{dR}(M, A)$ be a cohomology class, and choose a representative $\tilde{\varphi}$ of $\tilde{a}$. Suppose that $r(\tilde{a}) = 0$. That means that $\varphi = r(\tilde{\varphi}) = d\psi$ for some $(i - 1)$-form $\psi \in \Omega^{i-1}(M - U, A - U)$. By exactly the same logic, ψ extends to an $(i-1)$ form $\tilde{\psi} \in \Omega^i(M, A)$, and then $\tilde{\varphi} = d\tilde{\psi}$, so that $a = [\tilde{\varphi}] = [d\tilde{\psi}] = 0$. Hence r is injective. $\qquad\square$

We have the following almost trivial (but nevertheless important) observation.

LEMMA 8.2.18. *Let $M = \{p\}$ be a space consisting of a single point. Then $H^0_{dR}(M)$ is isomorphic to $\mathbb{R}$ and $H^i_{dR}(M) = 0$ for $i \neq 0$.*

Proof. $\Omega^0(M)$ is isomorphic to $\mathbb{R}$ by the isomorphism $f \mapsto f(p)$, and $\Omega^i(M) = 0$ for $i \neq 0$. $\qquad\square$

We now summarize the properties of de Rham cohomology that we have derived. There is nothing new in this theorem. Rather, it is simply convenient to have these properties all stated together. We assume in this statement that all manifolds/pairs are smooth, and that all maps/homotopies are smooth as well. We also implicitly restrict to situations where all maps are defined, and allow i to be arbitrary.

THEOREM 8.2.19. *de Rham cohomology satisfies the following properties:*

(1) If f is the identity map, then f^i is the identity. Also, if f and g are maps, then $(gf)^i = f^i g^i$.

(2) The following is a commutative diagram of long exact sequences:

$$\cdots \longrightarrow H^i_{dR}(N, B) \longrightarrow H^i_{dR}(N) \longrightarrow H^i_{dR}(B) \longrightarrow H^{i+1}_{dR}(N, B) \longrightarrow \cdots$$
$$\downarrow \qquad\qquad \downarrow \qquad\qquad \downarrow \qquad\qquad \downarrow$$
$$\cdots \longrightarrow H^i_{dR}(M, A) \longrightarrow H^i_{dR}(M) \longrightarrow H^i_{dR}(A) \longrightarrow H^{i+1}_{dR}(M, A) \longrightarrow \cdots$$

(3) If f and g are smoothly homotopic, then $f^i = g^i$.

(4) If $\overline{U} \subset \text{Int}(A)$, then the restriction map $H^i_{dR}(M, A) \longrightarrow$ $H^i_{dR}(M - U, A - U)$ is an isomorphism.

(5) If M is the space consisting of a single point, then $H^0_{dR}(M) \cong \mathbb{R}$ and $H^i_{dR}(M) = 0$ for $i \neq 0$.

Next we observe that we can define products in de Rham cohomology.

LEMMA 8.2.20. *Let M be a smooth manifold. Then, for any i and j, there is a product*

$$H^i_{dR}(M) \times H^j_{dR}(M) \longrightarrow H^{i+j}_{dR}(M)$$

given as follows:

$$\text{If } a = [\varphi] \text{ and } b = [\psi] \text{ then } ab = [\varphi\psi].$$

Proof. We merely have to check that this definition is independent of the choice of representative. Thus suppose we choose different representatives, so that $a = [\varphi']$ and $b = [\psi']$. Then $\varphi' = \varphi + d\sigma$ for some σ, and then $\psi' = \psi + d\tau$ for some τ, and then, recalling that $d\varphi = 0$ and $d\psi = 0$, since they are both cocycles,

$$\varphi'\psi' = (\varphi + d\sigma)(\psi + d\tau) = \varphi\psi + (d\sigma)\psi + (\varphi)d\tau + (d\sigma)(d\tau)$$
$$= \varphi\psi + d(\sigma\psi + (-1)^i\tau\varphi + \sigma d\tau)$$

so $[\varphi'\psi'] = [\varphi\psi]$. $\square$

The product gives the de Rham cohomology of a space the structure of a graded ring that is associative, commutative (in the graded, not ordinary, sense), and has an identity. We will not define this structure precisely here, but rather just list the key properties.

THEOREM 8.2.21. *The product on de Rham cohomology has the following properties:*

(a) If $a \in H^i_{dR}(M), b \in H^j_{dR}(M), c \in H^k_{dR}(M)$, then $a(bc) = (ab)c \in H^{i+j+k}_{dR}(M)$.

(b) If $a \in H_{dR}^i(M)$ and $b \in H_{dR}^j(M)$, then $ab = (-1)^{ij} ba \in H_{dR}^{i+j}(M)$.

(c) If $1 = [1] \in H_{dR}^0(M)$ is the de Rham cohomology class represented by the constant function with value 1 on M, and $a \in H_{dR}^i(M)$, then $1a = a1 = a \in H_{dR}^i(M)$.

Proof. Immediate from properties of the product of differential forms. □

REMARK 8.2.22. The product gives the de Rham cohomology of a smooth pair (M, A) a similar structure, except that if A is nonempty there is no identity element, as the constant function 1 does not vanish on A. ◇

8.3 Computations of cohomology groups

In this section we reinterpret some of our earlier results as results about de Rham cohomology groups, and prove some new results along these lines. We assume throughout this section that we are dealing with smooth manifolds with boundary/smooth pairs and smooth maps/smooth homotopies, and may not always state these conditions as part of the hypotheses of our results.

DEFINITION 8.3.1. Let $f : M \longrightarrow N$ be a smooth map between smooth manifolds with boundary. Then f is a *smooth homotopy equivalence* if there is a smooth map $g : N \longrightarrow M$ (which will also be a smooth homotopy equivalence) with the composition gf smoothly homotopic to the identity map on M, and with the composition fg smoothly homotopic to the identity map on N. ◇

LEMMA 8.3.2. *Let $f : M \longrightarrow N$ be a smooth homotopy equivalence. Then $f^i : H_{dR}^i(N) \longrightarrow H_{dR}^i(M)$ is an isomorphism, for each i.*

Proof. Let g be as in Definition 8.3.1. Then $(gf)^i = f^i g^i$ is the identity map on $H_{dR}^i(M)$, and $(fg)^i = f^i g^i$ is the identity map on $H_{dR}^i(N)$, i.e., $g^i = (f^i)^{-1}$, so f^i and g^i are both isomorphisms, for each i. □

We again restate the converse to Poincaré's Lemma (in its most general form).

COROLLARY 8.3.3. *Let M be a smoothly contractible manifold with boundary. Then $H_{dR}^0(M) \cong \mathbb{R}$ and $H_{dR}^i(M) = 0$ for $i \neq 0$.*

Proof. To say that M is smoothly contractible is the same as saying that the map from M to any point p of M is smoothly homotopic to the identity map on M, and so this is immediate from Lemma 8.2.18 and Lemma 8.3.2. $\qquad \square$

LEMMA 8.3.4. *Let M be a smooth n-manifold with boundary. Then $H_{dR}^i(M) = 0$ for $i > n$.*

Proof. In this case, $\Omega^i(M) = 0$ for $i > n$. $\qquad \square$

LEMMA 8.3.5. *Let M be the disjoint union of two components $M = M_1 \cup M_2$. Then $H_{dR}^i(M) = H_{dR}^i(M_1) \oplus H_{dR}^i(M_2)$ for each i.*

Proof. In this situation $\Omega^i(M) = \Omega^i(M_1) \oplus \Omega^i(M_2)$ for each i, and exterior differentiation preserves the summands. $\qquad \square$

Given this lemma, we will (almost always) restrict our attention to connected M.

LEMMA 8.3.6. *If M is connected and nonempty, then $H_{dR}^0(M) \cong \mathbb{R}$.*

Proof. As we have seen, in this case, the closed 0-forms are exactly the constant functions, and the only exact 0-form is the zero function, so there is an isomorphism given by $f \mapsto f(p)$, where f is a smooth function on M and p is any point of M. $\qquad \square$

DEFINITION 8.3.7. Let C be a piecewise smooth oriented closed loop in M, which we assume is parameterized by $f : S^1 \longrightarrow M$, where S^1 is the unit circle in the complex plane. Then for any integer n, nC is the loop parameterized by $g : S^1 \longrightarrow M$, where $g(z) = f(z^n)$. $\diamond$

REMARK 8.3.8. Geometrically, nC is the loop obtained by traversing C n times, where a negative multiple indicates reversal of orientation, and 0 times means a constant path (at the point $f(1)$). ◇

THEOREM 8.3.9. *Suppose that for every piecewise smooth closed loop C_0 in M, there is a nonzero integer n such that nC_0 is freely nullhomotopic. Then $H^1_{dR}(M) = 0$.*

Proof. Let φ be any closed 1-form on M. Then $\int_{nC_0} \varphi = 0$ by Corollary 6.4.20. But it is easy to check that $\int_{nC_0} \varphi = n \int_{C_0} \varphi$, so $\int_{C_0} \varphi = 0$. Since C_0 is arbitrary, that shows φ is exact, by the proof of Corollary 6.4.22. Hence every closed 1-form on M is exact, and so $H^1_{dR}(M) = 0$. □

THEOREM 8.3.10.

(a) Let M be a compact oriented smooth n-manifold. Then $H^n_{dR}(M) \cong \mathbb{R}$.

(b) Let M be a compact oriented smooth n-manifold with nonempty boundary. Then $H^n_{dR}(M) = 0$.

(c) Let M be a compact oriented smooth n-manifold with boundary. Then $H^n_{dR}(M, \partial M) \cong \mathbb{R}$.

Proof. We know that every n-form on M is closed.

(a) We have a map $i : \Omega^n(M) \longrightarrow \mathbb{R}$ by $i(\varphi) = \int_M \varphi$. By construction, this map is nonzero ($i(\varphi_0) \neq 0$ for φ_0 a form as in Lemma 7.7.5) and hence is onto $\mathbb{R}$ (by linearity), and by Theorem 7.7.6(2) φ is in the kernel of this map if and only if φ is exact, so i induces an isomorphism from $H^n_{dR}(M)$ to $\mathbb{R}$.

(b) We know from Theorem 7.7.6(1)(a) that every n-form on M is exact in this situation.

(c) This follows similarly from Theorem 7.7.6(1)(b). □

THEOREM 8.3.11 (*Mayer-Vietoris for de Rham cohomology*). *Let M be a smooth n-manifold with boundary. Let $M_1 \subset M$ and $M_2 \subset M$*

be smooth n-manifolds with boundary such that (M, M_1) *and* (M, M_2)
are both smooth pairs. Suppose that $M = M_1 \cup M_2$ *and that* $\partial M_1 \cap$
$\partial M_2 = \emptyset$. *Then there is a long exact sequence*

$$\cdots \longrightarrow H^i_{dR}(M) \longrightarrow H^i_{dR}(M_1) \oplus H^i_{dR}(M_2) \longrightarrow H^i_{dR}(M_1 \cap M_2)$$
$$\longrightarrow H^{i+1}_{dR}(M) \longrightarrow \cdots$$

Proof.   This situation is schematically illustrated as follows:

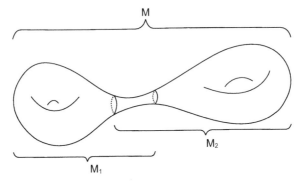

   In this situation, $(M_2, M_1 \cap M_2)$ is a smooth pair and the inclusion $h : (M_2, M_1 \cap M_2) \longrightarrow (M, M_1)$ is a map of smooth pairs.
Not only that, but, letting $X = M, A = M_1$, and letting U be
the complement of $M_1 \cap M_2$ in M_2, then $\overline{U} \subset \text{int}(A)$, and h is
the inclusion $h : (X - U, A - U) \longrightarrow (X, A)$, so by excision $h^i :$
$H^i_{dR}(M, M_1) \longrightarrow H^i_{dR}(M_2, M_1 \cap M_2)$ is an isomorphism for every i.
   Thus we have a commutative diagram of long exact sequences, with
the vertical maps induced by inclusion, and each h^i an isomorphism,

$$\cdots \longrightarrow H^i_{dR}(M) \longrightarrow H^i_{dR}(M_1) \longrightarrow H^{i+1}_{dR}(M, M_1) \longrightarrow H^{i+1}_{dR}(M) \longrightarrow \cdots$$
$$\downarrow \qquad\qquad \downarrow \qquad\qquad h^{i+1}\downarrow \qquad\qquad \downarrow$$
$$\cdots \longrightarrow H^i_{dR}(M_2) \longrightarrow H^i_{dR}(M_1 \cap M_2) \longrightarrow H^{i+1}_{dR}(M_2, M_1 \cap M_2) \longrightarrow H^{i+1}_{dR}(M_2) \longrightarrow \cdots$$

and so by Theorem 8.1.22 we have a long exact sequence as claimed
in the statement of the theorem. $\square$

THEOREM 8.3.12.   *Let* S^n *be the n-sphere,* $n \geq 1$. *Then the de
Rham cohomology of* S^n *is given by* $H^0_{dR}(S^n) \cong \mathbb{R}$, $H^i_{dR}(S^n) = 0$ *for*
$0 < i < n$, *and* $H^n_{dR}(S^n) \cong \mathbb{R}$.

Proof.   Since S^n is connected for $n \geq 1$, $H^0_{dR}(S^n) \cong \mathbb{R}$ for $n \geq 1$ by Lemma 8.3.6.

We prove the theorem for $i > 0$ by induction on n.
S^n is the unit sphere in $\mathbb{R}^{n+1}$,

$$S^n = \{p = (x_1, \ldots, x_{n+1}) | x_1^2 + \cdots + x_{n+1}^2 = 1\}.$$

Define subsets T (top) and B (bottom) of S^n by

$$T = \{p \in S^n | x_{n+1} \geq -1/2\}, \quad B = \{p \in S^n | x_{n+1} \leq 1/2\}$$

and let C (center) be their intersection $C = T \cap B$. We have the Mayer-Vietoris sequence Theorem 8.3.11 with S^n, T, B, and C playing the roles of M, M_1, M_2, and $M_1 \cap M_2$, respectively. We observe that each of T and B is smoothly contractible, so each has the de Rham cohomology of a point.

Let E (equator) be the subset

$$E = \{p \in S^n \mid x_{n+1} = 0\}.$$

We make two observations about E. First, the map $(x_1, \ldots, x_n, 0) \mapsto (x_1, \ldots, x_n)$ is a diffeomorphism from E to S^{n-1}, so induces an isomorphism on de Rham cohomology. Second, C is diffeomorphic to $E \times [-1, 1]$, and so the inclusion of E into C is a smooth homotopy equivalence, so also induces an isomorphism on de Rham cohomology. We use these isomorphisms to substitute $H^i_{dR}(S^{n-1})$ for $H^i_{dR}(C)$ in the Mayer-Vietoris sequence.

We begin the induction with $n = 1$. We observe that $S^{n-1} = S^0$ consists of two points. Then in the Mayer-Vietoris sequence we see the terms

$$H^{-1}_{dR}(S^0) \longrightarrow H^0_{dR}(S^1) \longrightarrow H^0_{dR}(T) \oplus H^0_{dR}(B) \longrightarrow H^0_{dR}(S^0)$$
$$\longrightarrow H^1_{dR}(S^1) \longrightarrow H^1_{dR}(T) \oplus H^1_{dR}(B)$$

which are

$$0 \longrightarrow \mathbb{R} \longrightarrow \mathbb{R} \oplus \mathbb{R} \longrightarrow \mathbb{R} \oplus \mathbb{R} \longrightarrow H^1_{dR}(S^1) \longrightarrow 0$$

and a simple dimension count shows $H^1_{dR}(S^1) \cong \mathbb{R}$.

For $i > 1$ we see the terms

$$H_{dR}^{i-1}(S^0) \longrightarrow H_{dR}^i(S^1) \longrightarrow H_{dR}^i(T) \oplus H_{dR}^i(B)$$

which are

$$0 \longrightarrow H_{dR}^i(S^1) \longrightarrow 0 \oplus 0$$

so $H_{dR}^i(S^1) = 0$ for $i > 1$.

Now suppose the theorem is true for S^{n-1}. In the Mayer-Vietoris sequence we see the terms

$$H_{dR}^{-1}(S^{n-1}) \longrightarrow H_{dR}^0(S^n) \longrightarrow H_{dR}^0(T) \oplus H_{dR}^0(B) \longrightarrow H_{dR}^0(S^{n-1})$$
$$\longrightarrow H_{dR}^1(S^n) \longrightarrow H_{dR}^1(T) \oplus H_{dR}^1(B)$$

which are

$$0 \longrightarrow \mathbb{R} \longrightarrow \mathbb{R} \oplus \mathbb{R} \longrightarrow \mathbb{R} \longrightarrow H_{dR}^1(S^n) \longrightarrow 0$$

and a simple dimension count shows $H_{dR}^1(S^n) = 0$.

For $i > 1$ we see the terms

$$H_{dR}^{i-1}(T) \oplus H_{dR}^{i-1}(B) \longrightarrow H_{dR}^{i-1}(S^{n-1})$$
$$\longrightarrow H_{dR}^i(S^n) \longrightarrow H_{dR}^i(T) \oplus H_{dR}^i(B)$$

which are

$$0 \oplus 0 \longrightarrow H_{dR}^{i-1}(S^{n-1}) \longrightarrow H_{dR}^i(S^n) \longrightarrow 0 \oplus 0$$

so $H_{dR}^i(S^n)$ is isomorphic to $H_{dR}^{i-1}(S^{n-1})$ for $i > 1$, which yields the truth of the theorem for S^n.   □

THEOREM 8.3.13. *Let M be a smooth manifold with boundary. Then:*

(a) $H_{dR}^i(M \times S^1) \cong H_{dR}^i(M) \oplus H_{dR}^{i-1}(M)$ *for every i, and*

(b) *If p is any point of S^1, $H_{dR}^i(M \times S^1, M \times \{p\}) \cong H_{dR}^{i-1}(M)$ for every i.*

Proof. Part (a) follows from a Mayer-Vietoris argument similar to the above, letting $M \times S^1 = (M \times T) \cup (M \times B)$, where T and B are the top and bottom of S^1. Then part (b) follows from the long exact de Rham cohomology sequence of the smooth pair $(M \times S^1, M \times \{p\})$.   □

8.4 Cohomology with compact supports

In this section we develop de Rham cohomology with compact supports. In the case of a compact manifold M with boundary, or a smooth pair (M, A) where both M and A are compact manifolds with boundary, this theory (trivially) agrees with (ordinary) de Rham cohomology, but in general, as we shall see, it differs. This theory is closely related to integration of differential forms on arbitrary oriented manifolds with boundary, since, as we have seen, we can always integrate forms with compact support, but not always arbitrary forms.

DEFINITION 8.4.1. Let M be a smooth manifold. Then

$$\Omega_c^i(M) = \{i\text{-forms on } M \text{ with compact support}\}$$

and the cochain complex $\Omega_c^*(M) = \{(\Omega_c^i(M), d)\}$ where d is exterior differentiation. If (M, A) is a smooth pair, then $\Omega_c^*(M, A) = \{(\Omega_c^i(M, A), d)\}$ where $\Omega_c^i(M, A) = \Omega_c^i(M) \cap \Omega^i(M, A)$.   ◇

Note that if φ has compact support, then $d\varphi$ has compact support as well, so $\Omega_c^*(M)$ is a indeed a cochain complex, as is $\Omega_c^*(M, A)$.

DEFINITION 8.4.2. Let M be a smooth manifold (resp. let (M, A) be a smooth pair). The ith *de Rham cohomology group with compact supports* of M (resp. of (M, A)) is the group

$$H_{c-dR}^i(M) = H^i(\Omega_c^*(M))$$
$$(\text{resp. } H_{c-dR}^i(M, A) = H^i(\Omega_c^*(M, A))).$$ ◇

The development of the basic theory of de Rham cohomology with compact supports entirely parallels the basic theory of de Rham cohomology, provided we suitably restrict our maps. Recall that, in general, if $f : X \longrightarrow Y$ is a continuous map, and A is any compact subset of X, then $f(A)$ is a compact subset of Y. The map f is said to be *proper* if it has the property that for any compact subset subset $B \subseteq Y$, the subset $f^{-1}(B) \subseteq X$ is compact.

In particular, let $f : M \longrightarrow N$ be a smooth map, and let φ be a differential form on N with compact support. If f is proper, then $f^*(\varphi)$ is a differential form on M with compact support, but if f is not proper, it may not be. Thus we need to restrict our attention to smooth proper maps (including restricting smooth homotopies to proper smooth homotopies). It turns out that is the only restriction we need.

Note that our definition of a smooth pair (M, A) includes the condition that A be closed as a subset of M. In this situation the inclusion $i : A \longrightarrow M$ is proper. (But if A is not closed, it will not be).

Note also that in our construction in the proof of Lemma 8.2.14, if we began with a form having compact support, we ended with a form having compact support as well.

THEOREM 8.4.3. *The analog of Theorem 8.2.19 holds for de Rham cohomology with compact support, where we require all maps and homotopies to be proper.*

THEOREM 8.4.4. *The analogs of the following results hold for de Rham cohomology with compact supports: Lemma 8.3.2, Corollary 8.3.3, Lemma 8.3.4, Lemma 8.3.5, Theorem 8.3.11.*

REMARK 8.4.5. We emphasize that the analog of Corollary 8.3.3 is that M be properly smoothly contractible, and that hypothesis implies that M must be compact (as if M is not compact, the constant map taking M to a point is not a proper map). Thus, for example, it applies to D^n (the unit disk in $\mathbb{R}^n$), but *not* to $\mathbb{R}^n$ itself.   ◇

LEMMA 8.4.6. *If M is connected but not compact, then $H^0_{c-dR}(M) = 0$.*

Proof. The only closed 0-forms on a connected smooth manifold with boundary M are the constant functions, but if M is not compact a nonzero constant function does not have compact support.   □

LEMMA 8.4.7. $H^n_{c-dR}(\mathbb{R}^n) \cong \mathbb{R}$.

Proof. Every n-form on $\mathbb{R}^n$ is closed. We have a map i : $\Omega_c^n(\mathbb{R}^n) \longrightarrow \mathbb{R}$ by $i(\varphi) = \int_{\mathbb{R}^n} \varphi$. Note that if we set $\varphi_0 = f_4(x_1) \cdots f_4(x_n) dx_1 \cdots dx_n$, with $f_4(x)$ as in Lemma 4.6.1, then $i(\varphi_0) \neq 0$. We claim that $\varphi = d\psi$ for some $(n-1)$-form ψ with compact support if and only if $i(\varphi) = 0$, and this implies that i induces an isomorphism from $H_{c-dR}^n(\mathbb{R}^n)$ to $\mathbb{R}$.

Suppose $i(\varphi) = 0$. Since φ has compact support, we may find a brick B with $\text{supp}(\varphi) \subset \text{int}(B)$. But then, by Lemma 7.7.4, $\varphi = d\psi$ for some $(n-1)$-form ψ with $\text{supp}(\psi) \subset \text{int}(B)$. But then ψ extends to an $(n-1)$-form with compact support on $\mathbb{R}^n$ by setting $\psi = 0$ outside of B, and $\varphi = d\psi$.

Conversely, suppose $\varphi = d\psi$ for some $(n-1)$-form ψ with compact support. Choose a brick B with $\text{supp}(\psi) \subset \text{int}(B)$. Then $\text{supp}(\varphi) = \text{supp}(d\psi) \subseteq \text{supp}(\psi) \subset \text{int}(B)$, so $\int_{\mathbb{R}^n} \varphi = \int_B \varphi$. But by the GST, $\int_B \varphi = \int_B d\psi = \int_{\partial B} \psi = 0$ as $\psi = 0$ on ∂B. $\square$

THEOREM 8.4.8.

(a) *Let M be an oriented smooth n-manifold. Then $H_{c-dR}^n(M) \cong \mathbb{R}$.*

(b) *Let M be an oriented smooth n-manifold with nonempty boundary. Then $H_{c-dR}^n(M) = 0$.*

(c) *Let M be an oriented smooth n-manifold with boundary. Then $H_{c-dR}^n(M, \partial M) \cong \mathbb{R}$.*

Proof. Part (a) follows from Lemma 8.4.7 by a partition of unity argument, similar to ones we have already done. (First show that any element of $H_{c-dR}^n(M)$ has a representative with support contained in a single coordinate patch.) Parts (b) and (c) are similar. $\square$

THEOREM 8.4.9. *Let M be a smooth manifold with boundary. Then*

$$H_{c-dR}^i(M \times \mathbb{R}) \cong H_{c-dR}^{i-1}(M)$$

for each i.

Proof. Let $S^1 = \{e^{i\theta} | 0 \leq \theta \leq 2\pi\}$ be the unit circle in the complex plane. We construct an isomorphism $\tilde{K} : H_{c-dR}^i(M \times \mathbb{R})$

$\longrightarrow H^i_{dR}(M \times S^1, M \times \{1\})$, which immediately implies the theorem, by Theorem 8.3.13(b).

Note that $S^1 - \{1\}$ is diffeomorphic to $\mathbb{R}$. For the sake of definitiveness, we choose the particular diffeomorphism $k(e^{i\theta}) = -\cot(\theta/2)$. Let $K : M \times (S^1 - \{1\}) \longrightarrow M \times \mathbb{R}$ by $K(m, e^{i\theta}) = (m, k(e^{i\theta}))$.

Set $M_0 = (M \times \{1\}) \subset (M \times S^1)$ and for any ε with $0 < \varepsilon < \pi$, set $M_\varepsilon = \{(m, e^{i\theta}) | m \in M, \ 0 \leq \theta \leq \varepsilon$ or $2\pi - \varepsilon \leq \theta \leq 0\} \subset (M \times S^1)$. The inclusion of $(M \times S^1, M_0)$ into $(M \times S^1, M_\varepsilon)$ induces isomorphisms from $H^i_{dR}(M \times S^1)$ to itself (obviously) and from $H^i_{dR}(M_\varepsilon)$ to $H^i_{dR}(M_0)$ (by Lemma 8.3.2) and so, by the five lemma (Lemma 8.1.12(1)), from $H^i_{dR}(M \times S^1, M_\varepsilon)$ to $H^i_{dR}(M \times S^1, M_0)$, for each i.

We now define $\tilde{K}$. Let $a \in H^i_{c-dR}(M \times \mathbb{R})$. Represent a by a form ψ on $M \times \mathbb{R}$ with compact support. Then $\varphi = K^*(\psi)$ is a form on $M \times (S^1 - \{1\})$ with support contained in $\{(m, e^{i\theta}) | \delta < \theta < 2\pi - \delta\}$ for some δ. Then φ extends to $M \times (S^1 - \{1\})$ by setting it equal to 0 elsewhere, and in fact to a form, which we still call φ, on $M \times S^1$ that is 0 on M_ε for some $\varepsilon > 0$, and in particular that is 0 on $M_0 = M \times 1$. We set $\tilde{K}(a) = b$ where $b = [\varphi] \in H^i_{dR}(M \times S^1, M \times \{1\})$. It is easy to check that $\tilde{K}(a)$ is independent of the choice of representative of a.

Now we must show that $\tilde{K}$ is an isomorphism.

Let $b \in H^i_{dR}(M \times S^1, M_0) \cong H^i_{dR}(M \times S^1, M_\varepsilon)$. Choose a representative φ of b with $\varphi = 0$ on M_ε. Restrict φ to $M \times (S^1 - \{1\})$, and let $\psi = (K^{-1})^*(\varphi)$. Then ψ is a form on $M \times \mathbb{R}$ with compact support. Let $a = [\psi] \in H_{c-dR}(M \times \mathbb{R})$. It is easy to check that $\tilde{K}(a) = b$. Thus $\tilde{K}$ is a surjection.

Suppose that $\tilde{K}(a) = 0$. Then, in the above notation, with $a = [\psi], [\varphi] = 0 \in H^i_{dR}(M \times S^1, M \times \{1\})$, so $[\varphi] = 0 \in H^i_{dR}(M \times S^1, M_\varepsilon)$. This means that $\varphi = d\rho$ for some ρ with $\rho = 0$ on M_ε. Restrict ρ to $M \times (S^1 - \{1\})$, and let $\tau = (K^{-1})^*(\rho)$. Then τ has compact support on $M \times \mathbb{R}$ and $\psi = d\tau$, so $a = [\psi] = 0 \in H^i_{c-dR}(M \times \mathbb{R})$. Thus $\tilde{K}$ is an injection as well. $\qquad \square$

COROLLARY 8.4.10. $\ H^n_{c-dR}(\mathbb{R}^n) \cong \mathbb{R}$ and $H^i_{c-dR}(\mathbb{R}^n) = 0$ for $i \neq n$.

Proof. This is true for $n = 0$ as $\mathbb{R}^0$ is a point, and then apply induction, using Theorem 8.4.9. □

We close with a few important results which we shall not prove, but which the reader should be aware of. First observe that, analogously to Lemma 8.2.20, there are products

$$H^i_{dR}(M) \times H^j_{c-dR}(M) \longrightarrow H^{i+j}_{c-dR}(M)$$

and

$$H^i_{c-dR}(M) \times H^j_{c-dR}(M) \longrightarrow H^{i+j}_{c-dR}(M)$$

given, on the form level, by multiplication of differential forms.

THEOREM 8.4.11 (*Poincaré duality*). *Let M be an oriented n-manifold. Then for any $0 \leq k \leq n$, $H^k_{dR}(M)$ is isomorphic to the dual of $H^{n-k}_{c-dR}(M)$ under the pairing*

$$([\varphi], [\psi]) \mapsto \int_M \varphi\psi,$$

i.e., given any linear transformation $T : H^{n-k}_{c-dR}(M) \longrightarrow \mathbb{R}$, there is a unique element $[\varphi] \in H^k_{dR}(M)$ such that $T([\psi]) = \int_M \varphi\psi$ for every element $[\psi] \in H^{n-k}_{c-dR}(M)$.

THEOREM 8.4.12. *If M is any compact n-manifold, $H^k_{c-dR}(M) = H^k_{dR}(M)$ is a finite-dimensional vector space, for every k.*

COROLLARY 8.4.13. *Let M be a compact oriented n-dimensional manifold. Then for any $0 \leq k \leq n$, $\dim H^{n-k}_{dR}(M) = \dim H^k_{dR}(M)$.*

8.5 Exercises

1. Let M be a compact oriented n-manifold. Then $[\varphi] \in H^n_{dR}(M)$ is defined to be the *fundamental cohomology class* of M if $\int_M \varphi = 1$. We denote this class by $\{M\}$. Let M and N be compact smooth n-manifolds and let $f : M \longrightarrow N$ be a smooth map. A point

$q \in N$ is called a *regular value* of f if for every $p \in f^{-1}(q)$, the induced map on tangent spaces $f_* : T_pM \longrightarrow T_qN$ is an isomorphism. (Note in this case that $\{f^{-1}(q)\}$ must be finite.) Let $\mathrm{ind}_p(f) = +1$ or -1 as $f_* : T_pM \longrightarrow T_qN$ is orientation-preserving or orientation-reversing. Let $d = \sum_{p \in f^{-1}(q)} \mathrm{ind}_p(f)$. Show that $f^*(\{N\}) = d\{M\}$. This integer d is called the *degree* of f. (Any smooth map f has a regular value, and in fact Sard's theorem states that the set of regular values of f is a dense subset of M.)

2. Let M be a smooth manifold with boundary. Show that the map $H^i_{dR}(M \times \{-1,0\}) \longrightarrow H^{i+1}_{dR}(M \times I, M \times \{-1,0\})$ in the exact sequence of the pair $(M \times I, M \times \{-1,0\})$ is given as follows: Coordinatize M locally by $(x_1, \ldots, x_n)$ and coordinatize $\mathbb{R}$ by t. A cohomology class in $H^i_{dR}(M \times \{-1,0\})$ is represented by a pair of i-forms $(\varphi_{-1}, \varphi_0)$ with φ_{-1} defined on $M \times \{-1\}$ and φ_0 defined on $M \times \{0\}$. Pull each of these back to $M \times I$ by the projections of $M \times I$ onto $M \times \{-1\}$ and $M \times \{0\}$, and (still) call their pullbacks φ_{-1} and φ_0. Write each of these as a sum of terms of the form $A_I dx_I$ and $B_I dx_I$ respectively, where I is a multi-index. For each multi-index I, let ψ_I be the $(i+1)$-form on $M \times I$ given by $\psi_I = d\big(A_I(1 - f_3(t)) + B_I(f_3(t))\big)$, where f_3 is as in Lemma 4.6.1 (and d is the exterior derivative). Then $[(\varphi_{-1}, \varphi_0)] \mapsto [\sum_I \psi_I]$.

3. Let M be a smooth manifold with boundary. Show that the isomorphism $H^{i+1}_{c-dR}(M \times \mathbb{R}) \longrightarrow H^i_{c-dR}(M)$ of Theorem 8.4.9 is given as follows: Coordinatize M locally by $(x_1, \ldots, x_n)$ and coordinatize $\mathbb{R}$ by t. Consider $[\varphi] \in H^{i+1}_{c-dR}(M \times \mathbb{R})$ and write φ locally as a sum of terms of two types, (i) $A_I dx_I$ where I is an $(i+1)$-element subset of $\{1, \ldots, n\}$, and (ii) $B_J dx_J dt$ where J is an i-element subset of $\{1, \ldots, n\}$. Then a term $A_I dx_I \mapsto 0$, and a term $B_J dx_J dt \mapsto \big(\int_{-\infty}^{\infty} B_J(x_1, \ldots, x_n, t)dt\big)dx_J$. Further, show that the inverse of this isomorphism is the map $H^i_{c-dR}(M) \longrightarrow H^{i+1}_{c-dR}(M \times \mathbb{R})$ given by $[\psi] \mapsto [\psi f_3'(t)dt]$ where f_3 is as in Lemma 4.6.1.

4. Carefully carry out the proof of Theorem 8.3.13.

5. Let p be a point of S^1 and choose a representative φ of the fundamental cohomology class $\{S^1\}$ that is 0 at p. Let $\pi_1 : M \times S^1 \longrightarrow M$

be projection on the first factor, and let $\pi_2 : M \times S^1 \longrightarrow S^1$ be projection on the second factor. Show that the map $[\psi] \mapsto [\pi_1^*(\psi)\pi_2^*(\varphi)]$ gives an isomorphism from $H_{dR}^{i-1}(M)$ to $H_{dR}^i(M \times S^1, M \times \{p\})$.

6. Formulate and prove an analog of Theorem 8.3.13, and the previous exercise, with S^1 replaced by S^j for any positive integer j.

7. Let $M = T^n = (S^1)^n$ be the n-torus, $M = \{(x_1, y_1, \ldots, x_n, y_n) \in (\mathbb{R}^2)^n | x_i^2 + y_i^2 = 1, i = 1, \ldots, n\}$. Let $\pi_i : M \longrightarrow S^1 = \{(x, y) | x^2 + y^2 = 1\}$ by $\pi_i((x_1, y_1, \ldots, x_n, y_n)) = (x_i, y_i)$ and let $\varphi_i = \pi_i^*(\varphi)$ where $\varphi = \{S^1\}$ is a fundamental cohomology class of S^1. If $S = \{i_1, \ldots, i_k\}$ is any k-element subset of $\{1, \ldots, n\}$, ordered so that $i_1 < \ldots < i_k$, let $\Phi_S = \varphi_{i_1} \cdots \varphi_{i_k}$. Show that $\{\Phi_S\}$ is a basis for $H_{dR}^k(M)$, where S ranges over all k-element subsets of $\{1, \ldots, n\}$.

8. Formulate and prove an analog of the previous exercise for $M = (S^j)^n$ for any positive integer j.

Index

Printed and bound by CPI Group (UK) Ltd, Croydon, CR0 4YY

08/05/2025

01864843-0001